Groundwater Pollution in Europe

WATER INFORMATION CENTER, Inc.

PERIODICALS

Water Newsletter
Research and Development News
Ground Water Newsletter

BOOKS

Geraghty and Miller *Water Atlas of the United States*
Todd *The Water Encyclopedia*
van der Leeden *Ground Water — A Selected Bibliography*
Giefer and Todd *Water Publications of State Agencies*
Soil Conservation Service *Drainage of Agricultural Land*
Gray *Handbook on the Principles of Hydrology*
National Water Commission *Water Policies for the Future*
Officials of NOAA *Climates of the States*
The Water Research Association *Groundwater Pollution in Europe*

Groundwater Pollution in Europe

Proceedings of a Conference
organized by
The Water Research Association
in Reading, England, September 1972

Edited by

John A. Cole
Chief Hydrologist
The Water Research Association
Medmenham, England

WATER INFORMATION CENTER, Inc.
Port Washington, New York

Library of Congress Catalog Card Number 74-84820

ISBN: 0-912394-12-9

Printed in the United States of America

Preface

Until recently relatively little attention has been focused on groundwater pollution, probably because the problem was out of sight and thus easily put out of mind. The causes of underground pollution are numerous and are as diverse as the activities of man. Recognizing the growing importance of the problem to the water supply industry, The Water Research Association organized a Conference on Groundwater Pollution. The purpose of the meeting was to provide a forum for the exchange of information on the causes and occurrence of groundwater pollution in the heavily populated and industrial nations of Europe, as well as on the administrative, legal and technical methods for its control. This book is a record of that Conference.

The Conference took place on September 25-27, 1972 on the campus of Reading University in Reading, England, The occasion represented the first time in Europe, I believe, when the chemical, hydrogeological, microbiological, and legal aspects of groundwater pollution were reviewed in a comprehensive manner.

Some 200 persons attended the conference. These participants represented a broad cross-section of professional backgrounds and affiliations. They came from water supply and river authorities, industries, government agencies, research institutes, and universities. A majority of those in attendance were from England; others came from several European countries and the United States.

The papers presented at the Conference, together with a summary of the discussions, a bibliography, and a list of attendees, were distributed in three separate volumes to the participants. Recognizing that the information presented at the Conference would be of interest and value to a much wider audience than just those who were able to attend, it was decided to publish the material in book form. In editing the original material I have regrouped the papers into a more logical sequence, have updated the bibliography, and have added an index to facilitate the usefulness of the book.

Within the time constraints of the Conference, it was not possible to cover all aspects of groundwater pollution. Thus, no claim to a complete coverage of the European situation can be made. But I do feel that the scope of the material is sufficiently broad to provide an insight to groundwater pollution causes, occurrences, measurements, and controls, which are typical in Europe today.

In preparing the manuscript for publication I have had the assistance and advice of many members of the Water Research Association staff. Particular acknowledgment is due Patti Dossett, Bob Odell, Leslie White, Dave Aston, Iris Rhodes, Julia Schoon, Fiona Chappell, Minna Fryman, Jill Wigmore and Isabelle Leese. And finally, the encouragement of Dr. R.G. Allen, Director of the Association, has been most helpful.

John Cole

Medmenham
June 1974

Contents

CHEMISTRY AND THE INTERACTION OF POLLUTANTS WITH AQUIFERS

MICROBIOLOGY OF GROUNDWATER POLLUTION

CASE HISTORIES OF POLLUTION OCCURRENCES

URBAN REFUSE DISPOSAL

MINING WASTE AND FLY ASH DISPOSAL

DEEP WELL WASTE DISPOSAL AND BOREHOLE TECHNOLOGY

GROUNDWATER POLLUTION CONTROL
by D. Mercer
Department of the Environment, London, England

SUMMARY

The effects of pollution of groundwater and surface water are compared, and distinction drawn between them. The control of groundwater pollution is by common law, statute law and by byelaws, and also by notification and consultation arrangements and similar semi-judicial procedures. The scope and application of the legislation are discussed.

1.   INTRODUCTION

Approximately one third of the public water supplies of the United Kingdom is obtained from underground sources and the need to protect the purity of this important source is obvious. Pollution of groundwater differs in some important ways from pollution of surface waters, rivers and streams, and the effects of pollution and measures to prevent and deal with them make necessary different approaches to the problems.

Pollution of groundwater is not normally easily noticed and the detection of the source or sources of pollution may present difficulty, particularly, for example, where the nature of the underground strata permits polluting material to travel considerable distances. Even when the sources of pollution have been traced and removed it may be that the resultant contamination of the aquifer will continue to pose problems of pollution of abstracted water. Equally important is the fact that groundwaters which are generally of good quality do not receive, or need, comprehensive treatment before passing into supply and in many cases simple, precautionary chlorination only is given.

Pollution of surface waters by distinction however, is easier to detect. The visual appearance of the stream, possibly its odour, the presence of dead fish, etc., could quickly lead to an investigation and discovery of the source of pollution. Indications of pollution could lead to a prompt decision to take appropriate action concerning water supply, i.e. temporary cessation of abstractions or modifications to treatment. The first indications of groundwater pollution, however, could in some cases be complaints by the consumers. Following actions taken to reduce or eliminate the pollution, the recovery of the stream to normal quality is often very rapid. (An example of such a case of quick recovery of a polluted stream concerned the River Cole (1), a tributary of the River Tame near Birmingham where, following the closure of the overloaded Yardley sewage treatment works of the Upper Tame Main Drainage Authority, from which poor quality effluent had been discharged, the recovery of the river was very substantial within a month). Unlike many abstracted sources of groundwater, surface water abstractions, particularly from rivers known to receive treated sewage effluent discharges, are given extensive treatment before passing into supply and this provides greater protection.

One of the most important differences, however, from the pollution aspect, is the great advantage conferred on groundwater generally by the purifying effect of passage of the water through soil and porous strata. This can effect very considerable purification by removal of bacteria and other undesirable organisms and also sub-stantially remove organic matter and oxidize ammonia which can interfere with disinfection by chlorination. The remarkable purifying effect of passage of waters through soil was well illustrated by a survey (2) carried out a few years ago by the Water Research Association to discover the amount of pesticides in rain water, rivers and groundwaters. It showed that the concentrations of organo-chlorine type pesticides were in some cases higher in rain water than in river water and were least in ground-waters. In fact, in samples from boreholes in fruit growing areas in the south of England where pesticides had been extensively used, pesticides of this type could not be detected.

The possibilities for pollution of groundwater are numerous. The well itself may be polluted by work during construction or by subsequent maintenance work. Contamination by polluted surface water may gain access to the water-bearing aquifers through fissures, cracks or swallow-holes, particularly in chalk or limestone. The vulnerability of sources is variable, depending on geological formations and depths of aquifers. Wherever an aquifer outcrops or approaches close to the ground surface there is a point of particular vulnerability. Near the coast, sea water may gain access to the water-bearing aquifer when a combination of high tide and/or excessive pumping permits sea water to percolate inland.

Pollution of groundwater can be caused by deposits on the surface of the land and also by discharges made directly to underground strata. Examples of the former include the application of fertilizers to farmland (of particular interest is the use of nitrogenous fertilizers which can result in an undesirable increase in the nitrate concentration of the groundwater, but in this connection the normal discharge of

2.

nitrate from cultivated land must be given due consideration). Percolate from rubbish tips can cause pollution in some circumstances and recommendations in this respect have been made (3). Recommendations have also been made on the operation of tips receiving toxic wastes to reduce pollution from them (4). Pollution by cesspit soakaways must be included in matters to be taken into consideration and also the pollution derived from burial grounds, leaking sewers and disposal of sewage and industrial wastes. A memorandum (5) has drawn attention to the need to protect public water supply resources from pollution of this nature.

Underground water is protected from pollution legally in a number of ways; by common law, statute law and by byelaws made under statute and also by notification arrangements and similar semi-judicial procedures.

2. COMMON LAW CONSIDERATIONS

Under Common Law a landowner has a right of action against anyone who pollutes water percolating through his land so that it reaches the owner's land in a polluted condition. The definition of pollution under common law is founded on case law and the effects can be onerous to the party causing the pollution. Legally, pollution means an act which changes the natural qualities of the water. The fact that the water may already be polluted is no defence, if what is added would pollute the water if it were otherwise pure. Theoretically, any change in quality from the "natural state" is pollution, but the "de minimus" rule acts to mitigate the severity of the results of this definition.

3. WATER ACT 1945

Section 21(1) of this Act states:

"If any person is guilty of an act or neglect whereby any spring, well or adit, the water from which it is used or likely to be used for human consumption or domestic purposes........,is polluted or likely to be polluted, he shall be guilty of an offence against this Act:

Provided that nothing in this section shall be construed as prohibiting or restricting -

(a)   any method of cultivation of land which is in accordance with principles of good husbandry; or

(b)   the reasonable use of oil or tar on any highway maintainable at the public expense so long as the highway authority take all reasonable steps for preventing the oil or tar.... from polluting any such spring, well or adit. "

From the viewpoint of the water authority seeking to prevent pollution of groundwater, the provisos (a) and (b) are of significance, being capable of liberal definitions which could be held to reduce the effectiveness of control under this provision of the Act. In addition, this section does not permit action to be taken to prevent the pollution of groundwater before it occurs but only allows the water authority to take action after it has established, probably with some difficulty, the source of and responsibility for, the pollution.

Section 18 of the Act permits water undertakers to make byelaws:
"...for the purpose of protecting against pollution any water, whether on the surface or underground, which belongs to them or which they are for the time being authorized to take, they may by byelaws -

(a)   define areas within which they deem it necessary to exercise control; and

(b)   prohibit or regulate the doing within that area of any act specified in the byelaws. "

If an owner or occupier considers that a requirement of him is unreasonable he may appeal to the Secretary of State who may determine the appeal himself or may refer it for determination by an arbitrator, and if it is decided that the requirement is unreasonable, it may be modified or disallowed.

An important provision of the Act requires the undertakers to pay compensation to persons within the area defined by byelaws for any curtailment or injurious affection of their legal rights, and for any expenses incurred by them in compliance of the byelaws. In this case also, in the event of dispute, an appeal may be made.

Since the passing of the 1945 Act, a number of statutory water undertakers have given consideration to the possibilities for the application of section 18 byelaws. Sixteen undertakers have actually made byelaws but most of these have lapsed and at the present time only six sets of byelaws are still in force. Of these, only three are for the purpose of protecting underground water; these are enforced by the Bucks Water Board, Brighton County Borough, and North Lindsey Water Board. As far as is known, no legal action has been taken by these water undertakers under the byelaws.

It is clear therefore that the byelaw making powers available to them are rarely used by statutory water undertakers, possibly because of the fact that compensation has to be paid to occupiers or owners for curtailment or injurious affection of legal rights.

4.   WATER RESOURCES ACT 1963

Section 72(1) of the Act states:

"...it shall not be lawful after the end of the initial period, by means of any well, borehole or pipe, to discharge into any underground strata within a river authority area -

(a)   any trade effluent or sewage effluent, or,

(b)   any poisonous, noxious or polluting matter not falling within the preceding paragraph

except with the consent of the river authority, which consent shall not be unreasonably withheld, and subject to any conditions imposed by the river authority in accordance with the next following subsection..."

The river authority may refuse consent or grant it unconditionally or subject to conditions specifying the nature, composition and volume of the effluent to be discharged; the strata into which it may be discharged; and subject to measures being taken to protect water contained in other underground strata. The provision and use of observation wells with facilities for inspection may be required. The potential discharger may exercise a right of appeal if he considers the conditions unreasonable.

The wording of section 72(1) indicates that there is no control over discharges to mine shafts, adits (horizontal entries to mines) and fissures. These have been used to dispose of various liquid wastes including coke-oven effluents, ammoniacal liquors, waste oil and distillery waste. (In 1967 a tank lorry was discovered to have discharged about 50 000 gallons of oils, triclorethylene, ketones, phenols and cyanide to a swallow hole in the Mendip Hills). Fortunately, however, cases in which underground waters have been polluted as a result of such activities are comparatively rare. It is not certain whether septic tank effluents are covered by this section, and this question has never been decided in the courts.

5.   DEPOSIT OF POISONOUS WASTE ACT 1972

This Act, which was passed too recently to permit observations being made on its operation, is another statute directed towards protecting groundwater supplies from pollution. It makes it an offence to deposit on land waste of a poisonous, noxious or polluting nature which could give rise to an environmental hazard. (It is to be noted that "land" for the purposes of the Act includes land covered with water). Such a hazard is defined so as to include circumstances where pollution or contamination of water (whether on the surface or underground) is threatened. It is to be noted that the threat of pollution is sufficient to enable the Act to be applied and it is not necessary to show that actual pollution has occurred. In this sense the Act remedies the possible difficulties of the Water Act 1945 where, under section 21, actual pollution of water must be shown to have taken place.

The Act also requires the responsible authorities to be notified before wastes, other than those exempted under the regulations, are removed or deposited. The responsible authorities concerned are the local authority and river authority, or river purification board for the area in which the premises are situated.

The principle statutes concerned in the control of groundwater pollution have now been described. There are, however, in addition, a number of official and semi-official procedures directed towards a similar purpose.

The construction of lagoons, used for refuse disposal and for tipping are controlled by planning acts. At the present time there are no statutory requirements for planning authorities to consult river authorities on these matters although consultation does take place. However, the procedure for control by river authorities via planning authorities creates many difficulties, especially once a proposal is approved and tipping of certain prohibited substances takes place.

Planning authorities have been advised by a circular of April 1971 from the Department of the Environment to consult river authorities on planning applications for developments which could cause inland water pollution. The type of operation envisaged include;

1. building or other operations or use of land for the purpose of refining, reception or storage of mineral oils and their derivatives;

2. the use of land for the depositing of all kinds of refuse, waste or other materials;

3. building or other operations or use of land for the retention or disposal of sewage, trade waste or sludge;

4. use of land for burial purposes.

The circular also stated that it was desirable that the river authority be consulted on applications for planning permission pertaining to the erection of buildings, etc., for the compounding, manufacture, storage or significant use of toxic chemicals, and proposals to change the use of existing buildings to use for that purpose. Authorities were asked to let river authorities know of any proposals which appeared to them likely to give rise to pollution.

Following these consultations, the river authorities are then in a position to advise whether some planning permissions should not be granted if they were likely to cause groundwater pollution. Alternatively, river authorities could ask that certain precautions were needed to be taken as a condition of granting planning permission.

The construction and siting of septic tanks is governed by building regulations (N. 17) - "... so as not to render liable to pollution any spring, stream, well, adit ..... used or likely to be used for drinking, domestic, etc, purposes". Thus, since 1966 when national building regulations came into force, new tanks should have been adequately covered by these regulations. Earlier constructions may in many cases have been subject to byelaw powers.

The views expressed in this paper are entirely those of the writer, and not necessarily those of the Department of the Environment.

REFERENCES

1. DEVEY, D.G., GREEN, M.B. and HARKNESS, N.

   Reclaiming the River Cole from pollution's clutches.
   Mun. Engng, 1972, 21 April.

2. CROLL, B.T.

   Organo-chlorine insecticides in water - Part 1.
   Wat. Treat. Exam., 1969, 18, (4), p 255.

3. HOUSING AND LOCAL GOVERNMENT, Ministry of

   Pollution of water by tipped refuse. Report of the Technical Committee on the Experimental Disposal of House Refuse in Wet and Dry Tips.
   London, HMSO, 1961.

4. HOUSING AND LOCAL GOVERNMENT, Ministry of

   Disposal of solid toxic wastes. Report of the Technical Committee on the Disposal of Toxic Solid Wastes.
   London, HMSO, 1970.

5. HOUSING AND LOCAL GOVERNMENT, Ministry of

   Safeguards to be adopted in the operation and management of waterworks.
   London, HMSO, 1967.

SAFEGUARDING THE WATER SUPPLIES IN UPPSALA, SWEDEN

by J. Sidenvall
Department of Public Works and University of Geology, Uppsala, Sweden

1. BACKGROUND

Water from the River Fyris was used in the early days for domestic purposes as it was the nearest water resource. Groundwater was only used as drinking water. It was then too far to transport several hundred metres.

During the last decades of the nineteenth century, people began to understand the increasing hygienic risks in using water from the river. The demand for groundwater increased. For this reason new water mains and wells were built. The population of Uppsala was doubled in the twenty years after World War II and the groundwater demand was more than doubled during the same time. This meant that the existing groundwater reserves began to shrink. Therefore, a system for infiltration of surface water was made.[*]

2. GEOLOGY

The area around Uppsala is rather flat. The plain is slightly broken by north-south and northwest-southeast faults. The western sides of these faults are seen as gentle slopes. The reason for this is that the rock has been broken up in large blocks, size around 5 x 10 km. The eastern parts of these blocks are generally uplifted. The western parts are on the contrary somewhat depressed.

The last glacial period (Weichsel) caused different sediments to fill these valleys. The author has shown that the faults in the area have caused the location of the Uppsala esker with its subsidiary eskers. The eskers are therefore deposited on the western slopes of the faults. The most coarse parts of the esker, (the central parts) are deposited on the lower part of the western side of the fault. Of course this means that there is plenty of groundwater here. The upper and more westerly parts of the esker are poor on this account though. These parts are on the other hand more used as gravel pits.

The rest of the valley is filled with till and the outlying parts of the glacial sediments such as sand, silt, clay etc. Above these layers there are layers of post-glacial clays.

Due to post-glacial emergence, the upper parts of all hills in this area have been strongly affected by sea wave erosion. Of course this also includes the upper parts of the eskers.

---

[*] See also 'Safeguarding Uppsala from oil pollution', pp. 328-331.

Therefore the stratification can be very complicated along the eskers. There are layers of redeposited clay, sand, silt etc. This is the reason why there can be several groundwater tables and artesian groundwater along the eskers.

3. NATURAL GROUNDWATER

The groundwater basin is unusually large for Sweden being around 1500 km^2 (Fig. 1). Through this basin the Uppsala esker with its subsidiary eskers are like drains. In some parts though, the eskers are leaking via springs. Under the plains, groundwater flows in the till. Groundwater streams here are directed towards the central parts of the eskers. Along the Uppsala esker the flow is towards the south. The termination of the groundwater flow along the esker of Uppsala is just to the south of Flottsund, some 8 km to the south of the city. An indication of the groundwater balance is given in Table 1.

It is understandable that people in the area took their water from springs and later from wells, especially as these parts of the Uppsala esker are distinguished by having sufficient groundwater resources in comparison with most Swedish eskers.

Among the springs, the most famous are St. Erikskällan and the Hospital Spring, the latter one yielded about 130 litres/sec a century ago. A water pipe was laid from the former as early as the middle of the seventeenth century. The pipe served the castle among other consumers. Unfortunately this municipal water main was in operation only about ten years. After that, no new water main was installed until 1875; St. Erikskällan was again used as a source. But this spring does not deliver groundwater from the primary parts of the esker. It comes from a secondary groundwater layer as do most springs in the area. The demand for groundwater was increasing all the time. New mains were built as well as a water reservoir. In the beginning of this century, J.G. Richert (1) warned against constructing new wells in the centre of this city, the reason being the risks of pollution of the drinking water.

TABLE 1

GROUNDWATER BALANCE;   UPPSALA ESKER (SIDENVALL 1971)

	litres/sec
Average yield:	
Municipal wells	475
Private wells	70
Test wells near the termination of groundwater flow in the esker	100
Spring flow	65
Total groundwater yield	710
Average volume recharged	210
Uppsala Esker yield	500

4. RECHARGING OF GROUNDWATER

During the 1950's the groundwater demand in the city was very large because of the increase in population. In fact the population doubled during the 1940's and 1950's. The water board wanted, therefore, to use recharge basins to improve the groundwater resources. It was important to have the basins in operation before a drawdown of the groundwater table became a reality. Luckily the city council decided to try to infiltrate surface water from the River Fyris into the Uppsala esker just as Richert had recommended fifty years earlier.

A suitable place for this purpose was found at a place called Tunaåsen, about 2 km north of the Galgbacken waterworks. This waterworks consists of a water well gallery where the wells are placed across the waterbearing layers of the eskers (2). Typically the well gallery is situated on the eastern side of the esker: the reason has been mentioned above. When the Municipal Services Division started the recharging, there were protests from some politicians and scientists who argued that all recharged water would flow back into the river. Other complaints were that all was pumped up at the Galgbacken well gallery; that the chemistry of groundwater at Galgbacken would change in a few years and that river water would pollute the groundwater. However, none of these things has been proved.

Considerable data have been collected during fifteen years of recharging and a few of the most important facts have been reported (3).

At Storvad, situated 2 km north of Tunaåsen, river water is abstracted through intakes provided with screens for arresting branches and floating debris. Water is then pumped up to the first part of the filtration plant; which consists of four rapid sand filter beds of the gravity type. The filters are cleaned by backwashing. The water is then pumped to the second part of the filtration plant. This consists of about ten recharge basins at Tunaåsen. These are constructed as slow sand filters right above the primary material of the esker. The sand filters consist of a sandlayer about 0.5 m thick. The construction of the filter plant is conventional, as well as the results.

The filtered water percolates about 20 m through the esker before reaching the groundwater table. The recharging causes a significant rise in the groundwater table, especially at Galgbacken waterworks, although there was a problem with the recharging. A few of the gravel pits, situated north of Storvad were flooded. To correct this and to get a reserve for the waterworks at Galgbacken a new groundwater well gallery was constructed at Storvad. Here the River Fyris crosses the esker but the top of the esker lies about 40 m beneath the river. The layers between consist of different sorts of clays. There is therefore no risk of pollution, even though the river water level lies about one metre higher than that of the groundwater. This means that groundwater is sub-artesian here.

Since the start of Storvad groundwater works no new springs have been observed in the particular area. No changes whatever have been noticed in groundwater chemistry at Galgbacken waterworks. This leads to the question, where does the recharged water disappear?

5. WATER CHEMISTRY OF THE RIVER FYRIS
First it is important to look at the chemistry of the water in the River Fyris. Certain people think that it is of standard composition (4). Experience has told us that such is not the case. During the floods in spring and autumn, water is milky with silt and clay. This is caused by autumn rains and by melting snow during springtime.

The river water originates then typically from surface flow. This has low ionic contents and the specific conductivity is consequently very low. The milky water is not recharged during these periods. If it were the filters would be completely clogged in less than a day. Immediately after the floods the specific conductivity of the river water rises and remains high until next flood.

Just before a flood the river water chemistry is almost like that of groundwater from these areas. The water in the River Fyris must originate then mainly from subsurface and groundwater flow. Several springs are noted along River Fyris and its subsidiaries. This is one of the reasons why groundwater chemistry at Galgbacken seems to be unaffected by the recharging.

6. WATER FLOW BETWEEN TUNAASEN AND GALGBACKEN
When water from the recharge basins reaches the groundwater table this will spread out on top of the natural groundwater, because of the laminar flow of groundwater under the recharging basins, (3)(5).

Between the recharge basins and Galgbacken waterworks there is a little natural gravitational water reaching the groundwater table. Of course this water will spread out on top of previously percolated water, for example, the recharged water (5). If the recharged water should reach the Galgbacken area it would, for the reasons given above, bypass high above the screens of the well gallery.

It is known that the recharged and the natural groundwater both flow southwards. This is the main direction of groundwater flow along the Uppsala esker.

Several tracer tests have been carried out to investigate what is happening to the recharged water.   Therefore, test wells have been driven into the esker between Tunaasen and Galgbacken.   Among these, Wells Nos. 1, 2 and 3 have attracted the most interest.

Uranin has been principally used as a tracer.   After many tests the results are that tracers put into the recharge basins can be detected in Wells Nos. 1 and 2.   But there has never been any sign of the tracers at Well No. 3.   But if the tracer was put into Well No. 2 it was detected at No. 3.   In any case, no tracers have been detected at the Galgbacken well gallery.   The average flow of groundwater between Wells Nos. 1 and 2 and between Nos. 2 and 3 has been 7. 3 m/day.

Further tracer tests were undertaken but with the same results.   On these occasions tracers were found in new test wells about 1 km to the west of Galgbacken but there was no explanation why they were found there.   However, the author's opinion why recharged water does not reach Galgbacken, is discussed below (3).

First we have mostly laminar flow in groundwater (5).   This means that a tracer must be put into the same groundwater bearing layer as that where the test well screen is located.   This is a difficult task especially when we are working with eskers of the so-called submarine type, such as obtained at Uppsala.

The second reason has been explained earlier (the superposition of the various infiltrating waters).   The third reason is also a geological one.   The location of the esker of Uppsala was caused by a north-south fault.   The lower parts of the esker are coarser and lie more easterly than the upper parts.   It is these parts which are seen as a ridge.   Also these upper parts appear to consist of finer material.

Rocks are seen in the bottom of the gravel pits between Tunaåsen and Galgbacken which must be parts of the very top of the western slope of the fault.   The top of the slope must be like a ridge which will most certainly divide groundwater flow in the esker into two different streams.

The recharge basins lie above and a little to the west of this ridge and most recharged water will be directed south-westwards.   Little flows directly southwards, i. e. on the eastern side of the ridge.   This means that probably little recharged water will reach the area of Galgbacken.

The author's opinion is that recharged water at Tunaåsen will pass west of Galgbacken unaffected by pumping at Galgbacken. After a few years the recharged water will reach the well galleries in the centre of the city. Here recharged water will be pumped up.

7. POLLUTION CAUSED BY AGRICULTURE

During the first years of recharging, samples for controlling the infiltrated water were collected very often. But the analysis of water samples (Figs. 2 and 3) that were taken during spring and autumn showed curious features. The sulphate ion concentration was then always very high. But during summer and winter sulphate concentration was normal again. It was also only the water samples from Test Wells Nos. 2 and 3 that showed these surprising results. It was obvious that sulphate ions came into the esker somewhere upstream of Well No. 2, but where? After investigations it was found that fields to the east of Tunaåsen were fertilized every spring and autumn with superphosphate, which consists of about 40% sulphate. All surface water from this area was drained by a ditch which ran into a gravel pit just to the north of Test Well No. 2. A rather surprising thing actually.

We have had several examples of pollution caused by agriculture in the Uppsala neighbourhood. But the above example did no harm whatever. In fact the sulphate ions worked as a tracer. Again it was proved that water passing Test Wells Nos. 2 and 3 never reach the well gallery at Galgbacken. In fact it is difficult to use sulphate ions as tracers in this case. The River Fyris has high concentrations of sulphate during spring and autumn. The ions come from the fertilized fields. Therefore the recharged water has small peaks of sulphate ions.

A few conclusions about groundwater flow in eskers are:

Generally laminar movement

A groundwater molecule follows in the first hand the same layer in an esker

There are different layers of groundwater with different densities caused by different chemistry

Youngest groundwater nearest the groundwater table.

8. COPING WITH ACCIDENTAL SPILLS

If an oil accident happens in Uppsala, special alarm lists will be followed. These lists tell how police, fire brigade and the Municipal Services Division must act. There is for example always one man detailed to handle oil accidents. If he needs help he follows the alarm lists.

During the first six months of 1972, only one serious accident with a chemical in transit happened in Uppsala. On that occasion it had a happy ending although this accident revealed how unprepared we were for such pollution.

Because of this accident we are now trying to map the transportations of all dangerous goods. We must know:

    (a)   Which different types of dangerous goods are transported in the area

    (b)   To whom and from which factory

    (c)   Are the goods transported by railway or by road

    (d)   Which railway or road

    (e)   How much dangerous substance is transported every time

    (f)   How often

    (g)   What day

    (h)   How are the goods stored

The first report on this subject revealed that very few could answer given questions like those above. Anyway the means of travel of the twenty most common and dangerous chemicals through the area of Uppsala are now known. A new and more detailed investigation on this subject will be made later.

In the meantime action lists are being prepared, one for every common and dangerous substance. If a chemical accident happens, this list will be used by the rescue personnel. Police, fire brigades, the different municipal boards, hospitals etc. must exactly know what to do and when.

Different people from these institutions have to be trained and taught how to handle their special equipment for such accidents. But the teachers must learn first, that is the problem and the necessary tools, materials etc. have already been obtained.

There are also questions like: what are we doing with chemically polluted soil, water or groundwater? All these questions must be answered soon.

Until these plans are ready special alarm lists have been made so that assistance is theoretically available anyway, if an accident happens.

REFERENCES

1. RICHERT, J.G.

   Om Sveriges grundvattenförhållanden. Stockholm, 1911.

2. WINQVIST, G.

   Ground water in Swedish eskers. Kungl. Tekn. Högskolans handl. nr 61, 1953.

3. SIDENVALL, J.

   Grundvatten i Uppsalatrakten. Uppsala, 1970.

4. ROHDE, W.

   The ionic composition of lake waters. Proc. Intern. Assoc. Limnol. 10, 1949.

5. GUSTAFSSON, Y.

   The influence of Topography on Ground Water Formation. In "Ground water problems", Proc. of the internat. symposium series. Oxford, Pergamon Press, 1967.

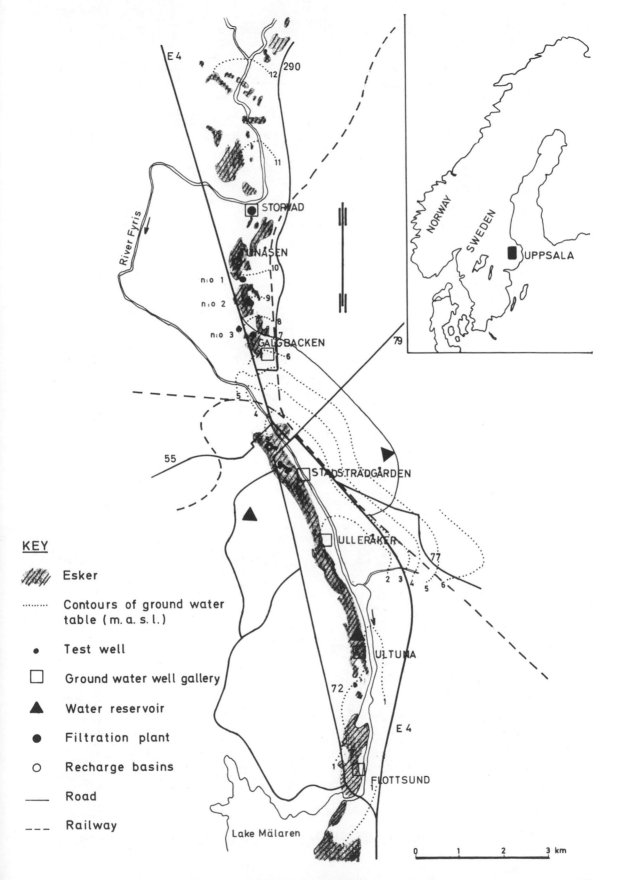

FIG. 1. THE WATER SUPPLY SCHEME AT UPPSALA

KEY

Esker

Contours of ground water
table (m. a. s. l.)

Test well

Ground water well gallery

Water reservoir

Filtration plant

Recharge basins

Road

Railway

E 4

River Fyris

STORVAD

LINÅSEN

n:o 1

n:o 2

n:o 3

GALGBACKEN

STADSTRÄDGÅRDEN

ULLERÅKER

ULTUNA

FLOTTSUND

Lake Mälaren

NORWAY

SWEDEN

UPPSALA

0   1   2   3 km

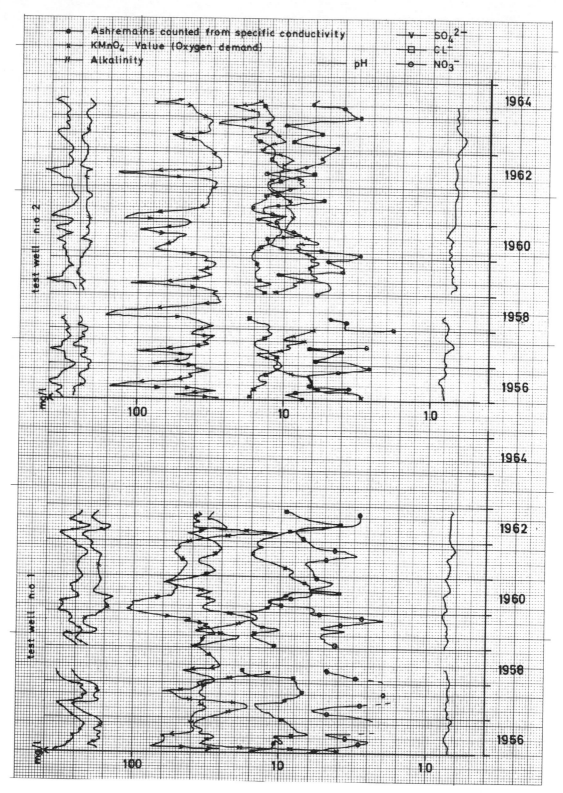

FIG. 2 <u>GROUNDWATER CHEMICAL ANALYSIS:TEST WELLS NOS 1 AND 2.</u>

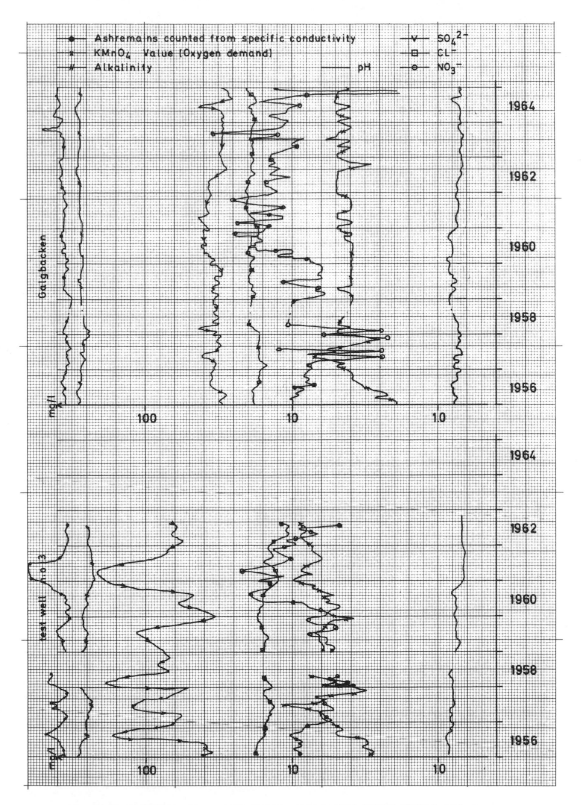

FIG. 3 GROUNDWATER CHEMICAL ANALYSIS: TEST WELL NO. 3.

LEGAL AND TECHNICAL ASPECTS OF GROUNDWATER POLLUTION CONTROL

by D.D. Young
Essex River Authority, Chelmsford, Essex, England

1. LEGAL CONTROLS IN RELATION TO DISPOSAL BY TIPPING IN THE ESSEX RIVER AUTHORITY AREA

1.1. Water Resources Act

This Act with its restriction to "well borehole or pipe", makes relatively little contribution to control.

1.2. Town and Country Planning Acts

Good liaison with the local authorities in Essex and East and West Suffolk county areas provides good control of new waste disposal, oil storage etc. proposals, provided that all the implications are clear at the time of granting of permission. Knowledge in the field is new and growing, so that some waste disposal permissions granted less than five years ago, with consultation with the Essex River Authority (ERA) have led to pollution by leachate. Also depends on liaison with intermediaries which is good at present, but what of future after reorganization?

1.3. Deposit of Poisonous Wastes Act, 1972

Applies severe penalties to creation of "environmental hazard". When proof "beyond reasonable doubt" is available, damage is done but deterrent may prove valuable. Notification of wastes only provides information and exemptions from notification are extensive. There is no provision for systematic control under this Act.

1.4. Essex County Council Act, 1967

This Act makes it an offence in Essex, "to form a deposit of refuse or continue to add refuse to an existing deposit or otherwise dispose of refuse in any place in the county" without the consent of Essex County Council (ECC) and local authority. There are a number of sensible exceptions (e.g. spreading of farm manure). ECC required to notify statutory water undertakers of applications. Consents may be withdrawn after six months. Usual appeal provision for applicant. ECC and local authorities liaise with ERA and system works well, but only applicable to Essex part of the area. Also at the Committee stages of the Bill, undertakings were given that there would be no "clamp down" on certain important sites within ten years. i.e. until 1977. Even with good liaison with local authorities now, changes of staff, local government reorganization could lead to loss of co-ordination in future.

1.5. Water Act 1945

Pollution prevention provisions not available to River Authorities but may not pass to Regional Water Authorities (RWA) outside water company areas. Powers have not apparently been used widely, possibly due to practical difficulties.

1.6. Essex River Authority

The Essex River Authority considers waste disposal far too crucial (34% of licensed resources are groundwaters) for anything but direct control and there is a need for control of discharges other than by well, borehole or pipe. Hence, Private Bill in 1972 session of Parliament, Clause 13, creates an offence of "act of neglect whereby water in any underground strata ...... is contaminated or likely to be contaminated by any poisonous noxious or polluting matter". There are sensible exceptions (e. g. acts in accordance with good agricultural practice). It also creates a requirement for consent for waste disposal, with usual appeal provisions for applicant. Compliance with consent exempts from offence of "act of neglect whereby ...".

There is provision for the statutory defence that the occurrence was due to an accident or some other cause beyond defendant's control and that defendant took "all reasonable precautions" to avoid the offence. This provision was thrown out by Parliament three years ago, but we are fairly confident this time. This should be a useful "trial run" for general legislation in this field.

2. HYDROGEOLOGICAL FACTORS IN THE PREVENTION OF POLLUTION BY DOMESTIC REFUSE

2.1. Groundwaters in Essex

Groundwaters contained in (a) chalk (24% of licences granted and 25% of licensed quantity) and (b) gravel (45% of licences granted and 9% of licensed quantity). Gravel often comes into contact with streams or discharges of spring water, so gravel ground-water pollution is frequently related to surface water pollution. All proposals should be considered from both aspects. Essence of control is close liaison between pollution inspectorate and hydrogeologist.

2.2. Classes of site in Essex

2.2.1. Clay workings.

Usually safe for domestic waste but regard must be paid to thickness and permeability of, and hydraulic gradient within, residual clay. Also, upper level of surrounding clay. Desirable to underdrain top cover.

2.2.2. Gravel workings.

(a)   Naturally dry. Under low rainfall and high evapotranspiration
conditions of Essex, downward percolation may be small and
acceptable.

(b)   Naturally wet (i.e. excavated below highest natural water
table level). Clay sealing to exclude clean water or contain
contaminated water has frequently been tried and usually failed.
Massive quantities of impermeable material essential for
success. Percolation through gravel does <u>not</u> effect adequate
purification (little effect on Cl^-, NH_4^+ and high residual BOD,
PV etc.) Groundwater levels, gradients, flows and effects of
any abstraction have to be assessed and provision made for
disposal of percolate usually required. No satisfactory method
for treatment to acceptable standards for inland water discharge.
Can be dealt with as a minor constituent of domestic sewage.
One case in Essex where percolate is passed to gravel washing
water supply. Polluted leachate known to continue to arise after
twenty years and may continue for much longer.

2.2.3. Chalk workings.

In continuity with vital aquifers. Depending on fissuring, transport of pollutants
may be very rapid. No waste disposal, other than inert, non-toxic material under closest
control.

2.2.4. Marshland.

Essex marshes have been used for London's refuse for many decades. Most
of surface drainage is via ditches to the River Thames. Gross surface pollution has
occurred, but most sites are relatively safe in relation to groundwater, due to low
permeability silty deposits overlying water-bearing strata*. The Essex River Authority
is examining this by borehole and permeability studies at one important site; the results
are not yet available.

---

* See Aspinwall's contribution on Case histories of some phenolic pollutions in Essex,
   p. 338-343.

NOTES ON THE DISPOSAL OF TOXIC SOLID WASTES
The Essex Water Company, Romford, Essex, England

1. POLLUTION OF UNDERGROUND WATER

Over the years there have been several instances where the Company have been concerned as to the dumping of solid wastes at points not far removed from their wells in South Essex. Here the cover over the chalk is often not great and there is a tendency for disposal in old chalk or gravel workings: the highly industrialized Thames estuary is nearby. Their concern has been accentuated by actual trouble which occurred many years ago with picric acid at Grays (Sect. 2) and with petroleum during World War II.

i. Petroleum (see also Sect. 3)

During 1939-45 there was a serious leakage of petrol or paraffin to the chalk at a point near the Thames estuary. As a result, certain private deep wells in the area acquired an oily taint, and this persisted for many years. In at least one case (Aveley) the smell was still evident in the water in 1954. While this case is not unique, it demonstrated the great danger of any form of petroleum or oily waste gaining access to the chalk.

ii. Metallic wastes

In 1956 the Company dealt with a proposal to dump non-ferrous foundry wastes in a disused gravel pit, 1560 yds from their Linford well: the waste had 21% of copper and 19% of zinc. It was finally agreed, provided selected earth filling was undertaken in the pits up to 2 ft above standing water level, and before dumping began.

iii. Gypsum wastes

In 1958 the Company raised objections to the dumping in a lagoon of waste gypsum from phosphoric acid manufacture, and containing traces of fluorine, at a site $1\frac{1}{2}$ to 2 miles from Linford well. Fears were expressed as to non-carbonate hardness increase and contamination with fluorine. The agreed solution was that, if necessary, the lagoon should be lined with impermeable clay to a depth approved by the planning authority.

iv. Clinker ash

In 1965 the South Essex Waterworks Company successfully opposed the storage of this ash in a disused quarry at Grays. While the Company were not greatly worried an objection on principle was made because the ash contained 10 mg/litre of arsenic and might have contained more.   It is noted that the Ministry Committee exclude pulverized fuel ash - do they consider there is any risk from traces of arsenic in this; this risk is admittedly remote?

v. General

One should make the point that though the risks from solid toxic wastes are small, great caution is warranted because if appreciable pollution occurs the waterbearing strata might become "poisoned" more or less in perpetuity, yielding traces of toxic material to the water.   It is well-known that soluble impurities may travel long distances underground.

2.   POLLUTION OF SURFACE WATER

The Company have had no trouble with toxic solid wastes and it is not unlikely that the greatest hazard is semi-solid oily wastes or tar wastes from garages etc. and in refuse dumps.   We have had recurrent trouble with river pollution by oil but mainly accidental spillages etc. of liquid.

In the South Essex Waterworks Act, 1935, there is a prohibition of the storage of manures or fertilizers within 50 yds of the nearest watercourse and such a provision may apply in other undertakings' legislation.   Conceivably some similar safeguard could be made to cover toxic wastes.

Technical discussion

1.   The Department of the Environment had commissioned a survey of waste tips or refuse dumps throughout the British Isles which was being carried out through the Institute of Geological Sciences.   That survey would include boreholes drilled through tips and in their vicinity to collect information on the migration of leachates.   Drilling investigations by River Authorities and others concerned with local pollution problems should be encouraged also.

2.   The Institute of Geological Sciences was maintaining a data store on the Permo-Triassic sandstones, their lithology, exposure and prevailing water quality.   It was very desirable to have standardization of measurement techniques with central coordination of data at a national level.

3.   The diversity of chalk aquifers should be recognized.   For instance, although puddling of a marly lower chalk was an effective way to obtain an impermeable base to a tip, such as had been used at Merstham, this would not be appropriate to upper and middle chalk having a negligible clay content.

4.   In Kent, one potential groundwater pollution problem was due to urea as a de-icing chemical for airfields, that led to abandonment of that method in the outcrop areas of the fissured chalk.

5.   Sources of pollution come broadly into the point source and distributed source categories.   Point sources can arise suddenly and unexpectedly, e.g. the accidental spillage of a liquid container, or may be continuous due to leaching of a solid deposit or leakage from a pipeline with strong dependency on percolation of rainfall.   Sources covering a large area are usually continuous in nature, and ipso facto involve very great volumes of water.

Research needs identified

6.   Water sampling in vadose (unsaturated) zone.

7.   How are the risks of contamination of groundwater to be evaluated?
The discussion here ranged from the attitude of 'one must leave it to the specialist', leaving site to be dealt with on its merits, to a completely structured rule for evaluation protection zones and risks (see addendum to A. Hunter Blair's paper regarding LeGrand's factors p. 65-67).   It was mentioned that for the Department of the Environment (DOE) survey the initial ranking was simply NO RISK/SOME RISK/SERIOUS RISK where the last

category clearly related to dumping or spillage over highly fissured strata, and appeared to include about 3% of the tipping sites in south east England. Hydrogeological criteria for assigning a ranking had not been explicitly formulated for the situation occurring in the UK with fissured sandstones or limestones and caution was expressed lest experience with intergranular deposits should be transferred unwisely to the other types of aquifer.

8.   Given that there was good experience of the migration of micro-organisms in filter beds and other sandy layers, there remained considerable variability of bacterial travel which would be a cause of worry until research had identified the cause.

Administrative comment

9.   The responsibility for avoiding groundwater pollution lay in two camps:
    (i)   at the water supply authorities who would take measures to protect their gathering grounds, and by regulation or enforcement prevent pollution occurring;
    (ii)   at the hand of potential polluters who would by waste treatment, by pre-processing of waste and by proper notification procedures, avoid dangerous situations.   Good liaison between these two groups is essential.

10.   The definition of criteria for landfill or other tipping operation to take place safely has been avoided in the existing British legislation in its application.   One River Authority has been careful to put the onus finally on the person wishing to deposit waste, although taking pains to explain their reasons for potential objections if wastes were to have been placed at the particular site.   A further issue not covered by the Disposal of Poisonous Wastes Act, is the question of who takes responsibility and therefore who would have to respond to any charge of environmental pollution e.g. due to persons having accidental contact with poisonous substances in the pit.

11.   The liaison with planning authorities was a continuing necessity because even if landfill took place in a safe manner, such as not to cause a risk of groundwater pollution, when it had been completed high buildings might be constructed with piles driven right through the sealing layer at the bottom of the landfill.   There also remains the danger of clandestine tipping of industrial wastes.   Stricter administrative controls by industry are now envisaged as recommended by the Institution of Chemical Engineers Working Party.

It was to be hoped that future laws for the prevention of groundwater pollution would also have 'real teeth'.

12.   The Rivers (Prevention of Pollution) Act, 1961 conferred strong powers on River Authorities to restrain potential and actual polluters.

1.   Here it was stated that liquid gases present a serious problem at the surface but because they are so extremely low in temperature they freeze the ground and are not a hazard to the aquifers below.

2.   There is a risk of overkill of an acid pollutant by use of too much alkaline and vice versa.

3.   Many spills were simply remedied by use of large quantities of dilution water.

4.   In the case of an oily spill, the ground was usually excavated, disposal of the excavated oil contaminated soil could be achieved by bacterial action when sub soil is mixed with urban refuse or with farmyard manure.

5.   According to experience in Scandinavia the best approach to countering damage from spills from tankers was to have from the factories concerned a listing of pollutants, their transport system and how the firms themselves cope with pollution incidents on their premises.

6.   Training in emergency procedures is given to fire brigades and will need to be co-ordinated with river authority requirements.

OPERATION OF THE DEPOSIT OF POISONOUS WASTES ACT, 1972

Discussion of this Act - here abbreviated 'DPW Act' - was called for by several participants, being the most recent legislation bearing on potential pollution of groundwater. Principal points made were:

1.   Some counties (e.g. Essex and Hertfordshire) had needed to pass their own legislation before the DPW Act, in order to regulate the quality of wastes accepted by District Councils, who were liable to sell landfill rights to waste disposal contractors.

2.   Applications under the Act involve Essex River Authority in up to 60 requests a week.   About once in 10 days a highly dangerous item is notified.   The burden of responsibility for avoiding groundwater pollution and surface hazards (e.g. to workmen and to the public) has to be put onto the contractor: the River Authority acts to warn of the dangers and to refer back unsuitable proposals.   One consultant warned of applications being resubmitted using alternative descriptions for the same substance.

3.   It was agreed that the Act had worked well in providing information, where little was available before, enabling action to become rational and effective.   All this was additional to the fact that the river authorities were required to give advice to local authorities under the Town & Country Planning Acts, as to where planning consents were acceptable for landfill operation.   An important weakness remained, however, in that no Inspectorate, with specific duties in the groundwater pollution area, had been set up.

4.   Surveillance of underground migration and dispersion was especially necessary for radioactive wastes and toxic metals.   Of the latter, electroplating residues containing chromium and cadmium and shotblasting residues containing lead were noted.

5.   A chemical engineer warned of the wastes that can arise from dismantled pilot plant, also from dry cleaning fluids.   By due insistence on the part of factory management, oily wastes could be cut by two-thirds quite readily.

6.   Waste incineration and waste reclamation are both contributors to reduction of waste bulk, thereby saving over-rapid fill of the safest sites.   A parallel danger was recognized in too fierce an application of the DPW Act, which would lead to undue deflection of wastes to the best tipping sites now, so causing problems for the next generation.

7.   Several speakers urged the need for a contractor's guide as to waste hazards: the Institution of Chemical Engineers Draft Code of Practice had been usefully specific on administrative practice, but the technology of waste disposal needed to be recognized as a process in itself, for which, as yet, there was no textbook procedure for avoiding groundwater pollution.

8.   The urgency of having technical guidelines for disposal contractors was stressed. These people had to cope with a giant problem which in the UK included 5 million cubic metres of liquid wastes per annum.   The disposal of such quantities could not be indefinitely delayed.

9.   The general philosophy of groundwater protection was questioned.   Was there any special privilege that underground strata could claim in being shielded from contamination? Any implied negative to this was swiftly countered by the water supply constraint of striving 'to keep decent water decent', i.e. not knowingly admitting a contamination whose effects on public health were unknown.   The irreversible nature of certain types of aquifer pollution, or the long time required to purge it, were further reasons to tread cautiously.   Only in aquifers never intended for public water supply, nor connected to those in such use, was it permissible to reckon the economic balance between deliberately causing surface or underground pollution.

10.   Several speakers made the point that groundwater pollution control will cost money and that a prompt expenditure in investigating the present problems would be recouped in terms of more efficient control and surveillance routines in future years.

HYDROGEOLOGICAL FACTORS IN GROUNDWATER POLLUTION

by A. Hunter Blair
The Water Research Association, Medmenham, Bucks, England

1. INTRODUCTION

Pollution of groundwater was defined by Walker (1) as an impairment of water quality by chemicals, heat or bacteria to a degree that does not necessarily create an actual public health hazard, but does adversely affect such waters for normal domestic, farm, municipal or industrial use. The term contamination denotes impairment of water quality by chemical or bacterial pollution to a degree that creates an actual health hazard to public health through poisoning or spread of disease.

Whilst groundwater is less vulnerable to pollution than surface water, the consequences of groundwater pollution last far longer than those of surface water pollution. This is because the time required to flush an aquifer is of the order of months or even years compared to a few days required to flush a river. Often groundwater pollution appears belatedly and is not recognized until a considerable extent of an aquifer has become polluted or contaminated. Once pollution has occurred, it may be very difficult and even economically unfeasible to reclaim the aquifer even when the source of pollution has been removed.

2. SOURCES OF GROUNDWATER POLLUTION

Pollution can originate from point or distributed sources within the recharge area of an aquifer. Generally the risk of groundwater pollution is highest in urban areas where large volumes of wastes are concentrated into relatively small areas. The risk is further increased where they are located on areas of permeable surface deposits or on highly creviced or fissured areas. By listing ways of groundwater pollution in terms of individual problems the outcome may only be the recognition of the more dramatic occurrences of pollution. Caution must be exercised when considering the "case history" approach which may only require short-term and local studies. This approach is inadequate for a study of broad regional problems occurring over large areas where a continuing slow deterioration of groundwater quality is proceeding. Groundwater pollution must therefore fall into two categories, first individual, short-term and local pollution, and secondly a continuing slow decline in groundwater quality over large areas requiring long-term and extensive investigations. This latter case is probably the more serious since the causes of pollution may be already well advanced by the time the effect is observed.

2.1. Contaminating sources at or essentially at the ground surface

Pollution of groundwater at or essentially at the ground surface can occur in the following ways. Some typical examples are shown in Figs. 4 to 13 (Figs. 4 to 11 from Deutsch (2) and Figs. 12 and 13 from Brown (3)).

(a) Individual sewage disposal systems such as septic tanks, cesspits, bored hole or pit latrines and other privy waste systems (Fig. 4). The potential for pollution from these sources is limited to areas not yet served by mains drainage, estimated in 1970 to be about 6% of the population of England and Wales (4).

(b) Refuse disposal sites such as landfills, open tips and dumps (Figs. 5 and 6). Due to convenience and the economic advantages of land recovery, abandoned gravel pits and quarries are often used as disposal sites. Rainfall infiltration, saturation of the refuse by groundwater, surface water flooding and surface water runoff will all produce potential leachates. Leachates can also be produced by decay and decomposition of vegetable matter and from the natural moisture content of the site.

Untreated refuse should not be tipped on sites where the percolate from it could pass direct through fissures to an aquifer without being effectively filtered (Fig. 7). Groundwater can be protected from pollution by a sensible choice of sites and where necessary the provision of protective measures (5) (6). These include for example:

  i)   waterproofing the base of a tip prior to refuse disposal,

  ii) collection of the percolate and its disposal into a foul sewer or into a stream or the ground following purification (such as has been carried out at Merstham, Surrey under the operation of the Greater London Council),

  iii) rapid tipping of successive layers of refuse in order to utilize its capacity to absorb rainfall,

  iv) covering the final layer with an earth cover and a cross fall to produce ready runoff of rainfall,

  v)   seeding of the cover material,

vi) provision of about 1 m depth of aerobic media such as gravel
or clinker at the base of the tip to purify the leachate
(see also Sect. 2.1. f),

vii) interception and diversion of all surface and groundwater before
it reaches the tip, and

viii) effective stabilization of the water table around the tip.
Apgar and Langmuir (7) show how highly contaminated leachates
from landfills can penetrate depths greater than 15 m, and that
groundwater pollution occurs by the leachate channelling down
fractures in bedrock or by the subsequent infiltration of surface
runoff contaminated by leachate.

(c)   Settling basins for disposal of sewage and liquid waste products.   In
industrialized regions the use of lagoons or basins for the disposal of
chemical and industrial wastes may be widespread and the use of the
oxidation basin for sewage treatment is frequently practised (8).   Where
such disposal lagoons are situated in a hydrogeologic environment which
readily permits infiltration, large quantities of these pollutants may
reach the groundwater.

(d)   Sewage and industrial wastes as discharged into water courses.   Large
quantities of industrial and sewage wastes are disposed of into streams
and rivers, the degree of dilution in dry periods being governed solely
by the base flow of that river.   Evidence has also been obtained (5) of
pollution by refuse tip percolates which either flow into them as surface
runoff, from discharges of porous land drains or from a leaky culvert
carrying a watercourse under a tip.   During prolonged dry periods the
base flow becomes minimal causing reduced dilution;   indeed the base
flow may cease completely and the river become influent to the aquifer.
In areas where groundwater is abstracted from aquifers in hydraulic
contact with the river bed, induced infiltration of the river water can
occur resulting from the establishment of influent conditions, and
polluted water can therefore be drawn into the aquifer from surface
streams (Fig. 8).   In areas where groundwater is obtained by induced

infiltration it would be necessary to stop the discharge of wastes within the reach of influent seepage caused by abstraction and to ensure at all times that the base flow is adequate to provide the required dilution.

(e) Agricultural sources. Pollution from agricultural sources can result from fertilizers, pesticides, cattle feed lots and cattle manure and wastes. Present groundwater pollution attributed to these causes is, for the most part, below the levels that have been demonstrated to cause disease, high reclamation cost or aesthetic nuisance. However in the future establishment of large agricultural operations the possibility of groundwater pollution should be thoroughly investigated. Under intensified farming large quantities of manure must be disposed of, and correlations have been noted between the number of livestock in the area and the nitrate concentration in the groundwater (9). Whilst no correlation was noted between the amount of fertilizer applied and the nitrate content of the groundwater, the Illinois Pollution Control Board have put together a detailed set of regulations concerning the rate and method of application of fertilizers and animal manure (10). For example for corn, the maximum nitrogen (as N) applied per hectare per crop is 180 kg/hectare to 250 kg/hectare (160 to 225 lb/acre).

(f) Artificial groundwater recharge as practised by surface spreading methods, must be considered as a potential form of groundwater pollution, with the recognition that a surface source of water is being deliberately introduced into the aquifer. Water sources include river waters which may be highly polluted, (for example the River Maun is recharged into the Bunter Sandstone,) also reclaimed sewage effluent and reclaimed industrial effluent. In this situation it is essential to ensure that the recharge water, subsequent to any required pretreatment and/or passage through filter media, is compatible with the aquifer media and groundwater already present. There should be no non-steady state quality changes occurring as a result of recharge in the aquifer, on the supposition that if there are, the aquifer may become blocked. It would be admissible to allow a quality change purely by the degree of dilution

afforded by the natural groundwater. Studies of water quality changes
that occur during recharge have been made for example at Flushing
Meadows, Arizona (11) where a treated sewage effluent is recharged,
and at Dortmund, Germany (12) where the River Ruhr is recharged.
Pilot scale studies of artificial groundwater recharge at Medmenham (13)
have demonstrated seasonal variations of the quality of a recharge water
entering the aquifer. During the winter months, with low water
temperatures, the water entering an aquifer, after passing through a filter
medium was aerobic and the nitrification and denitrification processes
were not observed. With increasing water temperatures the recharge
water tended to become anaerobic and the nitrification processes
commenced, followed at warmer temperatures by the denitrification
processes. During the course of a year therefore different qualities of
water would enter the aquifer, i.e. a non-steady state situation which
should be avoided.

(g)   Urban runoff resulting in subsequent infiltration from paved areas, towns,
roads and motorways. Pollution can occur, for example, from both the
application of road salt during severe winter conditions and from
unprotected salt stockpiles (Fig. 9). Consequent on a major thaw, the
salt may drain underground and cause a salt water problem some weeks
or even months after the thaw. This has already caused considerable
problems from motorway runoff, for example pollution by salt of
underground water in the Bedfordshire area. Whilst this pollution is of
a widespread nature, more severe and localized salt water pollution can
occur from runoff from any large unprotected salt storage piles. Appeals
have been made in the United States to prevent the use of salt on certain
roads (14). Soakaways from roads and motorways can provide a ready
access of pollutants resulting from oil and petrol spills on the road
surfaces and from accidents. In many cases it may be too late to prevent
the contents of a crashed tanker from entering such soakaways. It has
become necessary to provide legislation in the Dortmund City Waterworks
area, to ban tankers from the area of the waterworks.

(h)   Miscellaneous sources tend to be local problems and can be caused by,
for example, spillage of polluting materials, leakage from storage tanks,
leakage from pipelines, storage of unprotected chemical products
producing either a leachate or polluted runoff, etc. These sources are
not likely to result in problems of regional significance.

(i)   Acid water sources. Pollution by acid mine drainage water is described
in (15) and in the references quoted in that source. Water with a total
iron concentration in excess of 500 mg/litre and a pH level below 3 can
be pumped from coal mines at rates in excess of $0.05 \text{ m}^3/\text{sec}$.
Treatment of the iron and acidity generally occurs outside the mine in
lagoons and requires both oxidation and precipitation of the iron content
and neutralization of the acidity. To control pollution by this water,
mine sealing, refuse burying and groundwater diversion have all been
used with varying success (16).

2.2. Subsurface contaminating sources

The second major division of sources of contaminants is that introduced below the
aquifer surface either due to deliberate practice or induced by aquifer use.

(a)   Subsurface transportation and/or storage can occur both naturally and
artificially. For example an artificial recharge installation stores water
in an aquifer and can use the aquifer as a means of transport to the
abstraction point. Alternatively substances such as petroleum products
can be stored in underground tanks with subsequent abstraction via
pipelines. Steel tanks are affected by internal and external corrosion
and by the splitting of welded seams. In the case of tanks with a large
turnover the leak may not be noticed for some time. The quantity lost
depends on the size of opening, pressure and viscosity. For example a
leak of fuel oil from a hole 2.5 mm diameter under a pressure of 1 kg/cm^2
results in a loss of 25 000 litre/month (17). The number of pipeline
failures is low, but serious. They are generally caused not by pipeline
failures, but by carelessness on the part of others, e.g. during digging
operations. Dietz (18) divides the process of groundwater pollution by oil
into five phases:

    i)   flow of the oil phase,
    ii)   transfer of oil components into an aqueous solution,
    iii) transport of dissolved oil components with the water,
    iv)   underground oxidation of the oil phase,
    v)   anaerobic biochemical reactions of the oil components dissolved
          in the groundwater.

He concludes that pollution by oil spills is governed by a complicated set of
transport mechanisms and chemical reactions, and that an exact description
of the combined effects is neither necessary nor possible.

(b) <u>Subsurface waste disposal.</u>   Drainage wells provided to remove unwanted surface water provide ready access routes for groundwater pollution.   The subsurface waste disposal by deep well injection is becoming increasingly common, and it is questionable whether one is considering waste disposal or purely waste storage.   The hydrogeology of waste disposal systems is of paramount importance for successful waste disposal.   The formation must have sufficient porosity, permeability and areal extent to accept the wastes, should be below the freshwater level and confined horizontally by impermeable rocks to prevent pollution of fresh water aquifers and natural resources. Construction of the boreholes is also of great importance, for example correct lining of the borehole and suitable choice of material with regard to the injected waste material.   Failure of either hydrogeological or engineering aspects could lead to direct contamination of groundwater resources.

(c) <u>Interaquifer movement</u> can take place under natural conditions.   Polluted water can move downwards, or brackish water move upwards through leaky confining layers or through wells penetrating different aquifers. The main routes are through wells, for example unplugged and uncased test boreholes, or improperly or inadequately plugged and abandoned wells.   See also Sect. 2.2. (e).   The basic concept behind the proper sealing of any abandoned well is that of restoration as far as possible of the controlling geological conditions that existed before the well was drilled.   Abandoned wells should never be used for disposal of sewage or other wastes (Figs. 10 and 11).

(d) <u>Aquifer use</u> can cause lateral and vertical interaquifer movement.   This can be caused by boreholes producing a reversal of natural gradients and diverting polluted water into the aquifer.   It can also produce pressure reductions around pumping wells which causes encroachment upwards for example of saline or connate water.   Aquifer use can also cause salt water encroachment either by reducing the seaward flow of fresh water to such an extent that salt water encroaches landwards, or by causing "upconing" of an intruded salt water wedge underlying a fresh water aquifer (Fig. 12).   The subject of salt water encroachment was amply covered in the Haifa symposium (19).

(e)   <u>Borehole constructional and development methods.</u>   Dependent upon drilling methods, drilling fluids are frequently employed to assist the drilling operation.   During borehole development chemicals such as sodium hexametaphosphate, calcium hypochlorite and concentrated hydrochloric acid are frequently introduced directly into the borehole, and in the case of fissured media can travel considerable distances from the borehole.   The US Department of Health lists basic considerations for the protection of groundwater pollution by wells (20).

   i)   the annular space outside the well casing should be filled with a watertight cement grout from the surface to a minimum of 3.3 m below the ground surface.   This casing should be surrounded by a 12 cm thick concrete slab extending at least 60 cm in all directions,

   ii)   for artesian aquifers the casing should be sealed into the overlying impermeable layers so as to retain the artesian pressure,

   iii)   when a formation containing poor quality water is encountered, this formation should be sealed off,

   iv)   a well seal must be installed at the top of the well casing to prevent access of contaminated water,

   v)   all wells should be disinfected after construction or repair.

(f)   <u>Artificial groundwater recharge</u> by borehole injection.   As with surface methods of artificial recharge, it is essential that the water should be qualitatively compatible with the aquifer and groundwater already present.   However in the case of borehole recharge the water is injected directly into the aquifer and there is no filtering action (Fig. 13). In the majority of cases of recharge it would therefore be necessary to use either a water which has been pumped from a similar aquifer some distance away or to use at least a potable water.   Bize (21) conveniently summarizes processes of biological, chemical and physical clogging due to incompatability of the recharged water.

2.3. Distances to sources of contamination

Too many factors affect the determination of the safe distance between a groundwater supply and a source of contamination to enable a fixed distance to be set. Such factors include for example (20)(22)

a)   character of local geology - slope of ground surface,

b)   nature of topsoil and underlying porous strata,

c)   thickness of water-bearing strata, depth to water table, slope of water table, seasonal fluctuations of water table and direction and rate of movement of underground flow,

d)   the extent of the surface catchment area,

e)   well construction, screening, casing and depth, etc.

As a general rule the distance should be the maximum that economics, land ownership, geology and topography will allow. The US Department of Health recommends the following minimum distances (20).

TABLE 2
MINIMUM DISTANCES TO CONTAMINATION SOURCES

Contamination source	Distance to borehole (m)
Building sewer	15
Septic tank	15
Disposal field	30
Seepage pit	30
Dry well	15
Cesspit	45

The most important measure taken for the prevention of groundwater pollution is the establishment of protection zones. Within these zones certain activities are prohibited or restricted. These activities include building construction, establishment of cemetries or camping areas, the execution of drillings, excavations and the storing of gas and motor fuels, fertilizers, gravel, coal, refuse, waste matter or solid or liquid substances capable of polluting groundwaters (22)(23).

3. THE HYDROGEOLOGY OF GROUNDWATER POLLUTION

The hydrogeology of a given situation exercises a dominant control over the occurrence and movement of groundwater and so the hydrogeology must determine to a very considerable extent what happens to any pollutant that may be introduced into the groundwater regime. The hydrogeologic factors which control groundwater movement will control the movement of the pollutant since this pollutant is subject to the same physical laws that control groundwater movement.

Generally a polluted water moves downward through a zone of aeration to a saturated zone and then laterally to points of discharge. The natural direction and rate of such movements are generally dependent upon the geology, although artificial controls can alter the rate and direction.

3.1. Access to the aquifer

Surface and subsurface conditions that have created large catchments for underground water supply can just as easily create catchments for pollution. The pollution can enter the aquifer by percolation through a zone of aeration, by infiltration and migration in the zone of saturation, by vertical interaquifer leakage or by direct injection. Its movement and entry is influenced by a large number of geological conditions. (Fig. 14)(Fig. 15).

(a) Interstices, either primary dating back to the formation of the media, or secondary caused by joints, fissures, solution passages etc. The openings range in size from minute pores via fissures to, for example, limestone caverns.

(b) Geological movements such as dips, tilting, folding and warping, joints, faults. Fault zones all have a marked influence on the water movement. Stratigraphy, not only in the sense of the sequence between different formations, but also stratification within an individual formation can be helpful or harmful in the problems of groundwater contamination. Sedimentary formations often exhibit gradational changes, both horizontal and vertical, in the size and character of material deposited. Such changes determine the permeability which affects the occurrence and movement of water through the media. Especially important is the fact that permeability in the vertical direction is commonly much less than in the horizontal direction. This must be recognized when comparing the predominantly vertical fluid movement in the unsaturated zone and the horizontal movement in the saturated zone.

(c)   The topography of the land surface has an important influence. Generally groundwater table levels follow the land surface but tend to iron out the irregularities. For example in flat lying land bordered by higher land and not having an outlet, groundwater circulation will be slow. Under these conditions highly mineralized water can result. In hilly country water can percolate rapidly downwards and be discharged at lower elevations at springs etc. A pollutant introduced in the former case may remain for a long time, whilst in the latter case it may be rapidly flushed from the aquifer.

Shallow aquifers are those most susceptible to contamination since they may be either in direct contact with surface sources of pollution or because there is little or no natural treatment afforded by the overlying strata. The risk of pollution of deeply buried aquifers is less, but if dispersion and dilution of inorganic and non-degradable pollutants is insufficient, protection has to be provided. Confined aquifers are comparatively the safest since they are effectively sealed by impermeable layers.

Infiltration is the process by which a fluid is introduced into a medium through the medium surface, or through relatively shallow access points. It is caused by rainfall, irrigation, liquid waste disposal, groundwater recharge and by seepage and leakage from streams and rivers. A rigorous mathematical treatment of infiltration is beyond the scope of this paper since it is adequately covered in the literature (24)(25)(26)(27)(28)(29)(30).

If the intensity of fluid application is at all times lower than the saturated hydraulic conductivity of the medium, the medium will absorb the fluid without ever reaching saturation, and the wetted profile will attain a moisture content for which the conductivity is equal to the fluid application rate. The lower the rate, the lower the degree of saturation. If the intensity of application is less than the initial infiltrability of the medium but greater than the final value, the medium will first absorb the fluid at less than its potential rate and the flow will be unsaturated. However as the infiltrability decreases the medium surface will become saturated and infiltration will occur as under ponded conditions. If the intensity of application is greater than the soil infiltrability, the infiltration is also as for ponded applications. When the relation between the moisture content and the hydraulic conductivity is known the precise course of the moisture profile can be calculated.

Typical values of infiltration rate are given in Table 3. These values only state the order of magnitude.

TABLE 3

TYPICAL INFILTRATION RATES (24)

Soil type	Final infiltration rate (mm/hr)
Sands	> 20
Sandy and silty soil	10 - 20
Loams	5 - 10
Clayey soils	1 - 5
Clayey soils with sodium compounds	< 1

Infiltration into layered media in the case of a coarse layer overlying a finer textured layer is first controlled by the coarse layer but when the wetting front penetrates into the finer layer, the infiltration rate drops and tends towards that in the finer soil alone. Therefore in the long run it is the layer of lowest conductivity which controls the flow. If the infiltration continues for long enough positive heads develop above the boundary with the finer layer. In the opposite situation the initial infiltration rate is determined by the upper layer. However as the front reaches the coarser layer the infiltration rate can decrease. This is because water at the wetting front is normally under tension and this tension may be too high to permit entry into the relatively larger pores of the coarse medium. The lower layer can never become saturated since the restricted flow-rate through the less permeable layer cannot sustain the flow at the saturated hydraulic conductivity of the lower coarse layer (except in cases where the externally applied head is high enough, e.g. under deep ponding on the surface).

3.2. Dispersion

The situations described in Sect. 3.1. should be further enlarged to consider diffusion and dispersion of the pollutant.

Substances dissolved in a medium solution can move by molecular diffusion due to concentration gradients within the solution or by convection due to the mass flow of the solution. The process of diffusion plus convection results in a mixing process known as hydrodynamic dispersion.

As the flow in an aquifer goes through a network of interconnected pores or joints, the different paths in the network being traversed at various velocities, the aquifer causes a mechanical dispersion of any localized concentration of solutes.

Fried and Combarnous (31) describe the mathematical and physical principles of mechanical dispersion in porous media, distinguishing the tracer case, where a solute makes negligible addition to liquid density or viscosity, and the general case, where these effects are important.

Bear (32) has studied dispersion and shows how dispersion of a small volume of a pure tracer can be investigated experimentally.

3.3. Pollution travel in the saturated zone

The whole aspect of groundwater pollution must ultimately focus on the nature of fluid movement within a saturated zone. It is not intended to describe the analysis of saturated groundwater flow since it is amply covered in the literature already quoted. Simple calculations can be performed on the basis of the Darcy law and all its implicit assumptions, for example the case of a soluble pollutant entering the groundwater and migrating towards a pumped well. With assumptions of uniform depth and permeability of the aquifer, Fig. 16 shows the situation with a limiting stream-line flow, outside which any pollutant entering the groundwater flow would not reach the pumped well. The limiting breadth of the zone of flow is given by the formula

$$d = \frac{Q}{kbi} \qquad \ldots\ldots\ldots\ldots\ldots\ldots\ldots\ldots 1$$

where Q = discharge rate of the well

k = permeability of the aquifer

b = aquifer thickness

i = natural gradient of water table.

By integrating along the flow paths, times of travel of the pollutant along various stream-line paths can be obtained and are illustrated for a particular case. In the more general case where aquifer properties are not uniform and the natural flow system is modified by abstraction or recharge systems, one could resort to an analogue or digital model solution of the groundwater flow pattern. Integration along the stream lines will give travel times. These approaches can also be extended to derive concentration versus time graphs for pollutants reaching wells from refuse tips and other sources.

3.4. Hydrological aids

To assist the hydrogeologist in the evaluation of pollution hazards, he should have available information on

(a) direction and amount of surface runoff,

(b) rainfall and snowmelt, with any limits which these may encounter,

(c) amount of evaporation and transpiration,

(d) definition of the media and their geology,

(e) subsurface hydraulic properties,

(f) natural and artificial controls exerted upon aquifers,

(g) water table position, fluctuations, and water table map,

(h) bedrock configuration.

To assist the retrieval of such information, use should be made of existing hydrogeological surveys, pumping test analyses, borehole logging information, geophysical surveys, rainfall and evaporation details, river flow records etc. Given this information the hydrogeologist is able to design analogue models or digital models to examine the groundwater flow system, either as a design to, or backed up with, tracer experiments in the field to determine the groundwater movement and flow pattern, with the ultimate aim of assessing whether or not a given environmental situation affords adequate protection against sources of pollution.

In summary it can therefore be seen that predictions of where and how a polluted water may travel from its source can be based on relatively simple information, however if difficulties are envisaged specialist opinion may be sought regarding the physical characteristics, location and extent of all the pervious and impervious materials in the saturated and unsaturated zones.

4. THE QUESTION OF RESPONSIBILITY

4.1. The Engineer

With few exceptions, e.g. natural inter-aquifer movement, all the cases of groundwater pollution discussed in Sects. 2.1. and 2.2. are man-made. In the first instance therefore it must be the engineer's duty to consult with the hydrogeologist in all cases where engineering practice may result in groundwater pollution. If the natural environmental conditions do not provide adequate protection against groundwater pollution, the engineer must provide adequate protection against environmental hazards to the satisfaction of the hydrogeologist, chemist and biologist.

4.2. The Hydrogeologist

It is the responsibility of the hydrogeologist in consultation with the biologist and chemist to establish whether or not the natural environmental conditions provide adequate protection of aquifers and groundwaters against sources of pollution. It is also his responsibility to evaluate the hydrogeologic environment in terms of potential pollution from land and water use practices within the area. These criteria should serve as a basis for legislative and engineering specifications that would provide protection against groundwater pollution where natural safeguards are lacking. In this way the risk of the occurrence of groundwater pollution will be minimized.

Once groundwater pollution has occurred, it is once again the responsibility of the hydrogeologist to define clearly the pattern of groundwater movement in the affected area. Legislative steps may then be taken to eliminate the source of pollution and to protect water supply wells against pollutants which are already in the aquifer and groundwater. Indeed legislative steps have already been taken by the introduction of the Disposal of Poisonous Wastes Act (23) which creates an offence of depositing poisonous waste on land so as to give rise to an environmental hazard, and provides administrative machinery involving the service of notices in connection with removal and deposit of such waste.

4.3. The Chemist

The chemical composition of groundwater is strongly influenced by events that happened to the water before infiltration. Once the water has become groundwater it is subject to a fairly stable set of conditions. The equilibrium conditions that exist underground can be examined by the use of well known and established chemical principles. The effects of a polluted water arriving in an environment in which there is a natural equilibrium can be predicted and are better understood, if all the chemical factors involved, such as adsorption or desorption, solution or deposition and oxidation or reduction, are considered by the chemist.

Therefore if the hydrogeologist can provide the chemist with detailed geological information, the chemist can predict the likely qualitative effects of the introduction of a polluted or foreign water into the groundwater system.

4.4. The Biologist

The role of the biologist may indeed be more important than that of the chemist since many of the quality changes that occur within an aquifer are biologically-induced quality changes, for example the decay of organic matter due to bacterial action, or biologically-induced chemical reactions such as nitrification, denitrification and sulphate reduction.

More important however is the removal of bacteria and viruses during the movement of the pollutant. It must be the duty of the biologist to ensure that the aquifer remains bacteriologically and virally safe.

5. CONCLUSIONS

The paper has raised a number of fundamental issues which are summarized in the following questions:

**** Can groundwater pollution be studied by fundamental existing hydrogeological, biological and chemical techniques?

**** Are these basic concepts readily available to, and understood by those investigating groundwater pollution?

**** What extra knowledge must be made available to assist groundwater pollution studies?

**** Is groundwater pollution national or local?

**** Should the future emphasis on groundwater pollution studies concentrate on local case histories or gradual long-term effects?

**** Who should take the ultimate responsibility for groundwater pollution?

**** Is legislation necessary and, if so, how is it to be enacted?

**** Are we too late?

6. ADDENDUM

Refer to pp. 62-66.

REFERENCES

1. WALKER, W.H.

   Illinois Groundwater Pollution.
   J. Am. Wat. Wks Ass., 1969, 61, (1),
   pp. 31-40.

2. DEUTSCH, M.

   Natural controls involved in shallow
   aquifer contamination.
   Ground Wat., 1965, 3 (3), pp. 37-41.

3. BROWN, R.H.

   Hydrologic factors pertinent to
   groundwater contamination.
   Proceedings of Symposium on
   Groundwater Contamination, Cincinnati,
   Robert A. Taft Sanitary Engineering
   Center, Tech. Rep. W61-5, 1961, pp. 7-16.

4. HOUSING AND LOCAL GOVERNMENT,
   Ministry of

   Working party on sewage disposal.
   London, HMSO, 1970.

5. DEPARTMENT OF THE
   ENVIRONMENT

   Refuse Disposal. Report of the working
   party on refuse disposal.
   London, HMSO, 1971, 199p.

6. SALVATO, J.A., et al

   Sanitary landfill leaching: prevention
   and control.
   J. Wat. Pollut. Control Fed., 1971, 43,
   (10), pp. 2084-99.

7. APGAR, A., and LANGMUIR, D.

   Ground water pollution potential of a
   landfill above the water table.
   Ground Wat., 6, (9), 1971, pp. 76-96.

8. HACKETT, J.E.

   Groundwater contamination in an urban
   environment.
   Ground Wat., 1965, 3, (3), pp. 27-30.

9. LOEHR, R.C.

   Drainage and pollution from beef-cattle
   feedlots.
   Proc. Am. Soc. civ. Engrs, J. Sanit.
   Engng Div., 1970, 96, (SA6),
   pp. 1295-1311.

10. CHEMICAL AND ENGINEERING NEWS

    Chem. Engng. News, 1972, (1),
    pp. 17-18.

11. BOUWER, H.

    Water quality aspects of intermittent
    infiltration systems using secondary
    sewage effluent.
    Proceedings of Artificial Groundwater
    Recharge Conference, Reading, 1970,
    Medmenham, The Association, 1972,
    2 vol.

12. FRANK, W.H.

    Fundamental variations in the water
    quality with percolation in infiltration
    basins.
    Proceedings of Artificial Groundwater
    Recharge Conference, Reading, 1970,
    Medmenham, The Association, 1972, 2 vol.

13. HUNTER BLAIR, A.

The present position of artificial ground-water recharge research in the United Kingdom.
2nd National Conference on Artificial Recharge. Gottwaldov, Czechoslovakia, 1972.

14. SAINES, M.

News Notes.
Ground Wat., 1971, 9 (4), pp. 47-52.

15. WALSH, F., and MITCHELL, R.

Biological control of acid mine pollution.
J. Wat. Pollut. Control Fed., 1972, 44, (5), pp. 763-768.

16. MERRITT, G.C., and EMERICH, G.H.

The need for a hydrogeologic evaluation in a mine drainage abatement programme.
Proceedings of Third Symposium on Coal Mine Drainage Research, Pittsburgh, Pa., 1970.

17. GLEBIN, V.E.

Advanced techniques and methods of protection against the failure of storage tanks, pipelines and means of transport for crude oil and oil products. Proc. Seminar on protection of ground and surface water from crude oil and oil products, Economic Commission for Europe (Geneva 1-5 Dec 1969), New York, United Nations, 1970.

18. DIETZ, D.N.

Pollution of permeable strata by oil components in water pollution by oil.
Proceedings of Aviemore Seminar, May 1970.
London, Inst. of Petroleum, 1971.

19. SCHMORAK, S., et al

Salt Water Encroachment.
Proceedings of the Symposium of Haifa, Gentbrugge, 1967. Int. Ass. Sci. Hydrol. Publ No. 72, 1967.

20. US DEPARTMENT OF HEALTH, EDUCATION AND WELFARE

Manual of individual water supply systems.
Public Health Service Publ. No. 4, 1962.

21. BIZE, J., et al

Artificial Recharge of underground waters (in French).
Masson et Cie, Paris, 1972, 200p.

22. FOOD AND AGRICULTURE ORGANISATION

Groundwater legislation in Europe.
Rome, 1964, FAO Legislative Series No. 5.

23. GREAT BRITAIN: Statutes

Disposal of Poisonous Wastes Act.
London, HMSO, 1972.

24. HILLEL, D.

Soil and water; physical principles and processes.
New York, Academic Press, 1971.

25. CHILDS, E. C.

The physical basis of soil water phenomena.
London, Wiley, 1969.

26. SHEIDEGGER, A. E.

The physic of flow through porous media.
University of Toronto Press, Toronto, 1957.

27. BEAR, J., et al

Physical principles of water percolation and seepage.
Paris, UNESCO, 1968.

28. YOUNGS, E. G.

Two and three dimensional infiltration-Seepage from irrigation channels and infiltrometer rings.
J. Hydrol., 1972, 15, pp. 301-315.

29. HUISMAN, L.

The Hydraulics of Artificial Recharge.
Proceedings Artificial Groundwater Recharge Conference, University of Reading, 1970.
Medmenham, The Association, 1972, 2 vol.

30. MIKELS, F. C., and KLAER, F. H.

The application of groundwater hydraulics to the development of water supplies by induced infiltration.
Gentbrugge, Int. Ass. Sci. Hydrol., 1956, Publ. No. 4, pp. 232-242.

31. FRIED, J. J., and COMBARNOUS, M. A.

Dispersion in Porous Media.
Advances in Hydroscience, 1971, 7, pp. 169-284.

32. BEAR, J.

Some experiments in dispersion.
J. geophys Res., 1961, 66, (8), pp. 2455-2467.

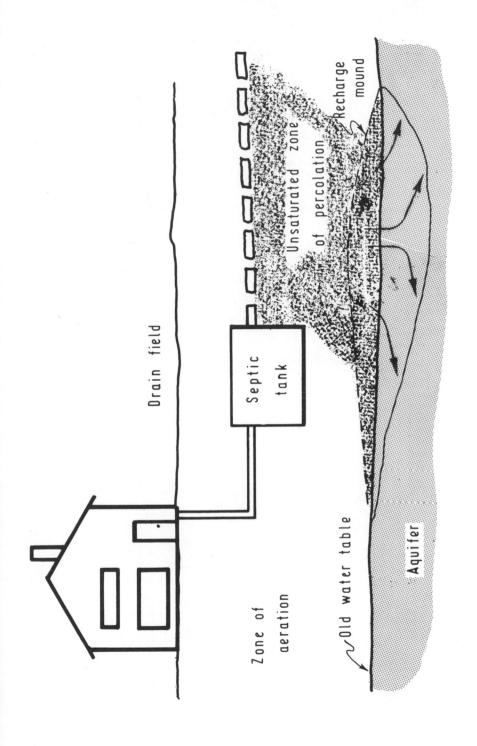

FIG. 4. PERCOLATION THROUGH ZONE OF AERATION. MOST OF THE
NATURAL REMOVAL OR DEGRADATION PROCESSES FUNCTION
UNDER THESE CONDITIONS (AFTER DEUTSCH)

49.

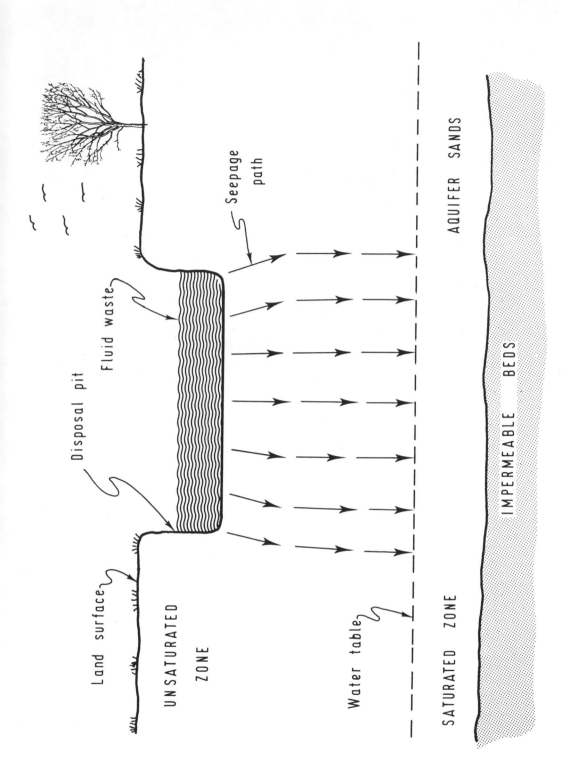

FIG. 5. CROSS SECTION OF DISPOSAL PIT, SHOWING SEEPAGE THROUGH AN IDEALISED UNSATURATED ZONE (AFTER DEUTSCH)

50.

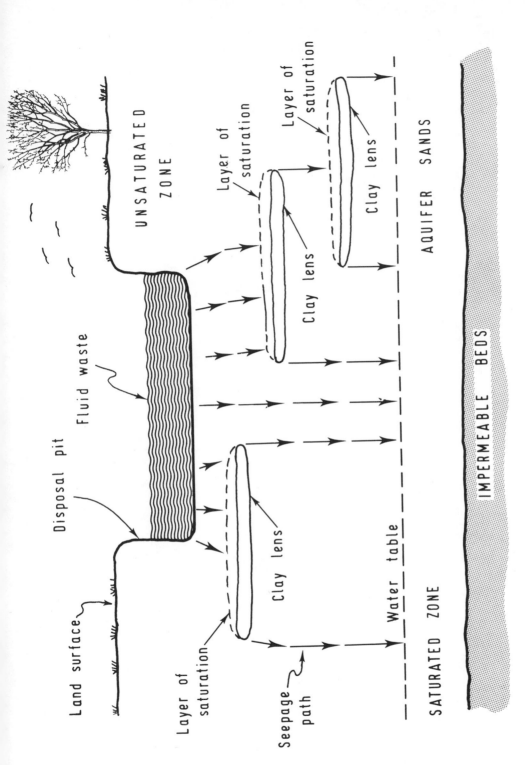

FIG. 6. CROSS SECTION OF DISPOSAL PIT, SHOWING SEEPAGE THROUGH AN UNSATURATED ZONE OCCUPIED BY CLAY LENSES (AFTER DEUTSCH)

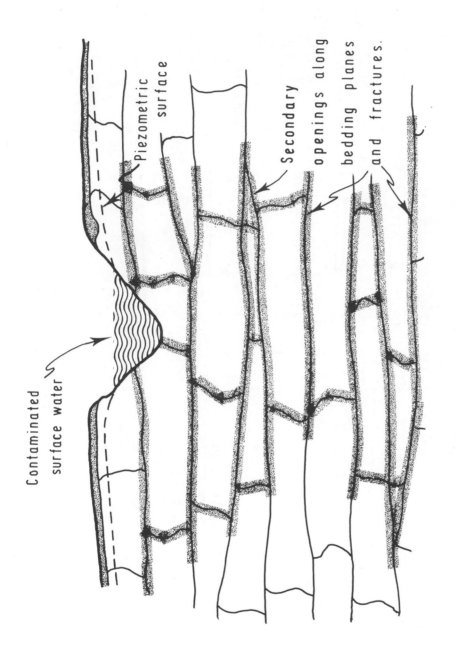

FIG. 7. DIRECT INJECTION INTO LIMESTONE AND DOLOMITE AQUIFERS. SEPTIC TANK EFFLUENTS CAN ENTER SUCH AQUIFERS IN THE SAME MANNER (AFTER DEUTSCH)

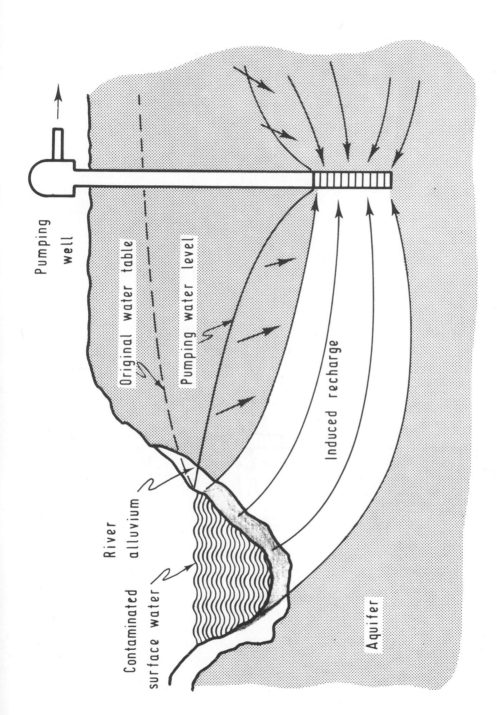

FIG. 8. INDUCED INFILTRATION FROM STREAM. STREAM POLLUTION CAN RESULT IN 'GROUNDWATER' POLLUTION. NATURAL PURIFICATION IS LIMITED TO THOSE PROCESSES REQUIRING NO OXYGEN (AFTER DEUTSCH)

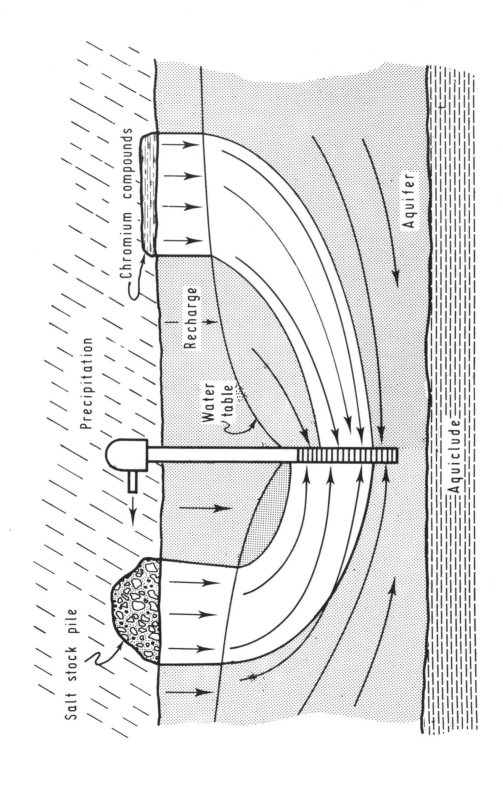

FIG. 9. LEACHING OF SOLIDS AT THE LAND SURFACE. THE POSSIBILITY OF GROUNDWATER CONTAMINATION UNDER THESE CONDITIONS IS RARELY ANTICIPATED (AFTER DEUTSCH)

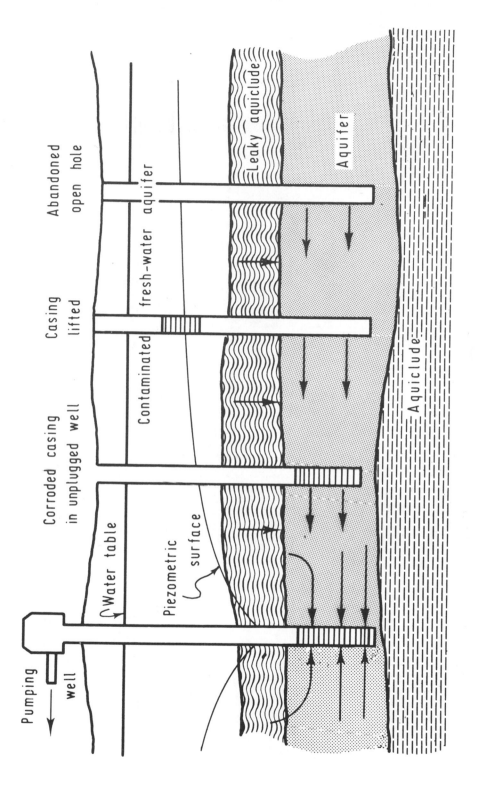

FIG. 10.   DOWNWARD LEAKAGE.   CONTAMINATION OF ONE AQUIFER CAN AFFECT OTHERS IN A MULTI-AQUIFER SYSTEM (AFTER DEUTSCH)

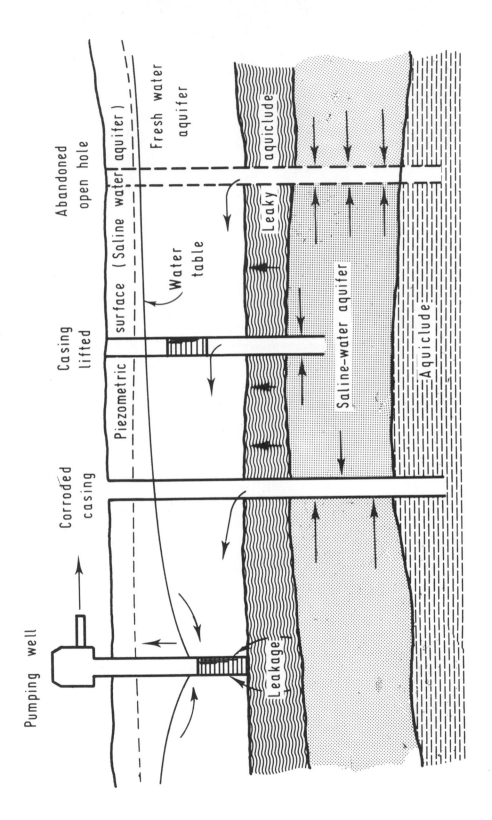

FIG. 11. UPWARD LEAKAGE AND FLOW THROUGH OPEN HOLES. SOME
IMPORTANT AQUIFERS HAVE BEEN RUINED BY IMPROPER
DRILLING PRACTICES (AFTER DEUTSCH)

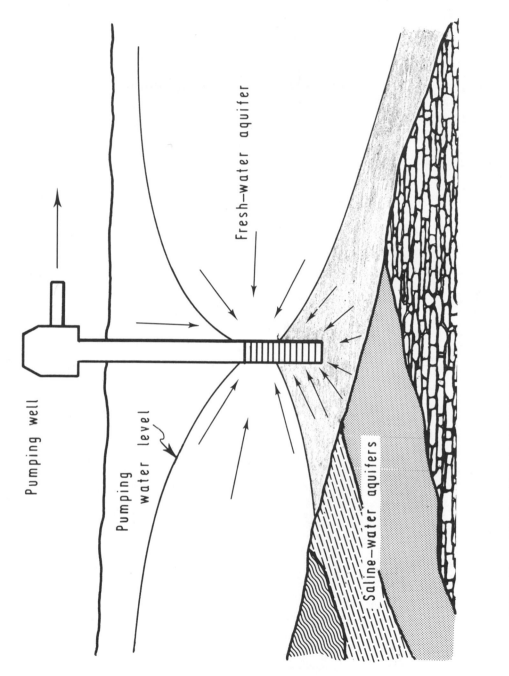

Pumping well

Pumping
water level

Fresh-water aquifer

Saline-water aquifers

FIG. 12. MIGRATION FROM HYDRAULICALLY-CONNECTED AQUIFER. OVERPUMPING MAY CAUSE CONTAMINANTS FROM NATURAL SOURCES TO ENTER A FRESH-WATER AQUIFER (AFTER BROWN)

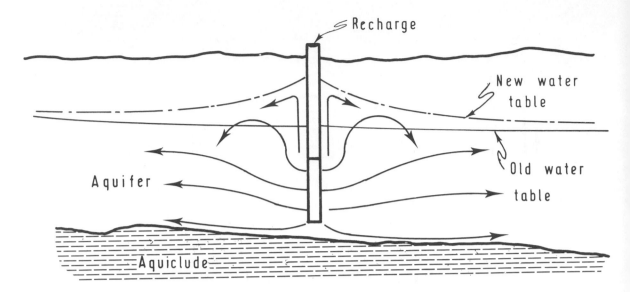

A. WATER—TABLE CONDITION

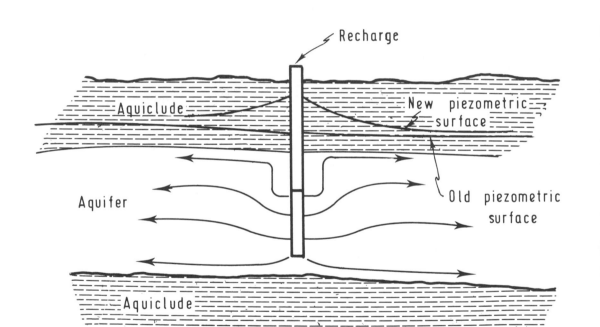

B. ARTESIAN CONDITION

FIG. 13. DIRECT INJECTION. THERE IS NO NATURAL TREATMENT
BEFORE THE WASTES ENTER THE AQUIFER
(AFTER BROWN)

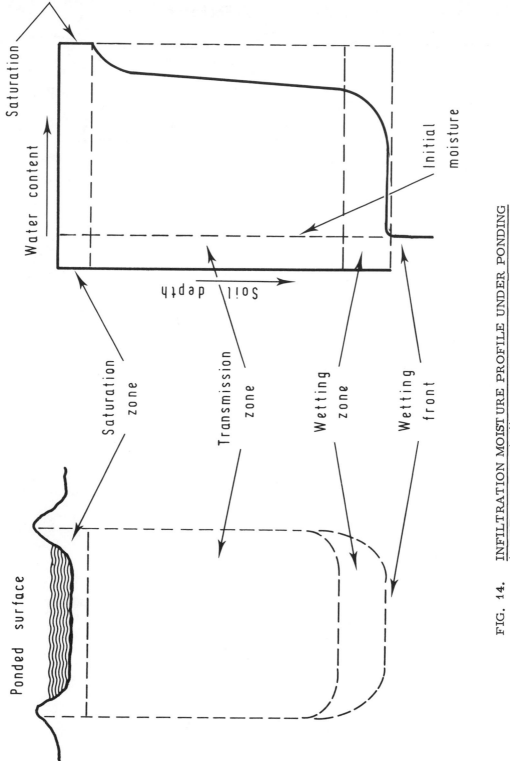

FIG. 14. INFILTRATION MOISTURE PROFILE UNDER PONDING
(AFTER HILLEL (16))

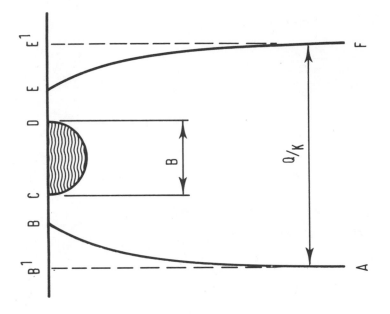

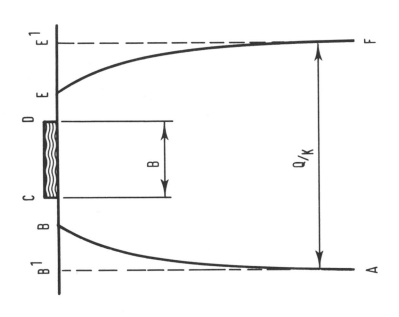

FIG. 15.   REGION OF FLOWS RESULTING FROM (a) A FLAT
BOTTOMED SHALLOW CHANNEL AND (b) A CHANNEL OF
SEMI-CIRCULAR CROSS-SECTION (AFTER YOUNGS (20))

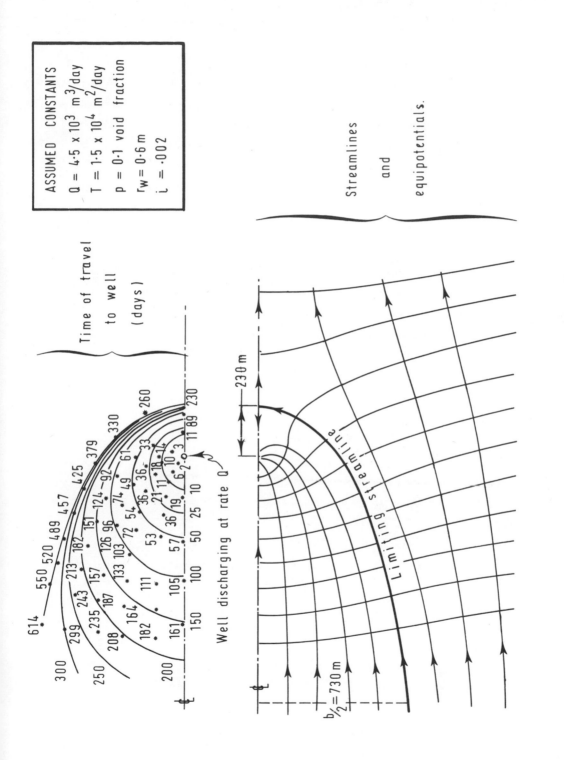

FIG. 16. FLOW DIRECTIONS AND TRAVEL TIMES OF GROUNDWATER NEAR A PUMPED WELL

61.

ADDENDUM TO THE PAPER 'HYDROGEOLOGICAL FACTORS IN GROUNDWATER POLLUTION'

by A. Hunter Blair

Sizes and shapes of contaminated zones

Contaminated areas of groundwater range in size from several square centimetres to several square kilometres. The majority however are no more than several tens of metres long and are elongated in the direction of the groundwater flow. Fig.17 after LeGrand (1) shows the shape in plan of some typical contaminated areas.

Not only is there an infinite variety of shapes of contaminated areas, but these change with time. A continuous waste discharge tends to create an ever expanding lobe of contamination, unless limited by degradation or dilution effects. Cessation of waste discharge does not immediately reduce the contaminated area, and in many cases, the contamination remains as an undegradable permanently adsorbed material. Soluble or degradable contaminants disappear eventually when waste discharge ceases, but may be conveyed some distance with the groundwater flow before vanishing completely. Water table fluctuations and associated changes in groundwater flow are another cause of varying distributions of contamination.

Assessment of pollution risk

Given the many factors governing the spread of a pollutant, LeGrand (2) has established a quantitative scale of assessing risk, in respect of water wells in the following formations:

      (a)   thick unconsolidated granular material

or (b)   unconsolidated granular material underlain at shallow depth by dense fissured media

or (c)   dense fissured media extending to the ground surface.

Weighted values of five factors, which can be evaluated from the field are used. These are:

      (i)   water table

    (ii)   adsorption

   (iii)   permeability

   (iv)   water table gradient and

    (v)   distance to the point of use.

Fig.18 shows a typical rating chart for sites in loose granular media. To apply the system to a given hydrological condition, one simply scans the lower scale for each factor on the appropriate chart, selects the approximate value for that factor in the given situation, and reads the point directly above . The sum of all the points for the five factors indicates the degree of probability that a site will become contaminated. For example figures of between 0-4 indicate imminent pollution, 4-8 probable, 8-12 possible but not likely, 12-25 very improbable, 25 impossible.

An example; a well penetrates coarse sand which extends to a septic tank 15 m distant. The water table can rise to within 1.5 m of the tank, and pumping the well results in 2% gradient from the septic tank to the well. The point values are water table 0.5, adsorption 1.5, permeability 0.5, gradient 2.5, distance 1, giving a total of 6 i.e. pollution at the well is probable.

Protection zones

One of the most important measures to be taken for the prevention of groundwater pollution is the establishment of protection zones. LeGrand's method of evaluation of contamination potential lends itself as a method to calculate the position of the boundaries of such zones.

In the USSR (3) the zone of sanitary protection of underground water sources is divided into three belts. The first belt is an area of strict regulation which embraces the source of the water supply. Within the boundaries of this belt all persons and buildings not directly concerned with the maintenance of the water supply installations are not permitted. This would correspond to the 0-4 category of LeGrand.

The second belt covers the area surrounding the source from which the water is abstracted, and the basin which feeds it. Within this belt, restrictions are imposed on water users and land users in order to preclude any deterioration of the groundwater quality. For example, all factories within this area must be equipped with appropriate facilities for the removal of their effluent and waste, application of manure is restricted, cattle are only allowed if the ground surface is impermeable etc. This would correspond to the 4-8 category of LeGrand.

The third belt, a control zone, is usually set up around the second belt in populated areas where an epidemic condition would, if it occurred, affect the sanitary condition of the second belt i.e. 8-25 on LeGrand's classification.

Such a legal organization of protection zones has worked well over many years in USSR and is an essential part of the legislation which integrates the use of water resources into a communal water supply scheme. It also provides legislation by which local councils must notify the public of the boundaries of the zones and of the conditions in force within these zones.

1. LeGRAND, H. E.   Patterns of contaminated zones of water in
                                      the ground.
                                      Water Resources Research, 1965, $\underline{1}$, p. 83.

2. LeGRAND, H. E.   System for evaluation of contamination
                                      potential of some waste disposal sites.
                                      JAWWA, 1964, $\underline{56}$, p. 959-972.

3. FOX, I. K.   Water resources, law and policy in the
                                      Soviet Union.
                                      Madison Wisconsin University Press, 1971,
                                      256 p.

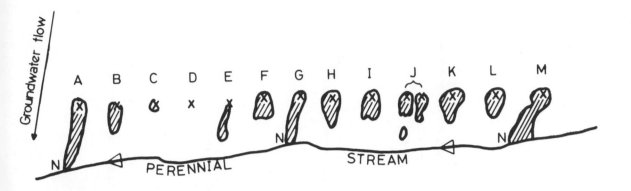

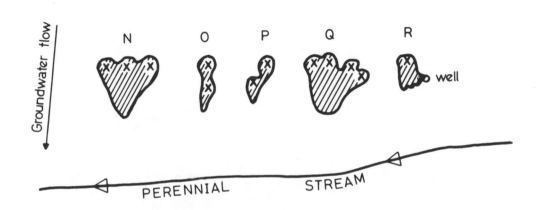

FIG. 17.   PATTERNS OF GROUNDWATER CONTAMINATION
AFTER LE GRAND

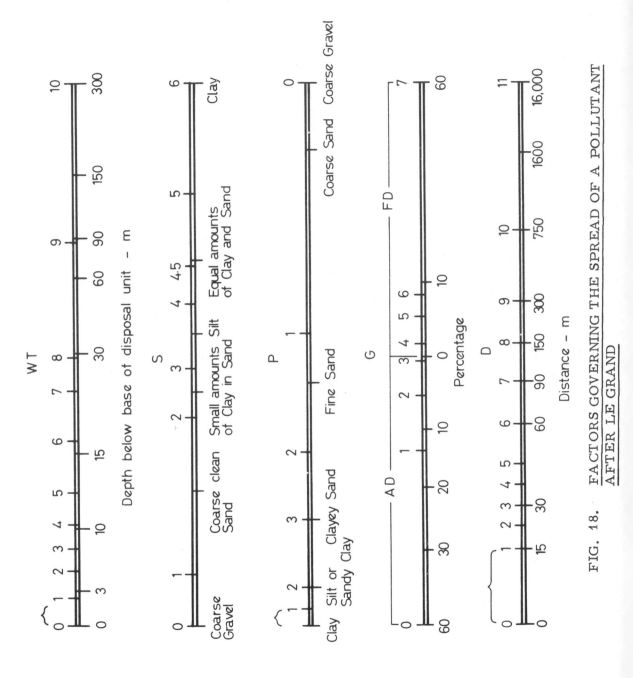

FIG. 18. FACTORS GOVERNING THE SPREAD OF A POLLUTANT AFTER LE GRAND

66.

FISSURE FLOW AND INTER GRANULAR FLOW

by S.S.D. Foster and P.E.R. Lovelock
Institute of Geological Sciences, London, England

It was pleasing to see the flag of hydrogeology waving in the field of groundwater pollution, which would be a fertile area of endeavour for the future. However, one should be critical of the context being restricted to unconsolidated formations with intergranular flow, as Hunter Blair's introductory remarks had seemed to suggest (p. 62-66).

Taking a general view of the major British aquifers, two of which are consolidated formations in which fissure flow predominates, Table 4 gives typical ranges of their hydraulic properties: it is these which quantitatively define 'the entry and movement of polluted water.... influenced by the interstices of the geological formation'. Note here the use of a 1/500 hydraulic gradient, purely to give realism in the comparison of aquifers. The fissure-flow in the Chalk and the Bunter Sandstone is an order of magnitude faster than that in the Alluvial Gravels, despite the high conductivity of the latter. These differing flow regimes also have a bearing on hydrodynamic dispersion and surface adsorption of pollutants.

Gaining knowledge of this sort is a slow process, partly because of the high cost of observation boreholes and of core-drilling. The locally great significance of heterogeneity in modifying the hydraulic properties was emphasized in the pollution context where one was more concerned to know the distribution of particulate travel times and distances rather than finding the average mass-movement through the formation as a whole. Heterogeneities were frequently responsible for the surprisingly early first-arrival times observed in many field tracer experiments.

The movement of polluted water in the unsaturated zone could sometimes be assessed from a knowledge of saturated vertical permeabilities found in laboratory tests, which allow for estimating maximum downward flow rates after making certain assumptions about the hydraulic conditions. Very little, however, is known, except by inference, about the influence of fissure-flow in the unsaturated zones of the Chalk and Bunter Sandstone, for example.

In summary, the significance of fissure-flow in the hydrogeology of the British aquifers and therefore in the pollution of their groundwaters should be recognized as fundamental, but has hardly been discussed.

TABLE 4

TYPICAL HYDRAULIC PROPERTIES AND INFERRED NATURAL FLOW RATES FOR SOME BRITISH AQUIFERS IN THE SATURATED ZONE OF THEIR INTAKE AREAS

| FORMATION | FLOW TYPE | HYDRAULIC PROPERTIES | | | REAL HORIZONTAL FLOW RATES FOR I = 1/500 V_H (km/year) | SOME SOURCES OF MAJOR HETEROGENEITY |
		HORIZONTAL PERMEABILITY K_H (metres/day)	POROSITY ϕ	SPECIFIC YIELD S_y		
CHALK East Yorks North Lincs	INTERGRANULAR	<0.0005	0.15-0.30	0.002	✕	✕
	FISSURE	40-80	0.005-	0.015	2-10	Zone of water table fluctuation K_H & $V_H \times 3$
ALLUVIAL GRAVELS Thames Valley Ulster	INTERGRANULAR	50-150	0.15-0.25	0.10-0.15	0.2-1	inter-bedded clean coarse-grained horizons K_H & $V_H \times 10$
	FISSURE	✕	✕	✕	✕	✕
BUNTER SANDSTONE Midlands Notts only	INTERGRANULAR	1-4	0.23-0.30	0.15-0.20	0.004-0.01	differential cementation
	FISSURE	10-20	<0.005		1.5-3	mining subsidence

COMPUTER SIMULATION OF POLLUTION FRONT MOVEMENT

by J.P. Sauty
Bureau des Recherches Géologiques et Minières, Orléans, Paris, France

1. PRINCIPLE OF THE CALCULATION

The aim is to compute the propagation of a pollution front in an aquifer, that is to say the displacement and the deformation in the course of time of a line of separation between pure water and polluted water having regard to the validity of the following hypotheses.

(a) Dispersion phenomena are negligible. That implies amongst other things that the porous media are locally homogeneous, so that digitation is limited.

(b) Molecular diffusion does not take place. This is the case at high Peclet numbers,[**] when the movement of water is sufficiently rapid.

(c) The pollutant is not adsorbed by the porous medium.

(d) The field of velocities is not disturbed by the advance of the front, i. e. the pollution does not modify the hydrodynamic characteristics of the water, such as the viscosity, density and capillary properties.

The computer program yields the streamlines for steady flow. Piezometric heads have been either observed in the field or computed by means of a classic steady state flow model. There may be a water table

---

[*] Originally this paper appeared as Présentation d'un modèle mathématique de la propagation d'un front de pollution dans une nappe d'eau souterraine' in La Houille Blanche 1971, 26, (8), pp 731-736

[**] The Peclet number (Pc) is a dimensionless group which may be defined as:

$$Pc = \frac{uk^{\frac{1}{2}}}{D_o} = \frac{v}{pD_o}\left(\frac{Kn}{\rho}\right)^{\frac{1}{2}}$$

where
- k = intrinsic permeability of the medium $[L^2]$
- p = effective porosity of the medium for fluid seepage
- u = pore velocity $[LT^{-1}]$
- v = seepage velocity $[LT^{-1}]$
- n = dynamic viscosity of the fluid $[ML^{-1}T^{-1}]$
- ρ = density of the fluid $[ML^{-3}]$
- D_o = molecular diffusivity of the fluid $[L^2T^{-1}]$
- K = hydraulic conductivity of particular medium/fluid combination $[LT^{-1}]$

or the aquifer may be confined. It may be homogeneous or not. The speeds of displacement along streamline trajectories allow the position of the pollution front to be defined at successive instants of time.

2. EXAMPLES

Three cases presented here concern steady state water table conditions established previously with the aid of an analogue model.

2.1.   The first example (Fig. 19) shows an unexploited alluvial aquifer such as might occur in the natural state. This aquifer is bounded by an impermeable barrier on the one side and by a fixed potential boundary, that is to say a river, on another. The latter controls the flows in the aquifer by recharging it on the upstream side and by draining it on the downstream side.

If there should be a pollution incident in the river and given the validity of the hypotheses stated in the first paragraph, the curves traced in Fig. 19 show the advance of the front with respect to the aquifer at 20, 50, 100, 200, 300 and 400 days after the onset of the pollution.

2.2.   The next example (Fig. 20) presents the same aquifer subjected to an intensive development by a cluster of pumping wells for a long enough time so that the flow is steady at the moment when the pollution occurs. The distribution of velocities demonstrated by the hydrodynamic model is modified in a very obvious fashion in comparison to that of the first example. It can be seen that the pollution front reaches the zone of capture from the eighth day onwards, but that the water pumped from the wells is not polluted totally until after a delay of about 100 days.

2.3.   The aquifer represented in Fig. 21 is fed in the north by a lake and bounded on the south-west by a river which feeds the aquifer or recharges the aquifer in the upstream zone and drains it in the downstream portion. Near the middle of the area there is a wellfield strongly drawing down the groundwater levels; further south a small pond is maintained artificially at constant level. It is assumed that after a certain date the pond gets polluted. Fig. 21 shows the progression of the front which can be expected at the wellfield on the one hand and the river on the other.

3. APPLICATIONS

The estimation given by this calculation of the pollution front's progress indicates the delays which can be counted on in order to set protection measures for the aquifer into action.

At successive instants of time, the flows entering a well can be identified, some being within the zone reached by the front and the rest coming from the zone not yet polluted, an estimation is obtained of the development of the concentration of the pollutant in the well discharge. Thus, examination of Fig. 20 shows a zero concentration during the first week, 40% after 10 days and 100% at the end of 100 days.

Similar considerations could be applied to a pollution which was limited in space and time, such as might have been produced if the river (Fig. 20) only suffered brief contamination upstream of the recharged zone of the aquifer. In this circumstance a band of pollution moves towards the pumps, so causing an arrival pulse that passes through a maximum. It should be observed that the simplified hypotheses adopted lead to a pessimistic evaluation of the maximum concentration of the pollutant (i.e. one that errs on the safe side). The phenomena which have not been taken into account (diffusion, dispersion, kinematic dispersion and adsorption by the terrain) would all tend to reduce the intensity of the concentration curves.

The work presented here is only a preliminary presentation of a more or less complete model which is being developed at the Bureau des Recherches Géologiques et Minieres (BRGM) in which the field of application is being extended in the following respects.

(i)   An auxiliary calculation of the reduction of the front by longitudinal kinematic dispersion, evaluated by semi-empirical formulae.

(ii) A mass balance to take account of the progressive dilution arising from the inflow of fresh water such as rainfall infiltration.

(iii) A version with transient flow allowing possible variations in the boundary conditions to be taken into account (e.g. fluctuations of levels or of discharges) and in the hydrodynamic characteristics as the front advances (e.g. due to modification of physical properties of the liquid).

The practical interest in this calculation procedure [*] resides in its simplicity.

Whenever the original hypotheses are acceptable a rapid and easy calculation can be done to evaluate the order of magnitude of the frontal movement.

---

* See the Appendix to this paper, Sections A2.1 and A2.2, p. 78-79.

The more advanced models which treat the problem more generally, incorporating dispersion and diffusion, are much more seductive from the scientific point of view but have the inconvenience for amateurs of requiring parameters such as the dispersion tensor whose real values are generally unknown.

4. APPENDIX

Refer to p. 76-85 giving the mathematics for calculating longitudinal dispersion and streamline directions, together with a computer programme outline.

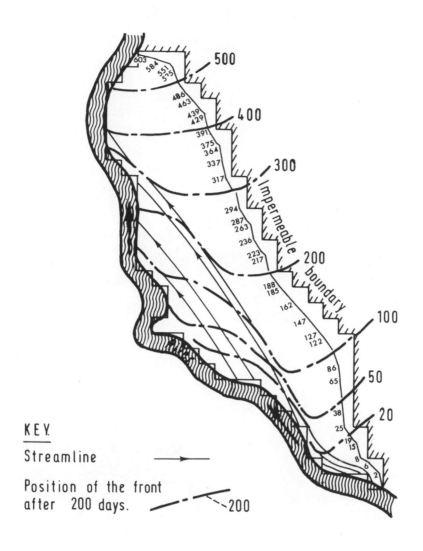

FIG. 19.   THE MOVEMENT OF POLLUTION THROUGH A
RIPARIAN AQUIFER (not pumped)

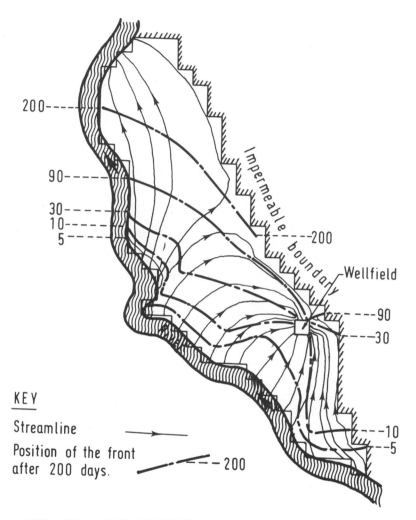

200----
90----
30----
10----
5----

Impermeable boundary
----200
Wellfield
----90
----30
----10
--5

KEY

Streamline

Position of the front
after 200 days.
----200

FIG. 20. THE MOVEMENT OF POLLUTION THROUGH A
RIPARIAN AQUIFER (with continuous pumping
at the wellfield)

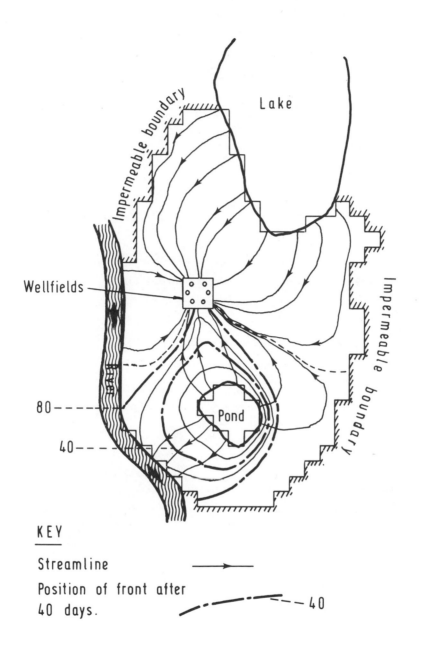

KEY

Streamline ———→

Position of front after
40 days. ━ ━ ━ ─ ─ — 40

FIG. 21.   POLLUTION MOVING FROM A POND TOWARDS
A WELLFIELD AND A RIVER

MATHEMATICAL APPENDIX TO 'COMPUTER SIMULATION OF POLLUTION FRONT MOVEMENT'
by J.P. Sauty

A1.   COMPUTATION OF LONGITUDINAL DISPERSION

A1.1. Introduction

The model presented in Volume I (conference papers) is designed to simulate the displacements of a sharp interface separating two immiscible fluids: pure water and polluted water.   In this piston type flow, dispersion is supposed to be completely negligible.

A recent extension to this model allows one to take into account the longitudinal dispersion when necessary and if one knows which numerical coefficients to use. Transverse dispersion is still neglected, but it is known to be much smaller than the longitudinal dispersion.   Then when transversal concentration gradients are not larger than longitudinal gradients, transverse dispersion has no really perceptible effects.

The method is based on an analytical technique respecting the variations of the dispersion coefficient with velocity and developed by Gelhar and Collins (1): it allows direct calculation of the concentration at any time and any position along a streamline; it is much quicker than a stepwise process and avoids the parasitic effects of numerical dispersion.

A1.2. Method*

In a locally homogeneous and isotropic porous medium, the concentration of a perfect tracer in the absence of absorption is governed by the following equation

$$\frac{\partial C}{\partial t} = \frac{\partial}{\partial x_i} \left[D_{ij} \frac{\partial C}{\partial x_j} - CU_i \right] + D_m \frac{\partial}{\partial x_i} \left(\frac{\partial C}{\partial x_i} \right) \quad \dots\dots\dots\dots\dots \quad 1.$$

where x_i $i = 1, 2, 3$ are cartesian coordinates,

t is the time,

C is the concentration of the tracer,

U_i is the velocity of the fluid,

D_{ij} is the dispersion tensor,

D_m is the coefficient of molecular diffusion,

---

*For more details see L. W. Gelhar and M.A. Collins (1).

If x_1 axis is parallel to the velocity vector (of length u), it is usually assumed (Bachmat and Bear (2)) that:

$$D_{11} = a_L \cdot u$$

$$D_{22} = a_T \cdot u$$

$$D_{ij} = o \ (i \neq j)$$

a_T being usually much smaller than a_L, transverse dispersion can be neglected as long as the transverse concentration gradients are not larger than the longitudinal gradients.

Then if one neglects the transverse dispersion, and the molecular diffusion, the equation of concentration along a streamline is the following:

$$\frac{\partial C}{\partial t} + u \frac{\partial C}{\partial s} = a_L \cdot u \cdot \frac{\partial^2 C}{\partial s^2} \qquad \dots\dots\dots\dots\dots\dots \quad 2.$$

where is the curvilinear distance along the streamline.

Gelhar and Collins (1) showed that, using the changes of variables:

$$X = \int_{s_o}^{s} \frac{ds}{u(s)} \qquad \dots\dots\dots\dots\dots\dots\dots\dots\dots \quad 3.$$

and

$$\omega = \int_{s_o}^{s} \frac{ds}{u^2(s)} \qquad \dots\dots\dots\dots\dots\dots\dots\dots \quad 4.$$

and matched asymptotic expansions, equation 2, reduces to:

$$\frac{\partial C}{\partial \omega} = a_L \frac{\partial^2 C}{\partial (X-t)^2} \qquad \dots\dots\dots\dots\dots\dots\dots\dots \quad 5.$$

This is the classical diffusion equation, for which there exist numerous analytical solutions (3).

For instance, the effect of a stepwise change of concentrations at the origin

$$t = o \qquad C = o \text{ for all} \qquad s \geqslant s_o$$

$$t > o \qquad \begin{cases} C = C_o \text{ constant at } s = s_o \\ C \rightarrow o \text{ as} \qquad s \rightarrow \infty \end{cases}$$

yields:

$$C(s,t) = \frac{C_o}{2} \ \text{Erfc} \left\{ \frac{X-t}{\sqrt{4 a_L \omega}} \right\}$$

A1.3. Results

Fig. 22, p. 82, shows the evolution of concentrations after a stepwise change C_o of concentration at the origin. The variations of concentration C/C_o with respect to time are plotted at various distances S from the source of pollution, and for 3 different values of the coefficient a_L.

For the considered watertable, the results are the following:

- When a_L = 1 m,

    at a distance of 229 m the front arrives after 50 days (C/C_o = 50%) and it takes 16 days for the concentration to grow from 5% to 95% of the concentration at the source

    It takes about 600 days for the front to arrive at approximately 3 km and the evolution of concentration between 5% and 95% takes 55 days.

- When a_L = 10 m,

    at the distance of 3 km this evolution takes 170 days.

- When a_L = 0.1 m,

    at the same distance, this evolution takes 16 days.

A2.   DETERMINATION OF STREAMLINES

A2.1. Computation of velocities (Fig. 23, p. 83).

Let us consider a square mesh, for which we know the potential H, the permeability K and the porosity P at every nodal point. If the subscript C corresponds to a nodal point, and the subscripts N, W, S, E correspond to the neighbouring nodal points, the pore velocity at node C is computed by the following finite difference formulae:

$$V_X = -\frac{1}{2}\left[K_{CE}\cdot\frac{H_E - H_C}{\Delta x} + K_{CW}\cdot\frac{H_C - H_W}{\Delta x}\right]\cdot\frac{1}{P_C}$$

$$V_Y = -\frac{1}{2}\left[K_{CN}\cdot\frac{H_N - H_C}{\Delta y} + K_{CS}\cdot\frac{H_C - H_S}{\Delta y}\right]\cdot\frac{1}{P_C}$$

This is the Darcy's apparent velocity divided by the porosity. The average permeabilities such as K_{CE} between C and E are obtained from the nodal permeabilities in the following fashion:

$$\frac{2}{K_{CE}} = \frac{1}{K_C} + \frac{1}{K_E}$$

A2.2. Construction of streamlines (Fig. 24, p. 84).

Streamlines are constructed from various preset starting points, situated at the source of pollution.

Each streamline is approximated by a polygonal succession of straight lines. In each square element the velocity vector is assumed to be constant and the streamline proceeds by straight segments parallel to the mean velocity of the element it crosses.

The transit times through each segment are computed (Fig. 24) and totalized giving the time necessary for the front to come from the source of pollution to the considered point.

A streamline stops when it reaches a boundary or a point of minimum potential at a well.

A3. DESCRIPTION OF THE PROGRAMME

This programme brings together finite difference and analytical methods of pollution movement prediction. It can be divided into four main blocks. (See General Flow Chart, Fig. 25, p. 85).

A3.1. Preliminary hydrodynamic model

The potential field is needed for the computation of velocities. This is why the first part of the programme is a classical steady-state two dimensional model (with Dupuit assumption) using a square mesh and finite differences.

INPUT : The data needed are the following:

either - the transmissivities T, the flow rates Q in the wells and on the boundaries with fixed flow, and the heads H_b on the boundaries with fixed head. This is in the case of fixed transmissivities: confined aquifers or unconfined with small variations of heads.

or - the permeabilities K, the elevation of the impervious bottom Z_B and top Z_T of the aquifer, and the flow rates Q and boundary heads H_b. These data are used for completely or only locally unconfined aquifers.

OUTPUT : The hydrodynamic model computes the heads at every node of the mesh.

In the case where these heads are already known this part of computation can be avoided. This occurs when heads have already been computed and punched, or when the field measurements are numerous and precise enough to yield an accurate contour drawing from which heads at every nodal point can be evaluated.

A3.2. Computation of velocities

This computation is described in Sect. 2, p. 78-79.

INPUT   : Permeabilities, heads and porosities are needed. It is necessary to give the values of porosity (node by node or only one uniform value for the whole mesh) plus the thicknesses TH of the aquifer when transmissivities have been used for the hydrodynamic model.

OUTPUT   : One gets then the two components V_X and V_Y of the velocity vector at each node of the mesh.

A3.3. Computation of streamlines

The method is described in Sect. 2.

INPUT   : coordinates x_o, y_o of the initial point for each streamline, and the value of a_L coefficient for longitudinal dispersion.

OUTPUT   : The programme yields a set of points (x, y) where the streamline crosses the elements, and the corresponding times t at which the piston flow front arrives.

           If a_L is non zero it computes also the values or the integral ω defined in Sect. 1.

A3.4. Computation of C(t)

This part is only used when dispersion is not negligible ($a_L \neq o$). Then the programme computes the concentration at any distance S from the origin of the streamline, and at any time.

The computation can be made for an initial step change of concentration at the origin, or an abrupt injection at time t = o, or an arbitrary law of concentration with respect to time at the origin, obtained by superposition of several step changes of concentration at different times.

INPUT   : The type of perturbation must be indicated, and the curvilinear coordinates S_i at which concentrations are desired for different times t_i^1, t_i^2, t_i^3 ....

OUTPUT   : Concentrations are printed.

A3.5. Conclusion

It must be emphasized that this treatment of dispersion takes very little computer time. The most costly operation is the evaluation of the potential field.

A complete treatment including hydrodynamic model with 600 nodes, and computation of concentrations only takes about 3 minutes of IBM 360-40 which is a relatively slow computer.

REFERENCES

1. GELHAR, L. W. and COLLINGS, M. A.

General analysis of longitudinal dispersion in nonuniform flow.
Water Resources Research 1971, $\underline{7}$ (6), 1511-1521.

2. BACHMAT, Y. and BEAR, J.

The general equations of hydrodynamic dispersion in homogeneous, isotropic, porous mediums.
Journal of Geophysical Research 1964, $\underline{69}$ (12), 2561-2567.

3. CARSLAW, H. S. and JAEGER, J. C.

Conduction of heat in solids, 2nd ed.
Oxford University Press, London, 1959.

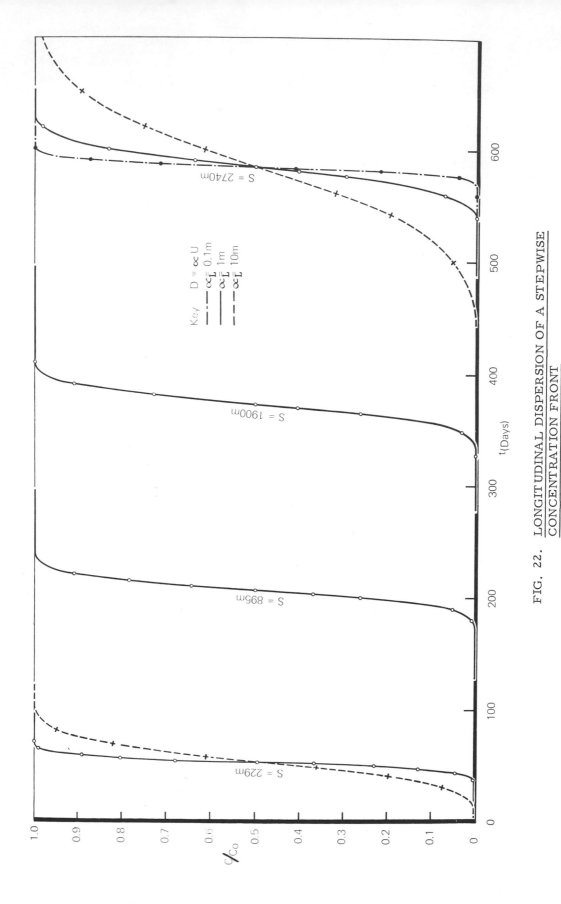

FIG. 22.   LONGITUDINAL DISPERSION OF A STEPWISE
CONCENTRATION FRONT

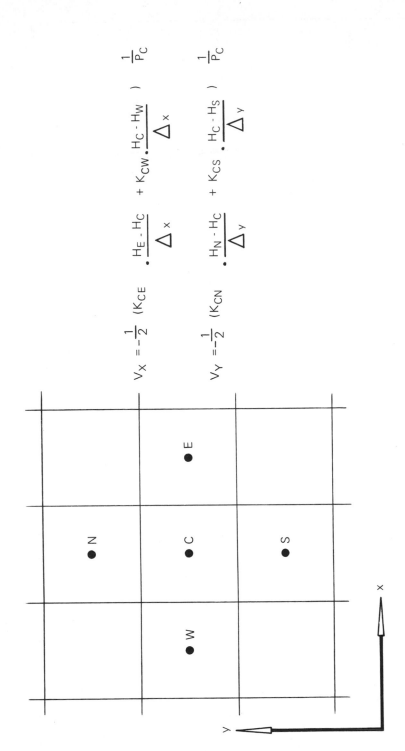

$$V_X = -\frac{1}{2} \left(K_{CE} \cdot \frac{H_E - H_C}{\Delta x} + K_{CW} \cdot \frac{H_C - H_W}{\Delta x} \right) \frac{1}{P_C}$$

$$V_Y = -\frac{1}{2} \left(K_{CN} \cdot \frac{H_N - H_C}{\Delta y} + K_{CS} \cdot \frac{H_C - H_S}{\Delta y} \right) \frac{1}{P_C}$$

FIG. 23. CALCULATING VELOCITY COMPONENTS

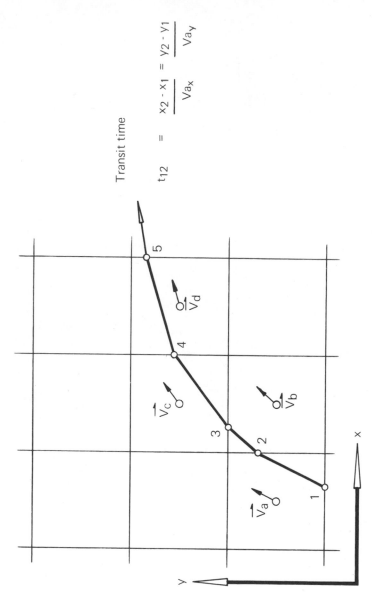

Transit time

$$t_{12} = \frac{x_2 - x_1}{V_{a_x}} = \frac{y_2 - y_1}{V_{a_y}}$$

FIG. 24.   CONSTRUCTION OF STREAMLINES

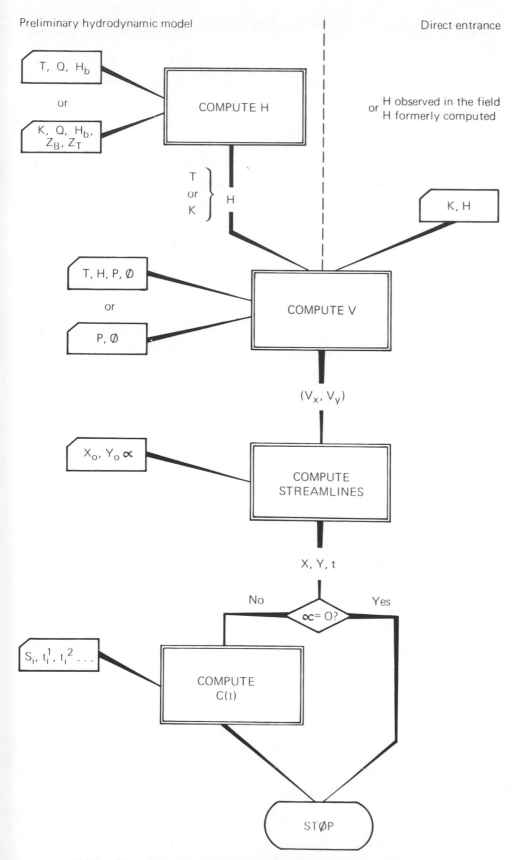

T, Q, H_b

or

K, Q, H_b, Z_B, Z_T

COMPUTE H

or H observed in the field
H formerly computed

T or K } H

K, H

T, H, P, Ø

or

P, Ø

COMPUTE V

(V_x, V_y)

$X_o, Y_o \propto$

COMPUTE STREAMLINES

X, Y, t

No $\propto = 0$?   Yes

$S_i, t_i^1, t_i^2 \ldots$

COMPUTE C(t)

STØP

FIG. 25. FLOW CHART OF POLLUTION FRONT CALCULATION

85.

SOME INTERPRETATIONS OF DISPERSION MEASUREMENTS IN AQUIFERS

by J.A. Cole
The Water Research Association, Medmenham, Bucks, England

Continuous and pulse inputs

The effect of these may be compared in Fig. 26, p. 90, where the lines of constant concentration relate to a tracer substance or soluble contaminant, entering the aquifer at points X and Y. The pulse input at X is carried with the regional groundwater flow to successive positions X', X", etc. with dispersion greater in the longitudinal than the lateral direction. See Fried and Combarnous's review (1) for explanations of the relationship between longitudinal and lateral dispersion.

The continuous input at Y will in time develop a steady-state plume of concentration, like that shown in Fig. 26. The plume shows the lateral dispersion effect primarily, the longitudinal dispersion tending to be concealed by the convection: in fact the continuous input can be thought of as a mathematical superposition of elementary pulse inputs, whose dispersed clouds are strung out in a chain and added together (cf. convolution of rainfall excess with a unit hydrograph to produce a river hydrograph).

Hanford reactor site fluorescein test

Theis (2) quoting Bierschenk (3) shows a diagram upon which Fig. 27, p. 91, is based. The two curves represent the concentration of fluorescein dye as a function of time, measured at observation wells downflow of an unused well, dosed with that dye over a period of about one day. Because the travel time is ~100 days in each case, the example is virtually one of pulse input.

The interest centres on the very different dispersion of dye arrival at the two wells. It is evident that both arrivals are multiple, but that the well at the SSE site receives most of the fluorescein in a single pulse, so that ramifications of dye path are minor in that case. In contrast the other well apparently received flows by very differing routes: it is the branching of flows that constitutes much of the dispersion phenomenon on a large scale.

To quantify the longitudinal dispersion one measures:-
Δt, the pulse width at 16% and 84% of pulse area $\left.\vphantom{\begin{matrix}a\\b\end{matrix}}\right\}$ Same time units
$\bar{t}$, the median time of the pulse, i.e. at 50% of pulse area
x, the distance from dye input point

and uses the formula
$$a = \left(\frac{\Delta t}{\bar{t}}\right)^2 \cdot \frac{x}{8}$$

a is a dispersion constant having dimensions of length. (Strictly the formula applies to homogeneous media of constant dispersion properties, such that the pulse spreads lengthwise in Gaussian fashion, without pronounced lateral dispersion. Also $\Delta t \ll \bar{t}$).

From Fig. 27, p. 91, the following are obtained:

Well number	Δt (days)	$\bar{t}$ (days)	x (metres)	a (metres)
699-14-27	16	137	3500	6
699-20-20	120	125	4000	460

The Hanford aquifer is 'very coarse and permeable', presumably a detrital deposit of intercalated gravels and sands.

Other measurements of longitudinal dispersion

It is appropriate to compare the Hanford a values with those obtained in other tracer experiments:

Aquifer or medium	Reference	a (metres)
Upper Chalk in Surrey, feeding the Addington well	Whitaker (4)	50
Middle Chalk of Cambridgeshire, near Wilbraham	Cole (5)	12
-	Fried and Ungemach (6)	11
Sands in laboratory columns	Lau and others (7)	< 0.001
0.47 mm glass ballotini in laboratory columns	Cole (5)&(8)	0.002

It seems reasonable to infer that the dispersion effects in real aquifers are dominated by permeability variations and/or fissure patterns on a scale of tens or hundreds of metres, in vast contrast to the tiny dispersion effects within well graded media.

Longitudinal dispersion in regularly stratified aquifers

Mercado (9) has treated a mechanism of dispersion in which plug flow within laminated strata of differing hydraulic conductivity produce a range of arrival times as shown in

Fig. 28(a), p. 92, adapted from his paper. When a tracer is introduced to the aquifer at a borehole, and flows towards an observation hole the latter shows a sigmoidal concentration rise of tracer as in Fig. 28(a), p. 92. Mercado effectively puts his results in the form:

$$\sigma_x = x \cdot \frac{\sigma_t}{\bar{t}} = x \cdot \frac{\sigma_k}{\bar{k}}$$

where $\bar{k}$ = mean hydraulic conductivity

σ_k^2 = variance of hydraulic conductivity

$\bar{t}$ = mean arrival time of tracer

σ_t^2 = variance of time of tracer arrival

x_2 = distance of tracer travel

σ_x^2 = spatial variance of tracer dispersion

Note that σ_x is here <u>proportional</u> to distance travelled, rather than to the square root of distance (as occurs in Fickian diffusion, in hydrodynamic dispersion and in dispersion of groundwater tracers traversing <u>branched</u> paths in the aquifer matrix). Mercado's interpretation rests on differing flow rates in laminae which do not inter-connect appreciably: it would however be inadmissible for lenticular deposits or others affording significant vertical transfers within the range of travel.

Dispersion from local sources of flow

Even when matrix and molecular dispersion effects are minimal, and even though the aquifer may have very uniform hydraulic conductivity, flow from localised sources inevitably causes dispersion to the extent to which flow lines spread from them. Often flow patterns are curved, the velocity profile along each being different: hence an impulse of tracer or contaminant in the flow will generally arrive at a sampling well by a variety of paths, and be spread in time accordingly. Brown (10) illustrates the flow paths for:

    (i)   an abstraction well near a stream,

    (ii) recharge and discharge wells in the same aquifer, the wells aligned to regional flow and

    (iii) as (ii), but the wells aligned at right angles to the regional flow.

See also the addendum to Sauty's paper to this Conference (p. 176-85).

Vertical dispersion

As well as matrix dispersion, which is a phenomenon independent of the tracer, one has to consider possible effects due to contaminant solution's density. Obviously an infiltrating liquor will tend to overlie the existing groundwater, if the former is is appreciably less dense; that a liquor will tend to sink through the groundwater if the opposite is true is really far less obvious. The conditions governing flow and any instability causing vertical mixing are spelt out by Fried and Combarnous (1).

The vertical dispersion phenomenon is conveniently visualized in model experiments by Zilliox and Muntzer (11), from which Fig. 29, p. 93, is derived. In practice, the sinking effect is of especial importance when considering brine residues from salt and potash mining and from the chemical and metallurgical industries.

References

1.	FRIED, J. J. and COMBARNOUS, M. A.	Dispersion in porous media. Advances in Hydroscience 1971, 7, 169-282.
2.	THEIS, C. V.	Aquifer and models. Proc. National Symp. on Groundwater Hydrology, American Water Resources Association, San Francisco, 1967, 138-148.
3.	BIERSCHENK, W. H.	Aquifer characteristics and groundwater movement at Hanford. US Atomic Energy Commission, 1969, Report H. W. 60601.
4.	WHITAKER, W.	The water supply of Surrey from underground sources. Memoirs of the Geological Survey, 1912, 95-96.
5.	COLE, J. A.	The flow of groundwater. University of Cambridge, MSc dissertation 1957, Table VI. 10 and Fig. 6. 15.
6.	FRIED, J. J. and UNGEMACH, P.	In situ determination of longitudinal dispersion coefficient in a natural medium. WRA Access 18571.
7.	LAU, L. K., KAUFMAN, W. J. and TODD, D. K.	Dispersion of water tracer in radial laminar flow through homogeneous porous media. University of California, Berkeley, 1959, Progr. Rept. No. 5, 93 p.
8.	COLE, J. A.	Discussion of a paper by de Josselin de Jong "Longitudinal and transverse diffusion in granular deposits". Trans. Amer. Geophysical Union, 1958, 39, 1160-1162.
9.	MERCADO, A.	The spreading pattern of injected water in permeably stratified aquifer. IASH Symposium of Haifa, 1967, Publn. No. 92, 23-36.
10.	BROWN, R. H.	Hydrologic factors pertinent to groundwater contamination. Ground Water, 1964, 2, (1), 5-12.
11.	ZILLIOX, L. and MUNTZER, P.	Physical model study of the mechanism of groundwater pollution by miscible liquids and non-miscible ones (in French). La Houille Blanche, 1971, 26, (8), 723-730.

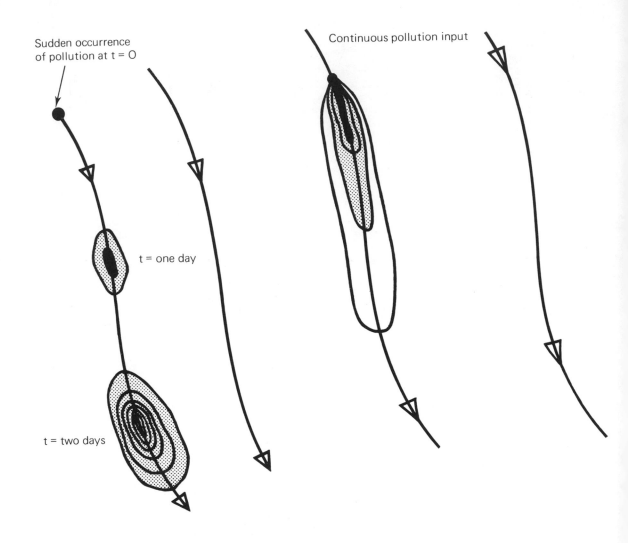

Sudden occurrence
of pollution at t = 0

Continuous pollution input

t = one day

t = two days

FIG. 26. SPREAD OF POLLUTION FROM SUDDEN AND
CONTINUOUS SOURCES

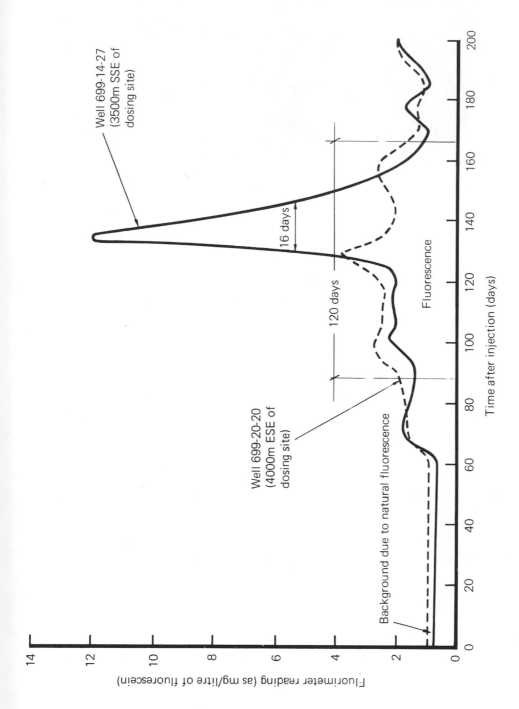

FIG. 27. DISPERSION OF ARRIVAL TIME OF FLUORESCEIN USED TO TRACE GROUNDWATER FLOW AT HANFORD, WASHINGTON (USA): AFTER THEIS

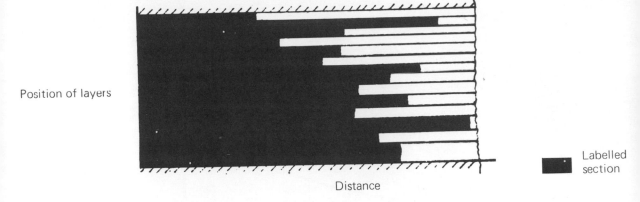

Position of layers

Distance

Labelled section

(a) Flow of labelled water—in a permeability stratified aquifer

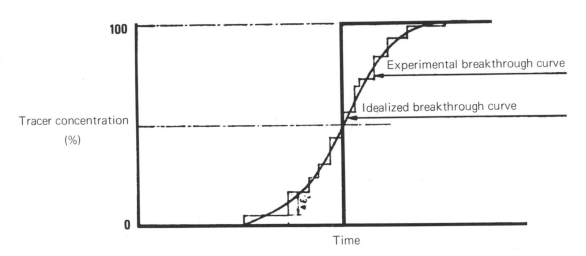

Theoretical breakthrough curve for the case of immiscible fluids in homogeneous aquifer

100

Experimental breakthrough curve

Idealized breakthrough curve

Tracer concentration (%)

0

Time

(b) Tracer breakthrough curve at an observation well

FIG. 28.   PLUG FLOW IN LAMINATED STRATA CAUSING LONGITUDINAL DISPERSION: AFTER MERCADO

(a) Entrainment with stable mix in flow direction and limited influence of density contrast

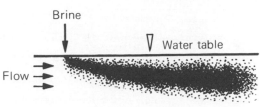

(b) Unstable regime: development of pockets at the lower interface

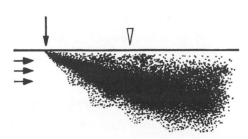

(c) Unstable regime: progressive lowering of upper interface

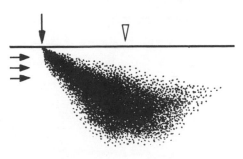

(d) Sinking with vertical infiltration predominant due to density contrast

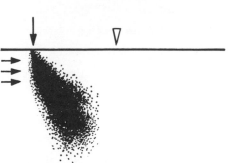

FIG. 29. VERTICAL DISPERSION. THE BRINE IS ENTERING A HORIZONTAL SEEPAGE FLOW OF WATER; DENSITY CONTRAST INCREASES (a) (d): AFTER ZILLIOX AND MUNTZER

DISCUSSION ON COMPUTING MOVEMENT AND DISPERSION OF DISSOLVED SUBSTANCES IN GROUNDWATER

1.   J. P. Sauty presented new material (reproduced on p. 76-85) on the extension of his original computer program to cases where longitudinal dispersion was included in the frontal migration of a pollutant.   He emphasized that the program could be run for only £2 for a 600-node solution if the head distribution was known.   Attention was also drawn to the assumptions of his model: these could cater for many practical cases in fact and the difficulties usually lay in knowing aquifer constants and boundary conditions rather than in accepting the model's approximate validity.   It could be converted to three dimensional flow and dispersion if need arose.   Other speakers drew attention to the costs of problem definition and of data assembly which would transcend Mr. Sauty's rather moderate computing costs.

2.   J. J. Fried and D. Poitrinal described their field equipment consisting of a gamma probe suspended on its cable, connecting to a single-channel energy-selective counter in the vehicle.   Depth is logged by a pulley-wheel operating switches.   The simplicity and safety of the techniques were emphasized, the recovery of much of the injected tracer being a special feature.   It was possible to reduce ('deconvolute') the field data into a layer-by-layer response, allowing for the probe's characteristic response ('impulse response') to a lamina of radioactivity.   The power of the technique lay in its diagnosis of high-risk zones, whose dispersion coefficients must be known for proper prediction of the movement of a pollutant.   A groundwater dispersion model for the urban area of Lyons had been made in this way.

3.   Supplementing Dr. Fried's remarks, P. Ungemach spoke of geophysical techniques used in France for the tracing of groundwater flow direction.   The resistivity anomaly created by a brine injection can be followed above ground; also the inferred contours of the brine concentration give a measure of the dispersion coefficients, once the brine has spread over distance ~100 m.

4.   Because the Sauty program was only of very recent development, its application to practical examples was only just beginning.   Apart from addressing real cases, Dudgeon made the suggestion that there was virtue in having a range of solutions precomputed for various longitudinal dispersion coefficient and flow geometries.   Others endorsed this on grounds of there being a good body of existing flow data on which to attach precomputed results.

5.   Reference was made to ongoing work in the US Geological Survey, where finite element models were preferred for the flow simulation, being easily adapted to a

changing boundary condition.   The USGS work involved cut-and-dry fitting of longitudinal dispersion coefficients to field data for a case of chromium contamination of a glacial aquifer.

SOME CHEMICAL PRINCIPLES OF GROUNDWATER POLLUTION
by E.S. Hall
The Water Research Association, Medmenham, Bucks, England

1. INTRODUCTION

Pollution of groundwater is usually attributable to some form of intensive activity that is a consequence of human civilization. The most obvious of these involve the disposal of various forms of waste or the spillage of chemicals and fuels. Also, the exploitation of groundwater contributes to its own pollution in systems where the water table is lowered to an extent that hydraulic opportunity for the flow of contaminants into the aquifer is intensified.

If the term 'pollution' is taken to refer to any undesirable contaminant acquired by water during its passage through the ground, consideration should be given to situations where pollution has an entirely natural origin. This is usually true, for example, of most of the iron, arsenic, nitrate and fluoride in natural water. On the other hand, obvious sources of artificial contamination do not always cause polluted groundwater. Irrespective of the source of contamination, what is important is the solubility and transport of contaminants in relation to the properties of soil and aquifer rock.

The solubility of many inorganic contaminants can be calculated from theoretical stability constants, but for such calculations to be valid, it is necessary to assume that chemical equilibrium is attained. This assumption is not justified for many reactions which require the introduction of considerable activation energy in the form of heat and without which proceed only very slowly. In the chemistry of groundwater however, equilibrium calculations usually give good indication of reality for two reasons:

(a)   Groundwater usually remains for many months or years in contact with the rock which governs the groundwater composition.

(b)   Most non-equilibrium oxidation-reduction systems in nature are used by various micro-organisms for respiration through the application of highly efficient enzyme catalysts. In this way the free energy associated with the non-equilibrium is used and equilibrium is thereby approached. Two examples of such systems are the oxidation of organic substances by oxygen and the oxidation of sulphur by nitrate. The dominant form of organism in any system is governed largely by the type and relative concentrations of oxidizable substance and oxidizing agent. Conditions in the ground, particularly in soil, are usually very favourable for the existence of many types of micro-organism.

Accordingly, this paper describes the more important systems of chemical equilibrium appropriate to groundwater quality under conditions both natural and subject to artificial influence. One artificial influence to which little or no attention has been given previously is the seemingly harmless introduction of air into the ground as a result of groundwater abstraction, agriculture and mining. The surface of the earth constitutes an interface between an oxidizing atmosphere containing oxygen and a reducing earth's crust containing organic substances and sulphides. Most of the latter occur as heavy metal sulphides in Precambrian rock, inaccessible to water except as results of mining, but sulphide (mostly iron pyrite) also occurs in sandstones used as groundwater aquifers. The division in oxidation state between this sedimentary sulphide and the oxidizing atmosphere has been generated by the sun's radiation through the process of photosynthesis, which has introduced oxygen into the atmosphere and organic substances into the ground. The latter have been the subsequent cause of sulphide formation by the bacterial reduction of sulphate. Photosynthesis continues to preserve the division by the generation of organic matter in soil and aquatic sediments. By its decay, this organic material uses up some of the oxygen which would otherwise pass into the ground, dissolved in rainwater. The natural oxidation balance in the ground is distorted by the opening up of the ground by mining and by the increased induction into the ground of aerated water as a result of water abstraction and agriculture. Two consequences of this to groundwater quality are described in the following section.

2.   POLLUTION BY AERATION

2.1. Nitrate in groundwater

Four oxidation states of nitrogen occur naturally in the following compounds, listed in order of increasing oxidation

NH_4^+ ammonium (or cell protein)

N_2 nitrogen gas

NO_2^- nitrite

NO_3^- nitrate

Their relative equilibrium stabilities at various oxygen tensions can be calculated from standard thermodynamic data (1) and the results are represented in Fig. 30 for the following oxidation-reduction equations:

Reaction	Standard Free Energy Change (kcal)	
$2NO_3^- + 2H^+ \rightarrow N_2 + \frac{5}{2}O_2 + H_2O$	- 3.83	 1
$NO_2^- + \frac{1}{2}O_2 \rightarrow NO_3^-$	- 18.18	 2
$NH_4^+ + \frac{3}{2}O_2 \rightarrow NO_2^- + H_2O + 2H^+$	- 45.95	 3
$2NH_4^+ + \frac{3}{2}O_2 \rightarrow N_2 + 3H_2O + 2H^+$	- 132.07	 4

The scheme of oxidation-reduction in the nitrogen system is complicated by the existence of two naturally-occurring reaction paths:

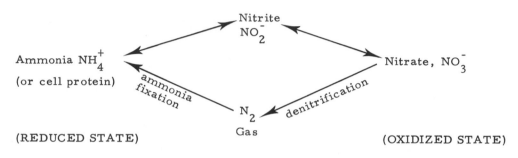

Nitrogen gas is stable over a very wide oxidation-reduction range which however, falls short of fully aerobic conditions. Under atmospheric conditions the ultimately stable form of nitrogen is nitrate, which is formed microbially at temperatures greater than $10^{\circ}C$. The existence of even very high concentrations of nitrate in water supplies should therefore be of no surprise; the wonder is that concentrations are usually so low. The explanation for this is that there are no naturally-occurring mechanisms by which ammonia is oxidized to nitrogen gas and nitrogen gas to nitrate (except by lightning discharge), whereas the reverse conversions do occur by the action of heterotrophic micro-organisms (2). In addition, nitrate and nitrite are used in the formation of plant and microbial cell protein, which is essentially the reduction to an ammonia derivative. Both the reduction of nitrate to nitrogen gas (denitrification) and the reduction of nitrogen gas to ammonia and proteins (nitrogen fixation) occur in the presence of organic matter (3). Nitrate is not produced naturally except when nitrogen fixation occurs in the presence of organic substances, which also allows some of the nitrate to disappear by denitrification. However, this is not true when ammonia is introduced artificially, such as by the application of ammonia fertilizers, when nitrate formation merely involves oxidation.

In soil, in which there is usually plenty of oxygen in the air spaces (4), the reduction of nitrate and nitrogen gas can only occur in micro-environments in which the appropriate low level of oxygen tension has been produced by the oxidation of organic matter. These regions occur either in soil crumbs and organic detritus or in the actual micro-organisms carrying out the reaction. The presence of oxygen, which is necessary for nitrate formation, does not prevent the occurrence of denitrification and nitrogen fixation provided that there is a sufficient concentration of organic matter in the system to enable anaerobic micro-regions to be formed (5)(6). A thorough account of the properties of nitrogen in soil is given in the complete publication listed under (3).

Nitrate enters groundwater by the flushing out of air spaces in soil by rainwater, the nitrate having been produced by the bacterial oxidation of ammonia and cell protein (7). The concentration of nitrate in soil depends upon the comparative rates of its production and consumption and hence upon the relative quantities of oxygen and ammonia on the one hand and non-nitrogenous organic matter on the other. Groundwaters containing little dissolved oxygen are also practically devoid of nitrate. Very high nitrate concentrations, however, are reported to occur in arid areas having poor soil cover (8).

Within the United Kingdom there is evidence of a relationship between nitrate in groundwater and the organic content of soil. Loss of soil humus has already occurred during the last twenty or so years and was the subject of a Ministry of Agriculture and Fisheries investigation in relation to an associated loss in soil mechanical stability (9). One area which was particularly badly affected was over the chalk outcrop in the Lindsey part of Lincolnshire, from which the North Lindsey Water Board also reported finding abnormally high nitrate concentrations in groundwater (10); a causal relationship was not proved.

The soil conditions required for high yield agriculture may not be those that give good groundwater. The deterioration of soil quality has been caused by intensive cereal production, involving heavy application of inorganic ammonia fertilizers. This requires that soil be well aerated by frequent ploughing which favours the oxidation of ammonia to nitrate. Such conditions also cause the inevitable loss of soil humus which, while diminishing the loss of nitrogen by denitrification, increases the overall concentrations of nitrate. The danger signal in this economy appears to agricultural interests as a loss in soil mechanical stability. It is a matter of urgent necessity to investigate whether the loss of soil humus is also a significant cause of high nitrate concentrations in groundwater.

2.2. Oxidation of iron sulphide

Iron sulphide occurs commonly as iron pyrite in sandstone, probably as a result of the bacterial reduction of sulphate in the presence of organic substances originally deposited with the sand. As organic substances are eventually oxidized in air, so also is the fate of iron sulphide. The fact that it still persists in an aquifer is evidence that the system has been confined and until it became exploited for water supply, did not receive a current of oxygenated rainwater.

The oxidation states of the iron and sulphur systems (1) are represented in Fig. 31. This shows that the reduction of sulphate to hydrogen sulphide occurs only under extremely strongly reducing conditions which could scarcely be achieved by the total oxidation of an organic substance, which is represented in the figure by the methane-carbon dioxide equilibrium. Formation of hydrogen sulphide requires the presence of fresh living matter, the decomposition of which involves an easy oxidation stage leading to only partial decay.

The presence of iron aids the reduction of sulphate by stabilizing the sulphide product as the insoluble iron sulphide. The reduction of sulphate need not signify therefore, the presence of significant concentrations of hydrogen sulphide.

The oxidation of iron sulphide by oxygen is represented by the following sequence of reactions which would occur if equilibrium oxygen tension were attained in all cases:

$$\underset{\text{iron pyrite}}{FeS_2} \xrightarrow{\text{(a)}} \underset{\text{ferrous sulphate}}{FeSO_4} \xrightarrow{\text{(b)}} \underset{\text{hydrated ferric oxide}}{Fe_2O_3 \cdot H_2O} \qquad \dots\dots\dots 5$$

However, because the stability of both iron sulphide and ferrous sulphate require the oxygen tension to be reduced to minute levels, reaction (a) does not in fact occur in the direction of oxidation. The presence of sufficient oxygen to cause any oxidation is enough to take the oxidation as far as the formation of ferric oxide.

Under natural conditions the oxidation of iron sulphide occurs through the action of sulphur-oxidizing bacteria (11)(12), which build their cellular substance autotrophically using the sulphur-oxidation as an energy source. They require also a sufficient concentration of carbon dioxide, which may be available either from the atmosphere or from dissolution of carbonate by the acid produced from sulphide-oxidation:

$$2FeS_2 + 5H_2O + 15\left[O\right] \rightarrow Fe_2O_3 \cdot H_2O + 4SO_4^{2-} + 8H^+ \qquad \dots\dots\dots 6$$

Under fully aerobic conditions Equation 6 has been shown to proceed by the agency of Thiobacillus thio-oxidans which requires dissolved oxygen as an oxygen source. At low

100.

oxygen concentrations <u>T. denitrificans</u> performs a similar process but by reducing nitrate to nitrogen gas (12). The latter process is probably responsible for the almost total absence of nitrate from waters associated with geological iron deposits. In formations containing sulphides not associated with carbonate rock, sulphur-oxidation may produce highly acidic water.

If conditions for sulphide oxidation are persistently favourable, the bacterial substance produced would in turn become oxidized and eventually disappear. When oxidizing conditions periodically cease and oxygen and nitrate concentrations fall to low values, sulphur oxidizers cease their activity, die and are replaced by anaerobic organisms. The formation of soluble iron then occurs by the reduction of iron oxide.

In the context of groundwater pollution, iron contamination was originally caused during the formation of an aquifer by the deposition of organic substance which subsequently gave way to iron sulphide. However, given that an aquifer contains iron sulphide, the factor which governs the amount of dissolved iron in groundwater is the quantity of dissolved oxygen which periodically reaches the sulphide. Sulphide will normally be more concentrated at the lower levels of an aquifer and therefore iron dissolution increases with exploitation of the aquifer.

Whereas the presence of iron dissolved in groundwater is inconvenient, it is undoubtedly preferable in the long term to the permanent precipitation of iron oxide in the aquifer rock that comes with unremitting pumping. Another possibility that should be avoided is the further introduction to the aquifer of soluble organic matter from surface contamination which is able to penetrate the aquifer to great depth. If this were to occur, the original undesirable composition of the aquifer rock would be perpetuated or made worse.

3. REMOVAL OF ORGANICS FROM GROUNDWATER

The principles governing the penetration of aquifers by organic substances are not well understood. In spite of the water-solubility of many naturally-occurring organic substances, their concentrations in groundwater are usually very small. For example, it is difficult to produce artificially water containing as little organic substance as occurs in chalk groundwater. This is all the more remarkable when the organic content of the overlying soil is considered. In this there is the risk of complacency with the belief that in the microbial processes affecting water during its passage into the ground, there exists the capacity for the removal of all manner of organic substances. This may not be the case.

Four types of natural process can be distinguished in the permanent removal of organic substances from water with its passage into the ground:

1. Filtration of insoluble matter. The process stops when the filtration capacity is filled, but the groundwater quality is thereby not impaired.

2. Consumption of substances for microbial respiration. This causes partial or total oxidation to carbon dioxide and water. Excretion products are likely to be more water-soluble than the food source because of the oxidation involved in their formation.

3. Precipitation of humus by microbial fermentation. In the absence of oxygen free energy for microbial metabolism may be obtained by the removal of carbon dioxide and water from organic substances. In the case of the large-molecular weight microbial excretion products known as 'humic substances', these processes generally cause a reduction in water solubility (13).

4. Consumption of substances for microbial cell synthesis.

The process of adsorption of dissolved organic substances onto mineral particles is not included as a natural removal process because its continuation is dependent upon the destruction of adsorbed material by microbial oxidation.

In the consideration of biodegradation of substances in water, a distinction should be drawn between conditions which favour the most rapid effect and those which bring about the most thorough removal of substance from water. It is undoubtedly true that oxidative degradation yields the greater energy for microbial growth and that microbial activity is therefore greater under aerobic conditions. However, there appear to exist biochemical 'dead-ends', as for example, the formation of the soluble humic substances in agricultural and moorland sub-surface water. These constitute most of the total organic content of river water (14), often at concentrations of several milligrams per litre, but with an associated bacterial concentration which is comparatively low. In comparison, a substance which is easily degraded may cause very high bacterial concentrations by waterworks standards when its own concentration is barely detectable by the most sensitive chemical methods.

The fact that humic substances occur in soil water, sometimes in high concentrations, but not in groundwater is not understood. These substances are known to be adsorbed by soil minerals to an extent depending upon the hardness of soil water, but this does not adequately account for the effect. The humic fraction in river water consists largely of the type known as 'fulvic acid' which depends for its solubility upon its possession of both carboxylic acid and other water-attracting units in its constitution. The reason why these substances do not rapidly decay is not known, but a contributing factor is possibly that they represent an 'oxidizing dead-end'. In physico-chemical terms this would signify that further oxidation requires an activation energy barrier which cannot be circumvented by any enzyme reaction. Substances having such an obstacle to decomposition may well, under suitably anaerobic conditions, undergo other biochemical reactions, including dehydration and decarboxylation, causing a reduction in water-solubility. A possible biochemical explanation for the presence of stable humic substances in surface water is therefore that they have been removed from the mixed aerobic-anaerobic environment of the soil in which alternative decay processes can occur. Another possibility is that under the consistently aerobic conditions of surface water, the decay of natural organic matter follows different biochemical pathways from those occurring in soil, yielding a product which can be neither rendered insoluble nor is susceptible to further rapid decay. With reference to organic pollution of the ground generally, it should be considered whether there are certain conditions in soil or in waste tips, etc., which either do not allow maximum removal of organic substances from water or favour the wrong sort of decay process, producing substances which cannot be removed subsequently by the normal processes of decay.

4. INORGANIC POLLUTANTS

For the prediction of the solubility of inorganic substances in the ground, it is necessary to assume the formation of well-defined insoluble compounds (1). Whereas the adsorption of ions onto mineral particles is bound to occur, this cannot be confidently taken into account because of the many uncertainties involved. Furthermore, for the removal of unnatural contaminants from groundwater, it is probably improper to depend upon adsorption by rock particles since this is essentially only a temporary effect.

The chemical equilibria prevailing in groundwater systems are largely governed by pH and bicarbonate concentration, which in turn are affected in most cases by the presence of calcium carbonate. Groundwater acquires carbon dioxide from microbial respiration and fermentation in soil and calcium carbonate is subsequently dissolved to form bicarbonate. The relevant equilibria equations for a temperature of $10^{\circ}C$ are as follows:

$$\left[Ca^{2+}\right]\left[CO_3^{2-}\right] = 10^{-8.01} \qquad \dots\dots\dots 5$$

$$\frac{\left[Ca^{2+}\right]\left[HCO_3^{-}\right]^2}{\left[CO_2\right]} = 10^{-3.98} \qquad \dots\dots\dots 6$$

$$\frac{\left[Ca^{2+}\right]\left[HCO_3^{-}\right]}{\left[H^{+}\right]} = 10^{2.48} \qquad \dots\dots\dots 7$$

the bracketed symbols indicating molar concentrations. When all of the calcium and bicarbonate are present as calcium bicarbonate,

$$\left[Ca^{2+}\right]^3 = 10^{-4.58}\left[CO_2\right] \qquad \ldots\ldots\ldots 8$$

$$\text{and}\ \frac{\left[Ca^{2+}\right]^2}{\left[H^+\right]} = 10^{2.18} \qquad \ldots\ldots\ldots 9$$

Equations 8 and 9 are represented in Fig. 32 which, when compared with values for actual groundwater composition, show that equilibrium with calcium carbonate is usually attained approximately. Many hard surface waters, however, are greatly supersaturated, which is due probably to the presence of impurities which prevent crystallization of calcium carbonate.

The solubility of other ions is affected accordingly:

4.1. Phosphate

The least soluble form of phosphate, and that which usually occurs in the ground is hydroxyapatite, $Ca_5OH(PO_4)_3$, the solubility product for which has been given as (15).

$$\left[Ca^{2+}\right]^5 \left[OH^-\right]\left[PO_4^{3-}\right]^3 = 10^{-58.3} \qquad \ldots\ldots\ldots 10$$

The relationship between $\left[PO_4^{3-}\right]$ and the total phosphate concentration depends upon the concentrations of the irons $H_2PO_4^-$, HPO_4^{2-} (16) and of their ion pairs with calcium, $CaH_2PO_4^+$, $CaHPO_4$ (17). The total phosphate concentration, $\left[P\right]$, is given by the equation,

$$\left[P\right] = \left\{10^{18.6}\left[H^+\right]^2 + 10^{11.5}\left[H^+\right] + 10^{19.7}\left[Ca^{2+}\right]\left[H^+\right]^2 + 10^{14.2}\left[Ca^{2+}\right]\left[H^+\right]\right\}\left[PO_4^{3-}\right] \ldots 11$$

From Equations 10 and 11 are obtained the plots of total phosphate concentration versus pH given in Fig. 33 for two calcium concentrations.

The equilibrium phosphorus concentration which occurs when water is also in equilibrium with calcium carbonate is less than $0.1\mu g/litre$. It is of interest that polluted waters which are supersaturated with respect to calcium carbonate are also supersaturated with phosphorus. It appears that the hindrance to the precipitation of calcium carbonate may also affect the precipitation of hydroxyapatite. In groundwaters which are in equilibrium with calcium carbonate, the concentration of phosphorus is usually less than the limit of detection by normal analytical methods.

4.2. Fluoride

Although artificial pollution by fluoride is uncommon, this ion occurs naturally in some waters at concentrations which are greater than recommended standards. It may therefore be considered as a pollutant derived naturally from the rock.

The least soluble naturally-occurring compound of fluorine is fluorapatite, which has the same structure and composition as hydroxyapatite, but with some of the OH^- replaced by F^-. No experimentally determined solubility data are available, but an estimate based upon the relative stabilities of hydroxyapatite, $Ca(OH)_2$ and CaF_2 corresponds to a solubility product,

$$\left[Ca^{2+}\right]^5 \left[F^-\right]\left[PO_4^{3-}\right]^3 = 10^{-80} \qquad \dots\dots\dots 12$$

This indicates that where there is sufficient phosphate and calcium and the formation of hydroxyapatite actually occurs, the concentrations of both fluoride and phosphate are reduced to extremely low values.

4.3. Chloride and sulphate

At the concentrations which these ions normally occur in water supplies, they form no insoluble salts in the ground and therefore travel freely with groundwater.

4.4. Iron

At marginal oxygen concentrations, the solubility of ferrous iron, Fe^{2+}, is limited by the formation of hydrated ferric oxide, as described by the equation (1),

$$- \log\left[Fe^{2+}\right] = 2pH + \tfrac{1}{4}\log\left[O_2\right] + 3.74 \qquad \dots\dots\dots 13$$

This condition is represented for pH 6.5 in Fig. 31, which shows the oxidation range of stability of soluble iron between its precipitation either as the hydrated oxide or the sulphide. Within this range, an upper limit to the solubility of Fe^{2+} occurs with the precipitation of ferrous carbonate, according to a solubility product (18),

$$\left[Fe^{2+}\right]\left[CO_3^{2-}\right] = 10^{-10.24} \qquad \dots\dots\dots 14$$

By comparison with Equation 5, it follows that when equilibrium with calcium carbonate is attained, the upper limit of $\left[Fe^{2+}\right]$ is given by

$$\left[Fe^{2+}\right] = \frac{\left[Ca^{2+}\right]}{167} \qquad \dots\dots\dots 15$$

4.5. Other heavy metals

Many heavy metals behave similarly to iron in forming highly insoluble sulphides under strongly reducing conditions, but not all of them form other very insoluble compounds under natural aerobic conditions. Table 5 lists the solubility of several metal sulphides calculated for equilibrium with elemental sulphur, which describes approximately the oxidation level in anaerobic deposits and geological formations where such metals are likely to occur. The relevant equilibrium equations are of the form,

$$3M^{2+} + 4S + 4H_2O \rightleftharpoons 3MS + SO_4^{2-} + 8H^+ \qquad \dots\dots\dots 16$$

and the solubilities are calculated for a sulphate concentration of $10^{-3}M$ and pH values of 2.5 and 6.5.

TABLE 5

SOLUBILITY OF SOME HEAVY METALS IN ANAEROBIC SULPHIDE DEPOSITS

Metal	Soluble form	Solubility of sulphide (molar)	
		pH 2.5	pH 6.5
Arsenic	$HAsO_2$	$10^{-10.2}$	$10^{-14.2}$
Cadmium	Cd^{2+}	$10^{-6.5}$	$10^{-17.2}$
Cobalt	Co^{2+}	0.2	$10^{-11.4}$
Copper	Cu^{2+}	$10^{-15.5}$	$10^{-26.2}$
Iron	Fe^{2+}	high	$10^{-7.0}$
Lead	Pb^{2+}	$10^{-7.5}$	$10^{-18.2}$
Mercury	Hg^{2+}	$10^{-32.4}$	$10^{-43.1}$
Nickel	Ni^{2+}	0.8	$10^{-10.8}$

The figures show that from an anaerobic deposit containing these metals, none of them would be liberated in soluble form if alkali were present to prevent the pH falling below the bicarbonate/carbon dioxide pK value of 6.5. If pH were to fall, as for example by the oxidation of mining waste tips, dissolution would occur in the order $Fe^{2+} > Ni^{2+} > Co^{2+}$. No significant concentration of other metals would occur while the deposit remained anaerobic.

In aerated water the solubility of metals is generally much greater. Divalent ions do not form very insoluble hydroxides; of the metals listed, nickel forms the least soluble hydroxide, with a solubility of 60 mg/litre of Ni at pH 6.5. Iron is insoluble in aerated water due to the formation of the trivalent hydrated ferric oxide. Another metal having an insoluble trivalent hydroxide is chromium. However, although $Cr(OH)_3$ has practical stability at neutral pH and normal oxygen concentrations, the thermodynamically stable form under these conditions is chromate CrO_4^{2-}, which is toxic to many organisms at very low concentrations. The reduction of chromate already present in a system to the insoluble hydroxide requires reducing conditions similar to those where nitrate reduction occurs. However, such reduction does not occur microbially because of the toxicity of chromate. Indeed, the reduction and consequent precipitation of $Cr(OH)_3$ on the outer microbial membranes may be the explanation for the toxicity.

A compound of arsenic which is stable in aerated water is the highly insoluble barium arsenate. The presence of this in the soil, in association with barium sulphate, may account for the very much greater arsenic concentrations in soil than can be accounted for in soil water. Arsenate is reduced to the soluble and more toxic trivalent form also under conditions similar to those for nitrate reduction. In the presence of arsenic compounds, groundwater would therefore contain arsenic in solution if its oxidation state were marginally anaerobic.

The only other inorganic forms in which metals may be precipitated in aerated water are the carbonates or basic carbonates. Of the metals listed in Table 5, cadmium and lead have least solubility in these forms. The solubility product for cadmium carbonate is (16).

$$\left[Cd^{2+}\right]\left[CO_3^{2-}\right] = 10^{-13} \qquad \dots\dots 17$$

When equilibrium with calcium carbonate is attained,

$$\left[Cd^{2+}\right] = \frac{\left[Ca^{2+}\right]}{10^5} \qquad \dots\dots 18$$

The solubility of lead as the basic carbonate is governed by the product (16),

$$\left[Pb^{2+}\right]^3\left[OH^-\right]^2\left[CO_3^{2-}\right]^2 = 10^{-45.46} \qquad \dots\dots 19$$

When equilibrium with calcium carbonate is attained,

$$\left[Pb^{2+}\right]^3 = 10^{-29.4}\frac{\left[Ca^{2+}\right]^2}{\left[OH^-\right]^2} \qquad \dots\dots 20$$

Equilibrium concentrations of Ca^{2+}, Fe^{2+}, Cd^{2+} and Pb^{2+} are given in Table 6 for a selection of pH values and the condition where $\left[HCO_3^-\right] = 2\left[Ca^{2+}\right]$

TABLE 6

CONCENTRATIONS (mg/litre) OF METAL IONS FORMING
CARBONATES OR BASIC CARBONATE WHEN EQUILIBRIUM
WITH $CaCO_3$ IS ATTAINED

pH	Ca^{2+}	Fe^{2+}	Cd^{2+}	Pb^{2+}
7.2	124	1.04	0.0035	0.0211
7.6	76	0.64	0.0021	0.0083
8.0	48	0.40	0.0013	0.0041
8.5	28	0.23	0.0008	0.0013

Under these conditions the concentrations of both cadmium and lead are well below World Health Organization standards for drinking water. Their concentrations could be much greater, however, if equilibrium with calcium carbonate were not attained. An instance of cadmium contamination in a sand and gravel aquifer was described in (19).

5. MOVEMENT OF POLLUTION

Polluted water may enter an aquifer by seepage through the substance of an aquifer or by flowing rapidly through comparatively wide channels in granular media and through fissures. The latter involve mainly hydrological factors, chemical considerations being of secondary importance because the polluted water is largely unaffected by chemical or microbial processes. If abstracted groundwater is affected by pollution through rapid-transit channels, the effects on the aquifer are equally rapidly overcome by the flushing out of the channels by clean water.

However, there is cause for concern in systems which are not fissured and which are apparently unaffected by obvious sources of pollution. The concern is that an aquifer rock removes pollution from the water and that the rock itself becomes contaminated. The question which should be answered is whether a pollution front can slowly move through an aquifer, eventually to fill it and with little chance of the pollution being removable.

For contaminants which are removed from water only by adsorption or precipitation onto the aquifer rock, both substances being in equilibrium, a front of the contaminant moves through the aquifer rock as described by the principles of chromatography (20).

The difference between adsorption and precipitation is merely in the extent to which a given bulk volume of aquifer rock provides either adsorptive capacity or chemical substance to allow the removal of the contaminant from water to occur. It would be reasonable to expect that an adsorption front would generally move faster than a precipitation front. In most hard water aquifers, for example, oils and other organic matter having surface active properties would be retained by adsorption, whereas phosphate, fluroride, lead, cadmium and arsenic would form precipitates. The capacity of an aquifer for retaining inorganic contaminants is likely to be much greater than for organic substances.

A factor which it is impossible to account for from theory is whether the transfer of contaminants from water to the aquifer rock would impede the flow of water. If the incidence of groundwater pollution is rare in comparison with the number of potential sources of pollution, it might be due to automatic sealing of the aquifer by the contaminant. It is important to know whether pollution fronts in aquifer rock originating from pollution sources are actually moving.

An explanation, other than by blocking, for a stationary front of an organic pollutant is that a steady state has occurred due to the oxidation of the substance by microbial action. For this to occur, without the microbial growths themselves causing blocking, is probably an indication that the pollutant is present in very low concentrations in the groundwater. Whether the microbial growths are themselves converted to insoluble humic substance and stop the continued flow of pollution may depend upon whether the flow is continuous or whether the aquifer periodically runs dry and the accumulated growths thus allowed to decay. Depending upon whether a pollution source is for the disposal of solid or liquid, there is reason for investigating conditions of operation. which favour either blockage or free water movement respectively. Whereas it may be feared, for example, that metal ions could be leached by water from mineral spoil, it may be preferable, nevertheless, to keep such a deposit totally water-logged. Under this condition microbial activity will be predominantly anaerobic, an organic component of the deposit will be preserved and heavy metals remain precipitated in the form of sulphides. For the disposal of sewage water from septic tanks, however, continual passage of water is required, for which it may be necessary to have occasional periods of unsaturated flow to permit the decay of microbial deposits. It is not necessarily true, however, that the best conditions for the disposal of waste water are also best for groundwater quality.

It is hoped that experience made known by this Conference will cast light on some of these matters.

REFERENCES

1. LATIMER, W. M.

The oxidation states of the elements and
their potential in aqueous solutions
2nd Edition.
Englewood Cliffs, N. J., Prentice-Hall Inc.
1961.

2. ROSE, A. H.

Chemical microbiology,
London, Butterworth, 1965.

3. GASSER, J. K. R.

Some processes affecting nitrogen in soil.
Min. of Ag., Fish. and Food.
HMSO, 1969, Tech. Bull. No. 15, pp 15-29

4. RUSSELL, Sir E. J.

Soil conditions and plant growth,
8th Edition.
London, Longmans, Green & Co., 1958.

5. WIJLER, J., and
DELWICHE, C. C.

Investigations on the denitrifying process
in soil.
Plant and Soil, 1954, 5 (2), pp 155-169.

6. SMITH, J. M., and others

Nitrogen removal from municipal waste
water by columnar denitrification.
Env. Sci & Tech., 1972, 6 (3), pp 260-7

7. COOKE, G. E., and
WILLIAMS, R. J. B.

Losses of nitrogen and phosphorus from
agricultural land.
Wat. Treat. & Exam., 1970, 19 (3), pp 253-274.

8. SALITERNIK, C.

Groundwater pollution by nitrogen compounds
in Israel. Proc. 6th Int. Wat. Poll. Res. Conf.,
1972.
To be published by Pergamon Press

9. AGRICULTURAL ADVISORY
COUNCIL ON SOIL STRUCTURE
AND SOIL FERTILITY

Modern farming and the soil.
HMSO, 1970.

10. NORTH LINDSEY WATER BOARD

An investigation into the nitrate pollution of
the chalk borehole water supplies.
Scunthorpe, The Board, 1970.

11. ZAJIC, J. E.

Microbial biogeochemistry.
London, Academic Press, 1969.

12. PURKISS, B. E.

Corrosion in industrial situations by mixed
microbial floras.
Chap. 4 of 'Microbial aspects of metallurgy',
J. D. A. Miller, (Ed),
Aylesbury, Medical and Technical Pub. Co.
Ltd., 1971.

13. KONONOVA, M. M.

Soil organic matter.
2nd English Ed.,
Oxford, Pergamon Press, 1966.

REFERENCES

14. PACKHAM, R.F.

Studies of organic colour in natural waters.
Proc. Soc. Wat. Treat. & Exam., 1964,
13 (4), pp 316-334.

15. WIER, D.R., and others

Solubility of hydroxyapatite.
Soil Science, 1971, 111 (2), pp 107-112.

16. SILLEN, L.G.

Stability constants.
London, The Chemical Society, 1964.

17. DAVIES, C.W., and
HOYLE, B.E.

The interaction of calcium ions with some
phosphate and nitrate buffers.
J. chem. Soc, 1953, pp 4134-6.

18. SINGER, P.C., and
STUMM, W.

Solubility of ferrous iron in carbonate-
bearing waters.
J. Am. Wat. Wks Ass., 1970, 62 (3),
pp 198-202.

19. LEIBER, M. and
WELSCH, W.F.

Contamination of groundwater by cadmium
J. Am. Wat. Wks. Assoc., 1954, 46, pp 541-547.

20. SIXMA, F.L.J. and
WYNBERG, H.

A manual of physical methods in organic
chemistry.
New York, John Wiley, Inc, 1964.

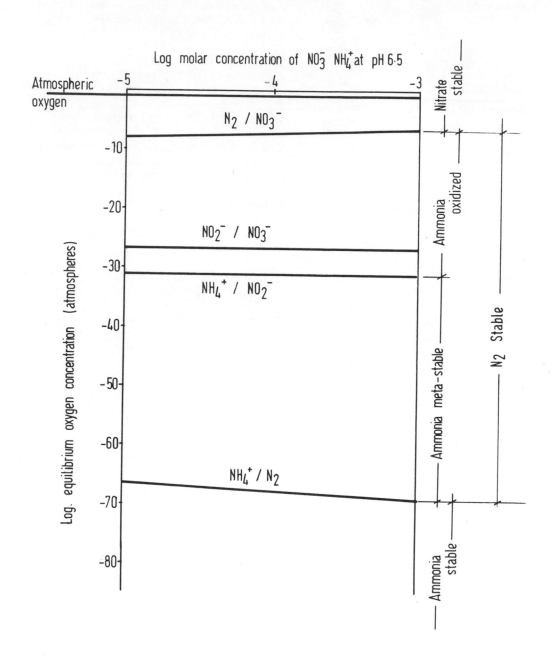

FIG. 30.   STABILITY DIAGRAM FOR THE NITROGEN SYSTEM

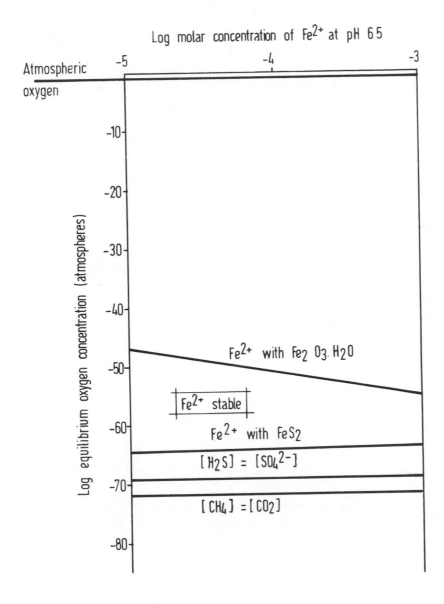

Log molar concentration of Fe^{2+} at pH 6 5

FIG. 31. STABILITY DIAGRAM FOR IRON, SULPHUR
AND CARBON SYSTEMS

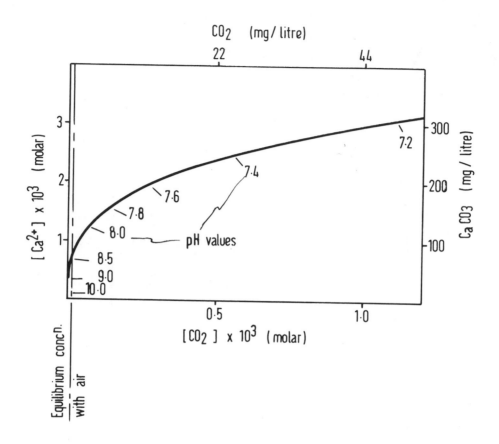

FIG. 32.   RELATION BETWEEN VARIABLES FOR WATER
IN EQUILIBRIUM WITH CALCIUM CARBONATE

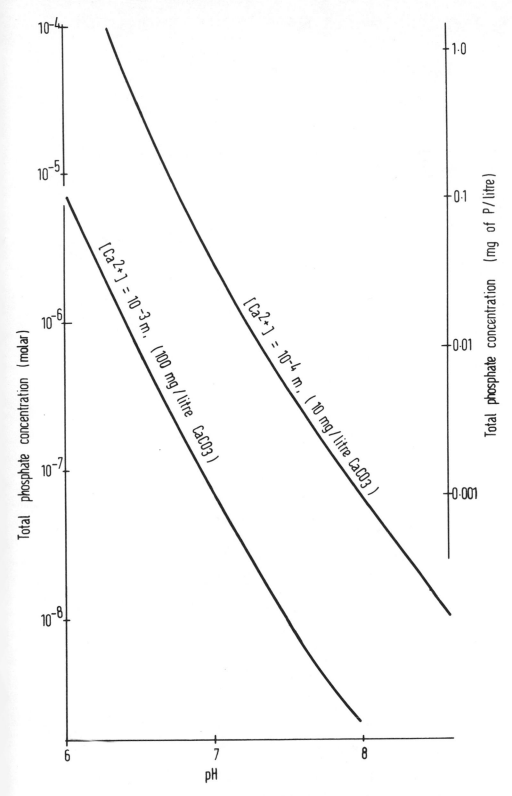

FIG. 33. PHOSPHATE CONCENTRATIONS IN EQUILIBRIUM
WITH HYDROXYAPATITE

POLLUTION AFFECTING WELLS IN THE BUNTER SANDSTONE

by G.D. Nicholls

Edgar Morton and Partners, Adlington, Cheshire, England

1. INTRODUCTION

Aquifers vary in their vulnerability to groundwater pollution, due in part to variation in their mineralogy and in part to variation in their water transmission characteristics. In general, those aquifers in which water transmission is largely along fissures rather than through the body of the rock (permeable but not porous) are more prone to pollution of well water than porous aquifers. Much of the Bunter Sandstone is a partially or wholly cemented sandstone through which water moves largely along fissures, though in some parts of the country and at some horizons in the geological succession, incomplete cementation has left some residual porosity.

The effects of groundwater pollution may be either direct or indirect. In cases of direct pollution the deterioration in water quality is of a character directly related to the nature of the pollutant, e.g. rise in nitrate content consequent upon the over-liberal use of nitrogenous fertilizer on soils overlying the aquifer or rise in chloride content consequent upon intrusion of saline waters into over-pumped near-coastal aquifers. With indirect pollution, the relationship between cause and effect may be concealed by a sequence of events and interactions spread over a time span of years, so that the pollution cause may, sometimes, not be recognized for the observed water quality deterioration. Detection and correction is obviously much more difficult in the case of indirect pollution.

Pollution may be widespread or local, dependent on the cause, and it is also possible to distinguish between the short-term pollution incident and long-term progressive pollution in which water quality deterioration is so slow that pollution may not be immediately recognized as the cause. Likewise it is possible to recognize in the history of water quality variation from some wells the superimposition of multiple incidents, the combined effect of which is to produce unsatisfactory, or even unacceptable water when any single pollution cause acting in isolation would not have done so.

The obligation placed upon officers of water undertakings to maintain the purity and wholesomeness of their supply necessitates constant vigilance over well catchments. In that well-known stimulant to intellectual strife, "The Limits of Growth", Professor Meadows and his associates have written, with penetrating insight into the heart of the pollution problem, that "any pollution control system based on instituting controls only when some

116.

harm is already detected will probably guarantee that the problem will get much worse before it gets better". We might simplify this into "pollution prevention is better than purification" and recognize that our true objective should be identification of potential pollution hazards before any pollution develops. Some guidance towards this identification may be found in the study of various examples of different types of pollution which have affected wells in the Bunter Sandstone. Some encountered by the author during the last decade are discussed in this paper.

2. LOCAL SHORT-TERM POLLUTION INCIDENTS

Accidental release of noxious liquids onto porous soils overlying well catchments is a continuing and constant pollution hazard. The transfer of much haulage business from rail to road increases the danger of pollution from this source. The movement about the country of large tankers carrying considerable volumes of noxious liquids of a wide variety of chemical character poses a threat to many wells located in reasonably close proximity to through or trunk roads. On 5 May 1972, the Gazette of the British Waterworks Association reported an accident involving a tanker carrying molten sulphur at the junction of the A18 and A614 trunk roads near the Hatfield Woodhouse borehole site of the Doncaster and District Joint Water Board. The tanker left the road and eventually crashed onto its side in a ploughed field, bursting open and spilling molten sulphur. Fortunately this borehole site is not at present in use for supply and, in any event, the escaping molten sulphur solidified on the ground surface. Although the Bunter Sandstone is here overlain by sands and gravels and percolation is relatively easy there should be no serious after-effects since the sulphur was excavated and removed from the site within a short time. Had the tanker load been a dangerous chemical, liquid at normal atmospheric temperatures, contamination of the groundwater would most likely have been almost immediate and unavoidable. Had the site been a pumping station it would have almost certainly been put out of commission for several weeks or months. Clearly there is little that a water undertaking can do to minimize this hazard if it is subject to it. The only practical action is to develop the supply position in such a way that wells whose catchments are traversed by major trunk roads become progressively less important in maintaining supply.

Local short-term pollution need not always be due to a sudden accident. One case in which it was not occurred in Shropshire in 1969-70. A farmer excavated an area of his land near Worfield in the Worfe valley, intending to use it as a reservoir for irrigation water and diverting for this purpose the waters of Merecot Brook, a tributary of the River Worfe. The excavations penetrated into the Bunter Sandstone over rather more than half the total area, breaching the river alluvium which seals the beds of the brook and the river from contact with the Bunter aquifer. A quarter of a mile to the north-east of the reservoir site the Stableford Pumping Station draws between 1.3 and 2.2 m.g.d.

from this same Bunter aquifer and prior to September 1969, the water was always of excellent bacteriological quality. Between January and September 1969, only one sample from 136 examined contained a single coliform organism. However, as the reservoir was filled with water the quality of water pumped from the station deteriorated, leaving no doubt that some of the diverted brook water was escaping underground into the aquifer and thus reaching the pumping boreholes. Of 13 samples examined in October 1969, five contained coliforms and in November, 15 samples were examined of which eight contained coliforms and three, E. Coli I. Bacterial examination of relevant waters then gave the following results:

Sample	Coliforms per 100 ml	
	Total	E. coli I.
Merecot Brook water	3033	791
Reservoir water (average)	3289	1253

Clearly, the time of passage of the water through the aquifer from the reservoir site to the pumping station was short enough for some of the coliforms to persist in the water until it was pumped. Deterioration of water quality in October and November 1969 necessitated taking the source out of supply. The stream diversion had been made without authority and in December 1969, the Severn River Authority required the farmer to return Merecot Brook to its normal course and original level. This was done and the reservoir drained during the first part of 1970. During 1970 the bacteriological quality of the water pumped at the pumping station improved, though in August and September of that year an occasional sample was still unsatisfactory.

A general feature of these short-term pollution incidents is that the source of pollution is at no great distance from the well, so that an immediate contribution of polluted water to pumped water is significant. Once the source of pollution is removed or dealt with the water quality improves and reverts to, or almost to, its pre-pollution quality.

3. LONG-TERM PROGRESSIVE POLLUTION

The existence at the surface above the catchment of unnatural accumulations (waste dumps) may result in a slow and progressive pollution of the groundwater that may eventually reach the stage where a well must be taken out of supply, but more generally results in some, though not excessive, deterioration in water quality during the lifetime of a well. Such deterioration, however, makes a well progressively more vulnerable to being rendered unusable by short-term local pollution incidents.

Colliery tips are examples of such waste dumps as are being considered here. Colliery tip material is a mixture of unweathered broken rock, subjected for the first time since its formation to the oxidative and destructive attack of rain water and the atmosphere, with various materials introduced in the course of mining operations. Reaction between the tip material and percolating waters takes place readily and, in consequence, the water seeping through colliery tips is very impure. For one Yorkshire colliery tip the chloride content of the tip seepage ranged from 1300 to 3000 mg/litre, sulphate from 2600 to 5500 mg/litre, sodium from 1750 to 3400 mg/litre, iron from 0.1 to 5.3 mg/litre and manganese from 0.1 to 2.7 mg/litre. Such seepage water is much richer in dissolved constituents than surface drainage from the tips which is more easily (and more usually) sampled. Entry of such seepage water into an aquifer will clearly cause deterioration in the quality of most groundwaters.

An example of long-term progressive pollution of this kind is afforded by the history of the Sandall Beat borehole and Nutwell Pumping Station east of Doncaster. Both the borehole and the pumping station are in relatively close proximity to a large working colliery in the concealed Yorkshire coalfield (Markham Main Colliery,Fig. 34). A very extensive colliery tip associated with this colliery has been deposited across an outcrop of the Bunter Sandstone aquifer and onto sand and gravel superficial deposits overlying the sandstone. There appears to be nothing to stop seepage from the tip percolating down into the aquifer and it is likely that pollution of the groundwater in the sandstone has been going on for a very long time. From the map (Fig. 34) it is clear that the Sandall Beat borehole is much more vulnerable than the Nutwell Pumping Station.

When the Sandall Beat borehole was commissioned in 1919 the total hardness of the pumped water was only 126 mg/litre and the chloride content 12.9 mg/litre. By June 1925 the total hardness had risen to 163 mg/litre and chloride to 25.7 mg/litre. In December 1932 the total hardness had reached 414 mg/litre and chloride, 337 mg/litre and the borehole was abandoned for public supply in 1933. During these and subsequent years, of course, the colliery tip continued to grow and the deterioration of the groundwater quality continued after 1933. In July 1965 the water had a total hardness of 2060 mg/litre and a chloride content of 2925 mg/litre. Since 1965 the chloride content of the water has dropped slightly and in 1971 it was 2600 mg/litre. Extension of the tip has brought the margin of it progressively nearer to the borehole, though it was never more than half a mile from the Sandall Beat site. The slow progressive replacement of unpolluted groundwater by seepage water from the tip leading to abandonment of the source after 14 years well illustrates long-term progressive pollution.

A further, more recent, possible source of pollution is the lagoon shown in Fig. 34, which receives effluent from a washery and briquetting plant at the colliery. This lagoon overlies porous superficial deposits, though its floor is now covered with black mud which probably protects the sands and gravels below. The chloride content of this effluent is in the range 2350 to 2750 mg/litre, the sodium content 1100 to 1300 mg/litre, sulphate 380 to 580 mg/litre and total hardness 980 to 1200 mg/litre. This lagoon drains into the Fore's Drain (shown in Fig. 34) via a complicated system of ditches in Sandall Beat Wood. On 6 July 1971 the water in this Drain where it passes the Nutwell Pumping Station was chloride-rich (2450 mg/litre). Along much of its course the Fore's Drain flows over a silty clay(Fig. 34) which probably protects the underlying aquifer from infiltration of this chlorine-rich water. In two places, however, the drainage from the lagoon flows over the outcrop of superficial sands overlying the Bunter aquifer. One is the Sandall Beat ditch area immediately south of the lagoon. The other is some 400 yards east of the Nutwell Pumping Station, where the Fore's Drain passes off the silty clay for a distance of about 300 yards.

Long-term changes in the total hardness, chloride and sulphate contents of water drawn from the Nutwell Pumping Station during the last twenty years are only slight:

	1951-3	1970
Total hardness	174 to 192	204 to 208 mg/litre
Chloride	21 to 30	33 to 34
Sulphate	18 to 20	22 to 25

These figures, however, conceal short-term fluctuations, especially during the last decade. Towards the end of 1966 the total hardness reached 286 mg/litre, chloride 39 mg/litre, sulphate 34 mg/litre and the iron content of the water became troublesome at 0.31 mg/litre, temporarily reaching 0.62 mg/litre on 1 November in water from Borehole 1. Comparable, though less striking, short-term changes occurred in 1963 and 1967. The iron content of the pumped water has become more troublesome in latter years as will be discussed shortly.

The most likely interpretation of these short-term changes in Nutwell water during the last decade is that water originating in the vicinity of Markham Main Colliery is temporarily but intermittently, invading the extraction zone of the Nutwell boreholes. As will be seen from Fig. 34 the colliery is over one mile from the pumping station and the possibility cannot be excluded that the Fore's Drain is transporting the pollutant water to a point of entry into the aquifer east of the pumping station. With the present pumping regime, the situation at Nutwell is reasonably stable against invasion by Markham Main waters, but if in future natural percolation over the present catchment should be reduced by industrial or urban development the water at Nutwell might be expected to deteriorate in quality in the same way but to a lesser degree as that in the Sandall Beat borehole.

In addition to the direct pollutant effect of Markham Main waters on the groundwater in the Bunter aquifer there is an indirect effect of equal or greater importance in determining the acceptability or otherwise of Nutwell water for supply. Both tip seepage and the lagoon water are de-oxygenated (negative oxidation potential). On mixing with normal percolate from the surface they will lower the oxidation potential of a very much larger volume of groundwater. The solubility of the iron oxide pellicles around the grains of Triassic sandstones in acid solutions is well-known, but their solubility in waters of low oxidation potential is less widely appreciated. In fact laboratory studies confirm thermodynamic calculations that a fall in the oxidation potential of groundwaters must result in considerable leaching of iron from any iron-bearing rock into the groundwater and if the water is drawn into a well while its oxidation potential is less than normal the pumped water will carry an abnormally high iron content. High iron contents at Nutwell are due to the de-oxygenating effects of some of the waters or drainage entering the aquifer. This has been developing progressively over the last twenty years as the colliery tip has extended and the volume of effluent discharged from ancillary plant has increased and illustrates long-term progressive pollution. Between 30 November 1970 and 28 May 1971 the iron content of Nutwell water varied within the range 0.12 to 0.68 mg/litre.

The high values for the iron content in Nutwell water are encountered rather spasmodically. For example, on 30 December 1970 the iron content was 0.32 mg/litre; on 31 December it had risen to 0.68 mg/litre and by 5 January it was back to 0.39 mg/litre. This suggests the intermittent and short-term discharge into the aquifer in relative close vicinity of the pumping station of either iron-rich or organic-rich liquids. There is no known source for iron-rich solutions, but there are two possible sources of organic-rich liquids, the entry of which into the aquifer would still further de-oxygenate the groundwater. These are a packing station (shown in Fig. 34) which, until restrained by the water undertaking, discharged rather obnoxious effluent from vegetable canning processes into soakaways on the catchment during the summer, and a piggery to the south of the pumping station which discharges its effluent into Black Dike (Fig. 34), which flows across the outcrop of sands overlying the aquifer. The problem of iron contents in Nutwell water illustrates the effect of multiple sources of pollution. The effluent from the vegetable packing station had maximum adverse effect on the quality of water pumped from the station because the groundwater was already adversely affected in respect of its oxidation potential by the influx of Markham Main water into the aquifer.

4. INDIRECT POLLUTION EFFECTS

In some respects the iron content in Nutwell water might be regarded as resulting from indirect pollution, since the iron is not present in the polluting effluents and drainage. Escape of hydrocarbons and oils, e.g. petroleum, within a catchment can, likewise, have indirect effects as well as the obvious direct ones. If such hydrocarbons are carried down into the aquifer by descending percolate, their slow oxidation by dissolved oxygen in the groundwater may increase the dissolved carbon dioxide content, lowering both the pH and the oxidation potential of the water. Likely delayed and indirect effects would be an increase in temporary hardness and of the iron and manganese contents of the polluted groundwater.

Even more indirect pollution of a well source may result from the discharge of organic chemicals over a catchment supporting a heavy plant cover and in this case the effects may not be apparent for some years. Aerial spraying of conifer plantations with emulsified concentrates of insecticides in organic bases such as xylene affords an example of the type of operation, the long-term effects of which require careful consideration. Soils developed on sandstone under conifer plantations are often acid (pH 3.0 to 4.0), implying almost total de-oxygenation of percolating rain water during its passage through the organic soil profile. The considerable quantities of solvent hydrocarbons that fall to the ground during aerial spraying operations are then likely to be carried down through the soil profile into the underlying aquifer before their oxidation has proceeded to any appreciable extent. Under the plantation they will then be in the presence of largely de-oxygenated water and oxygen-free soil gases as they descend to the water table. At the water table they will tend to concentrate in the uppermost part of the groundwater and move with it towards any pumping well drawing water from the polluted area. If the groundwater and associated hydrocarbons have a long passage, in both the temporal and the spatial sense, to the pumping station, they will, particularly as they move nearer to the pumping station, mix with more oxygenate water from elsewhere in the catchment. Oxidation of the hydrocarbons will then produce carbon dioxide in the water at a level in the strata well below that at which groundwater normally derives its dissolved carbon dioxide content. Since the hydrocarbons will be localized at the water table in a zone of no great thickness, carbon dioxide production will be similarly localized. Such carbon dioxide enriched groundwater would be corrosive to any calcium carbonate (calcite) cement or matrix in the sandstone at the level where the additional carbon dioxide was produced. Such corrosion would be particularly effective adjacent to joints and fissures along which the water is drawn to the pumping well.

Attack on the calcium carbonate cement, exposing chloride and sulphate films around the sand grains to solution, might be expected to lead to increase in the chloride and sulphate contents of the water. Such chemical changes in the water would not follow immediately after the aerial spraying occurred, since it will clearly take time for the hydrocarbons to pass down to the water table and laterally through the strata, to mix with more oxygenated water prior to their oxidation. A period of up to 2 to 5 years might elapse before the effects on the chemistry of the water pumped from the well became noticeable. Some increase in temporary hardness might be detected, but if the spraying area was at some distance (a mile or more) from the pumping well it is unlikely that the oxidation potential of the pumped water would be lowered sufficiently to cause increased quantities of iron to be leached from the aquifer and carried to the well.

Perhaps more striking, and in some cases more important, than the chemical changes in the groundwater could be the effects on the sandstone itself. Intrastratal solution of the calcite cement of the aquifer rock, especially adjacent to fissures and joints through which the groundwater moves, would liberate sand grains, which would then be drawn towards the pumping well. It is possible that, eventually, sand would appear in the pumped water. Development of a secondary porosity in an aquifer through solution of the cement can have a further complication if the water table fluctuates in the vicinity of the pumping well, either through natural causes or through inconstant pumping rates. At times of low water table level, air may penetrate this secondary porosity and some of it may be trapped therein when the water table rises again. By reaction with organics or hydrocarbons this air may become enriched in carbon dioxide and depleted in oxygen, after which it may be drawn into the groundwater, partially dissolving at the pressures obtaining at the water table level. When the water is brought to the surface by pumping, this carbon dioxide enriched air is likely to separate out from solution in a multitude of minute bubbles, giving to the water a milky appearance. Although such milky water might well be perfectly potable every water engineer will be aware of the response to be expected from the general public to such waters. It may not be polluted in the strict definition of that term, but it has been affected in such a way as to cause consumer concern. Moreover, the incident(s) responsible for the milky appearance of the water may have occurred years before the first indication of developing milkiness. By then preventive action is too late - all that is possible is remedial action, such as sleeving and grouting the borehole with possible consequent reduction of yield.

Where indirect pollution is being considered the quotation from Meadows and his associates given in the introduction to this paper is particularly apt. Once the effects of such pollution are detected they will probably get much worse before they get better. If the water supply profession is to advance its understanding of groundwater pollution, free interchange of information, at least within the confines of the profession, is vital and necessary. On the other hand it may be argued that a wider dissemination of knowledge is not desirable since it may cause undue and unnecessary consumer alarm.

5. ACKNOWLEDGEMENTS

The author desires to express his thanks to Mr. M. Cawley and other officers of the Doncaster and District Joint Water Board and to Mr. W. C. Johnson and other officers of the Wolverhampton Water Undertaking for their help and willingness to allow him to describe examples from their areas.

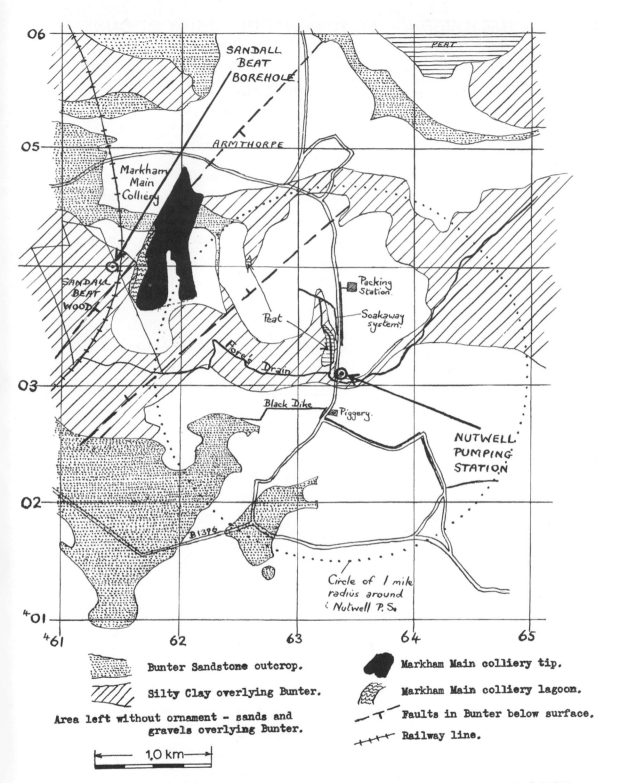

Legend

(stippled)	Bunter Sandstone outcrop.
(hatched)	Silty Clay overlying Bunter.
Area left without ornament	sands and gravels overlying Bunter.
(black)	Markham Main colliery tip.
(scaled)	Markham Main colliery lagoon.
—T—	Faults in Bunter below surface.
+++++	Railway line.

1.0 km

FIG. 34.   SANDALL BEAT BOREHOLE AND NUTWELL PUMPING STATION
IN RELATION TO MARKHAM MAIN COLLIERY.

MONITORING RIVER WATER TO CONTROL PURITY OF DUNE RECHARGE WATER

by G. Drost
Duinwaterleiding van Den Haag, The Netherlands

SUMMARY

The large waterworks of the western half of the Netherlands are practically all situated in the dunes and are running short of raw water. Because of this they are now recharged by infiltration of prefiltered Rhine water. This very polluted water is partly purified during its passage through the ground. In this respect the composition of the 'sand' is very important. Much attention is given to the results of analyses of the infiltrate including heavy metals and pesticides.

1. INTRODUCTION

During the last fifteen years the Dutch have been continuously recharging the ground-water of the whole dune area of the provinces of South and North Holland, that is from the Hook of Holland in the south to den Helder in the north (this area is practically the only part of the western half of the Netherlands that is used by the larger groundwater works).

In order to understand this situation we have to go back in history for several thousands of years. In those days the seas were considerably higher and therefore the western coast of the Netherlands was to be found in what is now the middle of the country, as shown in Fig. 35. Places like Amsterdam, The Hague, Rotterdam and Vlissingen are situated in what was once the North Sea.

As the seas of the world became lower because of the formation of ice caps at the Poles, the North Sea retreated from the shallow coast of the Netherlands. After some time, sand banks were formed in the sea where the coast is now situated. As the sea became still lower, sand dunes were formed from the sand banks by the wind.

To the east of the dunes, the enclosed shallow sea became brackish and was gradually filled by sedimentation and vegetable growth, the surface of the resulting swamp never rising much above sea level. This part of the country is of marine origin and therefore the groundwater had to be sea or brackish water. The rain falling upon the land could not penetrate the ground because its surface was at about sea level and the groundwater was at the same level. Practically all the soft rainwater therefore had to flow away as surface run-off.

Only in the higher dunes rising up to about 30 m could the rain penetrate into the dry sand until it reached the salt groundwater.   It did not mix with this heavier water but floated on it and replaced it.   This process continued for about a thousand years until some hundred years ago man started to look for groundwater fit for drinking purposes. In the western part of the Netherlands it was only found in considerable amounts in the dunes as explained above.   For this reason the large waterworks of today, for instance those of Amsterdam, Leiden, the Hague and the waterworks of the whole province of North Holland (Provincial Waterworks of North Holland - PWN) take their raw water out of the dunes.

The waterworks were started a hundred years ago without much idea of the amount of soft water floating on the old sea water in the dunes.   Nowadays this is known and the situation is illustrated in Fig. 36.   The differences in vertical and horizontal scales of Fig. 36 should be noted.   The real form of the lens shaped soft water body in the dunes is very flat, but the figure gives a good idea of the dynamic equilibrium between the soft water body and its surroundings.   There must be a constant flow of soft water along the surface and along the boundary with the salt water to either side, the western, sea side and the eastern, polder side.   The volume of the soft water is nevertheless constant due to replenishment by the rain falling upon the dunes.

The dynamic equilibrium was disturbed, however, by the waterworks taking water without replenishing it; this was done for about 100 years.   The effect was that the water table in the dunes was lowered several metres, at the same time the fresh water/ salt water boundary below the wells was brought up at some places to only 20 m below those wells.   In 1955 the waterworks of The Hague pumped 15 million m^3 of water out of its dune area, while the normal effective rain upon that area is only 5 million m^3 a year. It was then that replenishment started and since that time River Rhine water has been infiltrated artificially.   The amount of infiltration water is slightly more than the amount taken out in order to restore the predatory cultivation of the past.

The prefiltered river water is pumped to the dunes and flows into canals or half-natural ponds which are surrounded by shallow wells or open canals in which the water potential is kept some metres lower than in the infiltration ponds.   A scheme of the process is given in Fig. 37.   The horizontal distance between ponds and wells varies but is about 70 m; the water stays in the ground for one to two months or more, flowing from pond to wells.   During its passage through the ground there is a considerable change of composition.   This has always been the case.

2. COMPOSITION OF THE ORIGINAL GROUNDWATER

The composition of the old dune water differed in principle from that of the natural infiltration water, the rain. Table 7 gives a composition of the old dune water.

TABLE 7

COMPOSITION OF OLD DUNE WATER

Colour		mg Pt/litre
Permanganate number	15	mg/litre
Chloride	80	mg/litre
Nitrate	2.7	mg/litre
Bicarbonate	254	mg/litre
Silicon dioxide	21	mg/litre
Oxygen	0	mg/litre
Ammonia	0.90	mg/litre
Org. ammonia	0.13	mg/litre
Iron	0.50	mg/litre
Manganese	0.20	mg/litre
Total hardness	3.2	m.aeq/litre

The composition differed from place to place, the reason being that dune sand as a marine sand consists not only of silicon dioxide but also contains shell fragments composed of calcium carbonate together with organic matter containing nitrogen. As explained earlier, the sand also contains sediments from the inner sea that was once at the east of the dunes and peat from the old vegetable growth in it. The organic matter in the dunes was slowly consumed by micro-organisms using up the oxygen out of the percolating rainwater. The water thus made anaerobic contained iron and manganese in solution while hardness was dissolved from the shells. Ammonia was liberated by microbiological action from the peat which also caused colour. The level of permanganate number indicated organic substances in general. Sulphate reduction caused hydrogen sulphide and methane to be present.

The same phenomena could be expected after changing from rain to Rhine water was as infiltrate water, together with other changes caused by the extreme difference in composition.

3. COMPOSITION OF THE RIVER WATER

It is commonly known that the Rhine is particularly polluted. Rhine water is nevertheless used as infiltration water for two reasons. First it is the only water source in the Netherlands that always contains a sufficient quantity of water; secondly, at the time the Rhine was chosen (before 1950) its water was not as polluted as it is today. Some waterworks are now changing from the Rhine to the Maas, a river which makes the use of reservoirs necessary under certain circumstances, but Rhine water will flow through Dutch dune sand for many many years.

Naturally the composition of the Rhine water is not constant. Not only is there a slow but continuous deterioration from year to year, but within this trend there are differences caused by yearly fluctuations (wet or dry years) and within the years caused by seasonal differences. Flow differences within the year are not so extreme as in England because the Rhine is a part rain part glacial river. (The Maas is only a rain river.) A glacial river contains a lot of glacial water during the warm, mostly dry season.

Table 8 gives the data of the routine analysis of the river water before the pretreatment (biological rapid sand filtration) that takes place before transport to the dunes. The table contains the results for 1971, a relatively dry year. The mean flow for that year was 1451 m^3/sec, while the mean flow over 50 years is 2200 m^3/sec. The frequence used for each analysis is also indicated in the table.

TABLE 8

COMPOSITION RHINE WATER 1971

	Min	Mean	Max	Analysis frequency per week
Riverflow m^3/sec	760	1451	3633	7
Conductivity $20^{\circ}C$, 10^{-6} recipr. Ohms	760	1085	1400	2
Total solids $180^{\circ}C$, mg/litre	497	679	858	7
Chloride mg/litre	126	221	295	* 7
Bicarbonate mg/litre	134	163	193	2
Sulphate mg/litre	72	101	127	1
Calcium mg/litre	72	91	103	1
Magnesium mg/litre	10.6	13.9	20.0	1
Sodium mg/litre	66	123	156	1
Potassium mg/litre	6.5	9.5	11.9	1
Total hardness m. aeq/l	4.3	5.5	6.8	1
Bicarbonate hardness m. aeq/l	2.5	2.7	3.2	2
Permanganate number (10 min, $100^{\circ}C$)	19	27	35	* 1
T.O.C. mg/litre	6.2	8.7	10.8	* 1
Colour mg Pt/l	20	33	50	* 1
Taste number	7	32	127	* 1
Phenol $\mu g/l$	1.0	12	80	* 1
Ammonia mg/litre	0.13	2.9	6.0	* 5
Albuminoid nitrogen mg/litre	0	0.71	5.8	5
Nitrite mg/litre	0.06	0.37	1.0	* 2
Nitrate mg/litre	7.0	14	20	* 2
Phosphate mg/litre	0.125	0.490	1.200	* 1
Polyphosphate mg/litre	0.030	0.190	0.630	1
Silicon dioxide mg/litre	1.1	5.0	10.0	1
Temperature $^{\circ}C$	0.5	11	22	5
Oxygen mg/litre	2.6	6.9	10.3	2
Oxygen saturation %	28	63	84	* 2
BOD_5, not filtered, mg/litre	2.3	6.4	10.2	* 1
Carbon dioxide mg/litre	2.0	6.9	14.0	2
Ph	7.3	7.6	7.9	5
Iron mg/litre	0.3	0.97	·2.5	2
Manganese mg/litre	0	0.09	0.33	2
Sediment $110^{\circ}C$, mg/litre	10.0	26.3	174.4	5
Detergents Manoxol OT mg/litre	0.04	0.11	0.22	* 1
Fluoride mg/litre	0.24	0.37	0.62	* 1
β - Radioactivity, not filtered, pCi/litre	4	8	13	1
" " " "	4	7	10	1
" , sediment/litre water pCi/litre	< 1	< 1	6	1
Radioactivity, potassium present pCi/litre	5	7	9	1
Bacterial count, 3 days 22 $^{\circ}C$ per ml		90.000		1
" " , 3 " 37 $^{\circ}C$ "		190.000		1
Coli-count (MPN/ml)		18		* 1

These analyses are made by the waterworks using Rhine water, each sampling at its intake. Samples are also taken at Robith, the village where the Rhine enters the Netherlands. All data are brought together in a report by the Dutch water chemists and presented to the International Rhine Commission to assist their fight against pollution. In spite of this struggle the Rhine deteriorates from year to year. In Table 9 some data from 1971 are compared with those from 1959 and 1964 which years showed the same mean flow as 1971.

TABLE 9

COMPARISON OF WATER QUALITY DATA 1959, 1964 and 1971

		1959	1964	1971
Colour	(Hazen)		34	30
Chloride	(mg/litre)	179	204	232
Sulphate	(mg/litre)		100	107
Hardness	(degrees)		5.6	5.7
Nitrite	(mg/litre)			
Nitrate	(mg/litre)		5.5	12.0
Ammonium	(mg/litre)	2.4	3.6	3.6
Albuminoid nitrogen	(mg/litre)		0.97	0.00
Orthophosphate	(mg/litre)		0.14	0.67
Oxygen	(mg/litre)	6.9	5.6	4.5

Fig. 38 gives an impression of the variations within a year, entirely due to differences in flow as they show themselves in the chloride content. Variations mainly due to seasonal effects (microbiological oxidation of ammonia to nitrate) are shown in Fig. 39.

All the figures indicate the poor condition of the Rhine water which, after a simple pretreatment, is recharged to the groundwater of the dunes and rendered into drinking water. The most important constituents which indicate pollution are given in Table 8 and marked thus *. Pollution with organic matter is shown by permanganate number, TOC*, colour, taste number[+], phenol concentration, BOD_5. The MPN for coliform

*TOC = Total organic carbon.

[+]The taste number is the number of times a volume of river water has to be diluted with good tasteless water to bring the original taste to the limit of detection.

organisms indicates that the water is unfit for swimming. The ammonia, nitrate and phosphate contents indicate trouble with algal growth in reservoirs and in infiltration ponds. Half of the chloride content is due to the waste salts of the potassium mining of France, which are obtained in solid state, brought into solution and dumped into the Rhine.

In the last two or three years, there is more and more evidence of specified organic pollutants in relatively high concentrations. Special research is undertaken to prove their presence and to estimate their concentrations. This research is carried out with the co-operation of the waterworks, KIWA and institutes of the government.

This special research deals with heavy metals, oil (aliphatic hydrocarbons), aromatic hydrocarbons and naphthenes, volatile organic substances causing taste and odour, pesticides (chlorinated hydrocarbons, organic phosphate esters and carbamates) polychlorobiphenyls.

The most alarming heavy metals present are mercury, chromium and zinc. The greater part of the heavy metals is absorbed on the silt in the water. The analysis of heavy metals is made by atomic spectrophotometry.

Aliphatic and aromatic hydrocarbons are 'fingerprinted' by gas liquid chromatography (GLC) as illustrated in Figs. 40 and 41. Concentrations are calculated by comparison with chromatograms of samples of oils (fuel, diesel oil, lubricating oil) of known composition. As the figures show, the 'oils' are for a considerable part removed by rapid sand filtration. There are indications that these substances are also adsorbed onto clay particles. Concentrations of aliphatic oil seem to point to >0.5 mg/litre. The main problem of oil is its biological transformation to substances giving taste and odour problems.

By a combination of GLC and mass spectrometry a more heterogenous group of organic substances, partly aromatic, mostly containing chlorine are determined. Only a low percentage of these substances are removed by the treatment before infiltration (rapid sand filtration). Fig. 42 in combination with Table 10 gives a typical example after rapid sand filtration. Because so many substances contain chlorine and such substances cannot be of natural origin it is planned to develop an analysis for total organic chlorine. The sanitary institute of the government (RID) now seems to have succeeded in this.

TABLE 10

AROMATIC SUBSTANCES IN RHINE WATER (RAPID FILTRATE)

Peak number *	Substance
5	monochlorotoluene
6	1.3 dichlorobenzene
7	1.4 dichlorobenzene
8	butylbenzene
9	1.2 dichlorobenzene
12	dichlorotoluene
13	0-nitrotoluene
14	trichlorobenzene
15a	m-nitrotoluene
16	trichlorobenzene
18	p-nitrotoluene
19	chloronitrobenzene
19a	dichlorodiisopropylether
20	methylnaphthalene
21	trichlorotoluene
23	tetrachlorobenzene
24a	trichlorodiisopropylether
25	tetrachlorobenzene
26a	dimethylnaphthalene
27a	trichlorodiisopropylether
31	dichlorobiphenyl
32	alkylbenzoate
33a	dichloro-N-methylpyrrol
34a	(di-or tri-)chloro-N-methylpyrrol
35a	trichlorobiphenyl
36a 40 t/m 45	dialkylphthalates

* The peak numbers refer to the gas chromatograph (Fig. 42)

Many pesticides are also found in the river water. These substances are not adsorbed onto sediment so that rapid sand filtration does not reduce their concentration. Some of them are to be found even in the drinking water. The highest concentrations are of αBHC and γBHC (Lindane), 0.20 to 0.25 mg/litre. The actual concentration of Endosulphan (Thiodan) is low in comparison with some years ago when, during a period of mass fish deaths, concentrations were found of tenfold, i.e. 1.5 μg/litre.

The river water shows a marked cholinesterase reducing activity partly due to parathion, malathion and carbaryl (a carbamate). Aldrin, Heptachlor, Heptachlor-epoxide, pp'DDE, pp'DDe, pp'DDT, Dieldrin and Endrin are all present at the limit of detection (0.01 mg/litre).

There is a certain anxiety about the polychlorbiphenyls and also about carcinogenic polycyclic aromatics such as 3,4 benzpyrene, but investigations still have to be started.

4. CHEMICAL CHANGES DURING PURIFICATION

4.1. Before infiltration

The first step in the purification of river water to drinking quality is the pretreatment before infiltration, which consists, up till now, of a simple rapid sand filtration without coagulation. The purpose of this stage is to make the water fit for transport and for infiltration. In any case the sediment which causes clogging of transport lines or infiltration ponds has to be removed. About 90% is removed leaving 10% (2 to 3 mg/litre) of the silt and 10% of the absorbed substances in the filtrate. Because no chlorine is used before filtration there is considerable microbiological action in the filters. This, together with normal chemical reactions also brings about a change in the composition of the dissolved substances. About 1 mg/litre ammonia is oxidized to nitrate (partly to nitrite) reducing the oxygen content sometimes to zero. The taste is reduced by about 20%, the permanganate number by about 10%.

In order to prevent biological scaling of the transport pipes, the filtrate is chlorinated (3 mg/litre). This chlorine is mostly consumed when the water arrives at the dunes but some combined chlorine still remains.

In the infiltration ponds new changes can be observed, mainly due to biological activity, primary production of algae and especially higher water plants. This means that, depending on circumstances, the pH may rise sometimes to 9.9, phosphate is reduced to ≤0.03 so is ammonia, nitrate and even silicate, while there are no changes in permanganate number. The colour may, however, become higher. These phenomena are to be seen during the spring and summer seasons giving rise to the organic substances of living organisms. Most of them die in autumn and winter and then the organic materials come (partly) into the water. Mineralization takes place and the mineral substances originally removed by growth are back adding to the concentration present in the newly arrived infiltration water.

This makes it difficult to compare the composition of the recovered water with the 'infiltration water' but this is necessary in order to understand what the infiltration of polluted water does underground.

4.2. During infiltration

In order to be able to follow the alterations in chemical composition of the infiltrated water during its journey through the sand, two rows of shallow wells were constructed in two parts of the infiltration area of The Hague. The wells of either row are situated so that they are in the same streamline from pond to draining pipe. The two parts of the infiltration area show two differences. The first is that there is a difference in time, the water taking respectively 6 and 20 weeks to reach the draining pipes. The other difference is that in the second part, peat layers of limited area are present, in the first part they are not. Some results are given in Figs. 43 to 46.

Fig. 43 shows the effect upon the organic content in general. In pure sand there is a steep fall in the first week (= first 10 m of sand or less) followed by a small gradual decline during the rest of the journey. In sand with peat there is first a considerable rise followed by a gradual decline, without reaching the value of the pure sand even after 20 weeks (= 120 m). The same phenomenon is shown by the colour (Fig. 44); with peat there in no improvement at all. It seems that there is a strong influence of the 'sand' on the quality of the passing water. Organic substances are taken out of it into the water, whether or not organic substances are taken out of the water by the 'sand' is not to be decided.

Fig. 45 gives the fate of the taste causing substances. Here the peat layers are of significant value. The explanation is that in pure sand bacteriological activity only is responsible for the reduction, but in peat containing sand, biological activity and adsorption are possible. The bacteria deal exclusively with the 'soft' substances, while 'hard' substances only are removed by absorption.

A dramatic development, especially in the last years, is that taste reduction by infiltration is getting worse and worse; this phenomenon being stronger in the area of pure sand than in that of peat containing sand. This is shown in Table 11.

TABLE 11

TASTE REDUCTION DURING 6-7 WEEKS

	pure sand	peat sand
1961 - 1968	61%	80%
1968	53%	55%
1969	37%	59%
1970	20%	60%

An explanation is that the 'hard' substances among the taste and odour producing substances gradually become greater as the result of a gradually enlarged (biological) purification of the waste waters. In the peat sand relatively more can be removed (by absorption) than in the pure sand. If this is true the situation becomes alarming.

Fig.46 shows the influence of the sand upon the ammonia content. At present there is no obvious explanation for the decline in peat sand after eight weeks.

Finally, in Table 12 is given a summary of the changes in concentration of a number of particulate substances during the different purification steps (mean values during about half a year).

TABLE 12

HEAVY METALS AND PESTICIDES REDUCTION DURING PURIFICATION

		River μg/litre	After pretreatment (rapid sand filtration) μg/litre	After infiltration μg/litre	Drinking water μg/litre
Lead	Pb	10*		2.3	2
Copper	Cu	20*		12	15
Chromium	Cr	15**	4	<2	<2
Zinc	Zn	63*		3.5	8
Cadmium	Cd	1*		<1	<1
Mercury	Hg	0.7**	0.11	<0.1	<0.1
HCB		0.05			0.02
α BHC		0.25			0.03
γ BHC (Lindane)		0.20			0.03
Endosulphan		0.15	0.05	0.04	0.04
Parathion		0.04	0.04	<0.01	<0.01
Malathion		6.01	0.01	<0.01	<0.01
Carbaryl		0.045	0.045	0.005	0.005

* after centrifugation (sediment removed)
** centrifugation brings these figures down to 4 (Cr) and 0.11 (Hg)

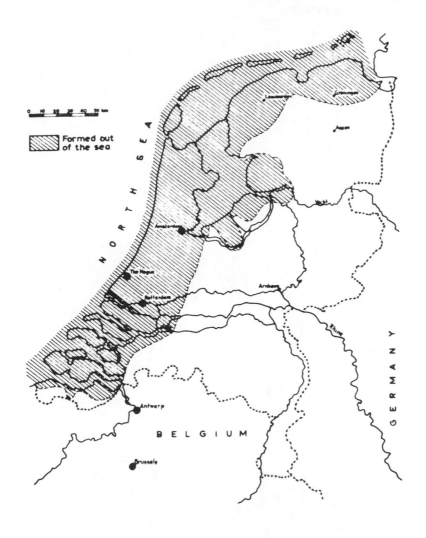

FIG. 35 THE NETHERLANDS SHOWING LAND AREAS
FORMED OUT OF THE SEA

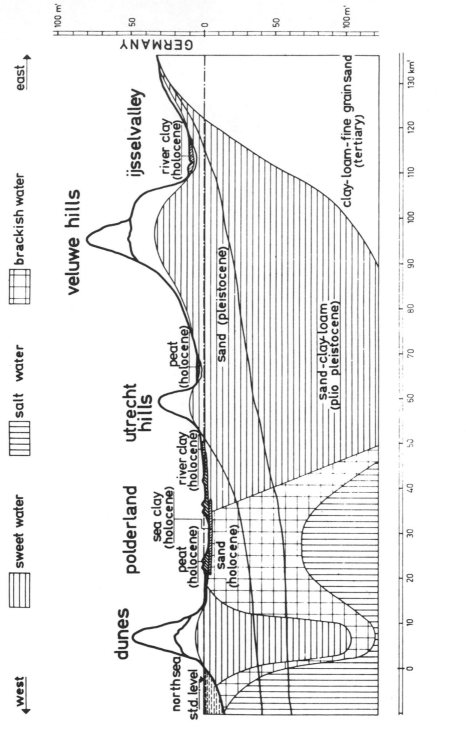

FIG. 36 GEO - HYDROLOGIC CROSS-SECTION

OVER THE NETHERLANDS

138.

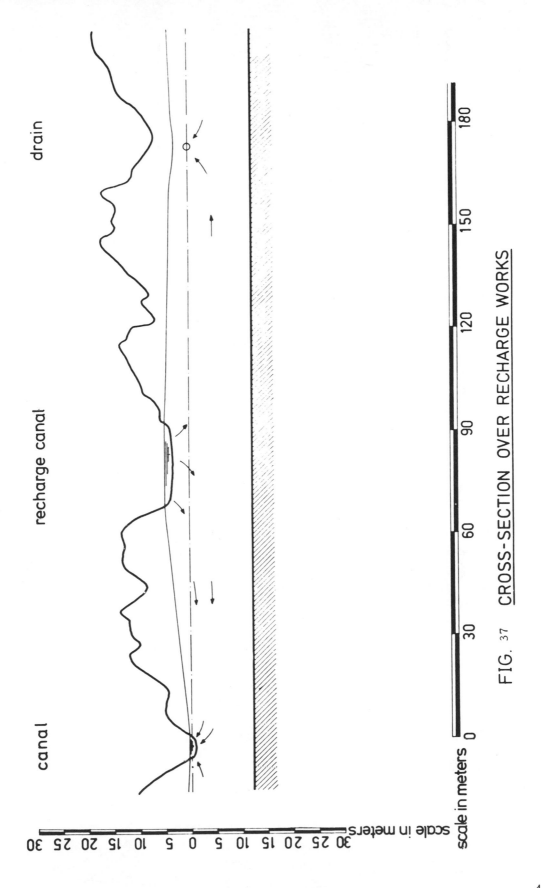

canal   recharge canal   drain

scale in meters
30 25 20 15 10 5 0 5 10 15 20 25 30

scale in meters
0   30   60   90   120   150   180

FIG. 37   CROSS-SECTION OVER RECHARGE WORKS

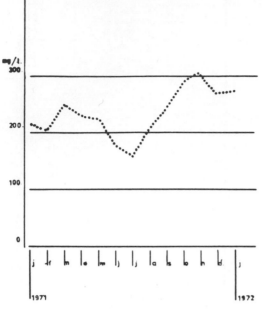

FIG. 38 RHINE WATER CHLORIDE
CONTENT AT LOBITH

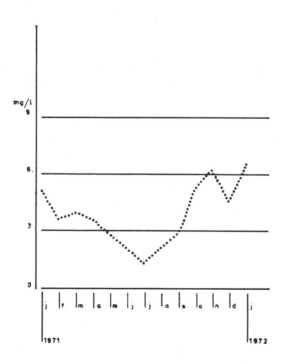

FIG. 39 RHINE WATER AMMONIA
CONTENT AT LOBITH

140.

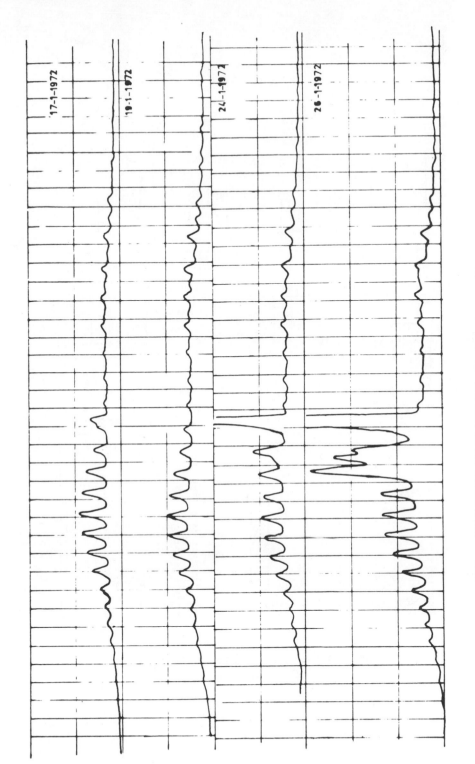

FIG. 40 HIGH SPEED FILTRATION EFFLUENT OF RHINE WATER:
OIL:ALIPHATIC HYDROCARBONS

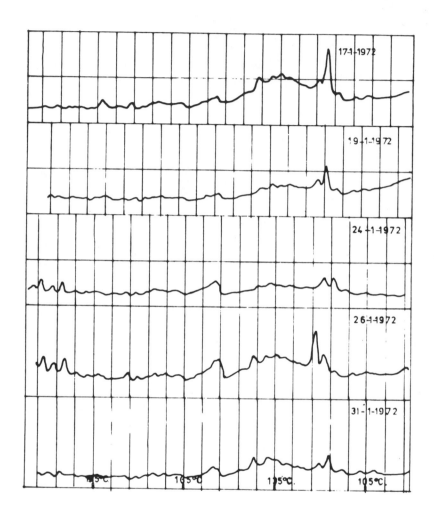

FIG.41 <u>HIGH SPEED FILTRATION EFFLUENT OF RHINE WATER:</u>
<u>AROMATIC HYDROCARBONS</u>

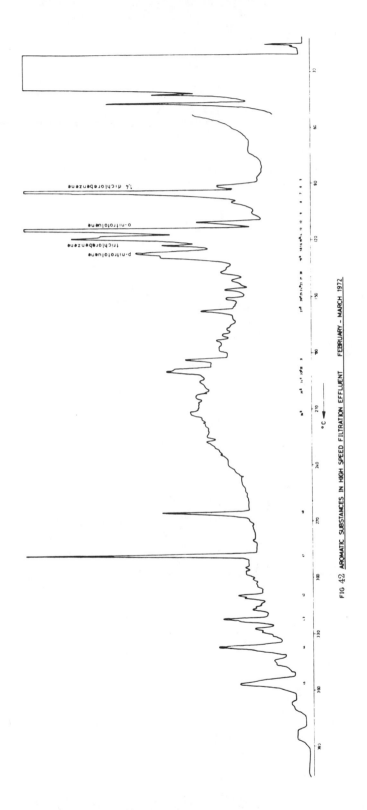

FIG 42 <u>AROMATIC SUBSTANCES IN HIGH SPEED FILTRATION EFFLUENT</u>   FEBRUARY – MARCH 1972

143.

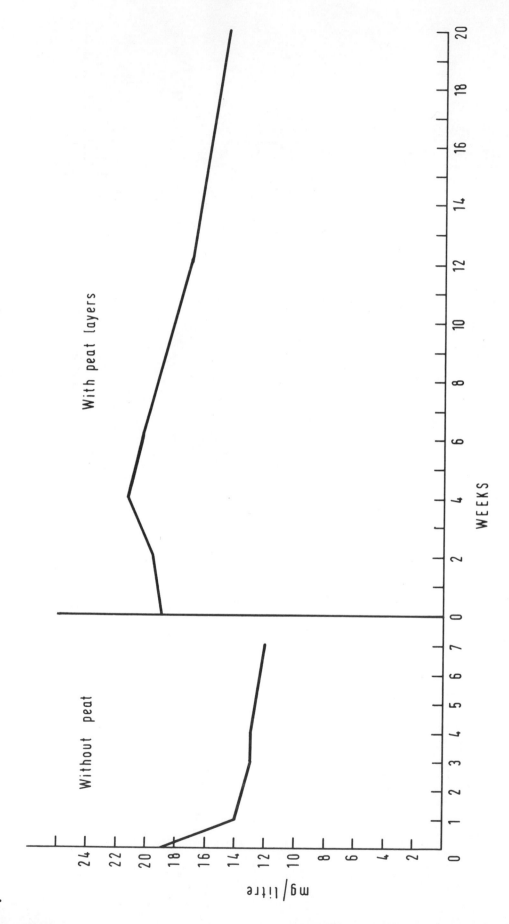

FIG. 43.   MEAN PERMANGANATE VALUES
(The Hague, 1965–1970)

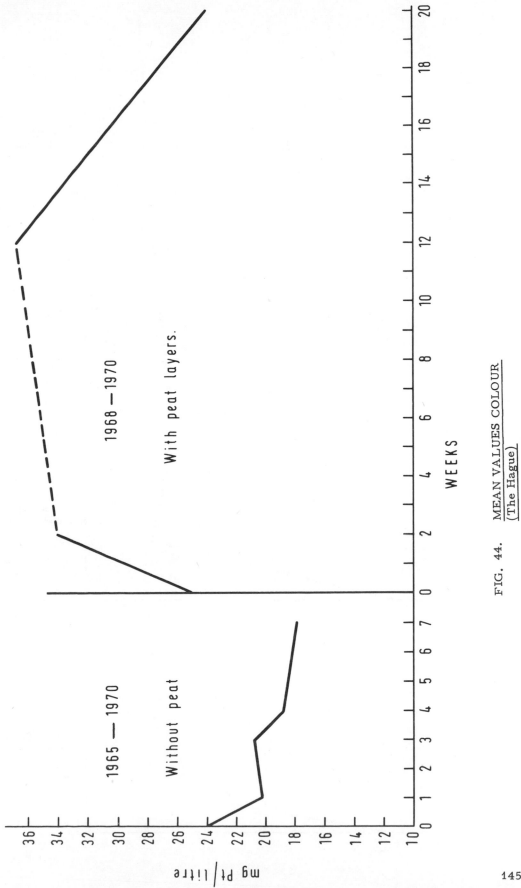

FIG. 44.   MEAN VALUES COLOUR
(The Hague)

145.

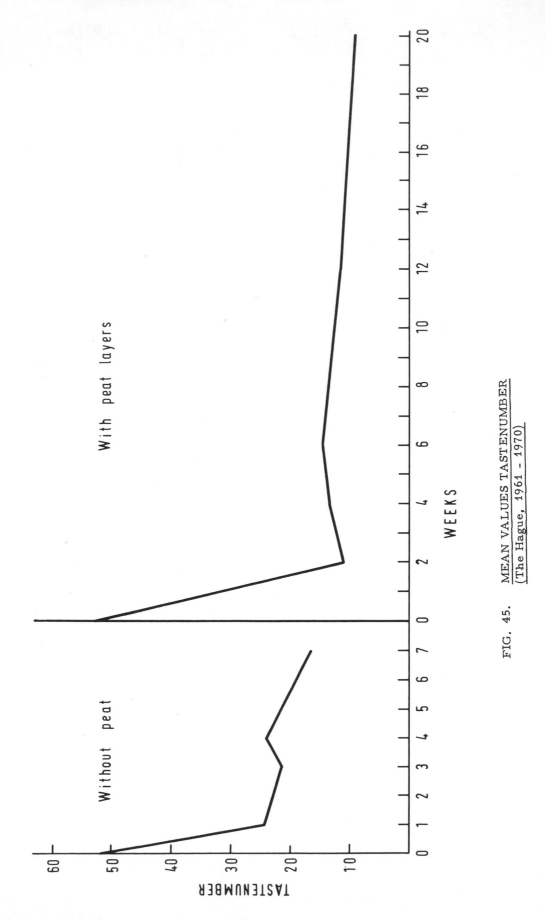

FIG. 45. MEAN VALUES TASTENUMBER
(The Hague, 1961 - 1970)

146.

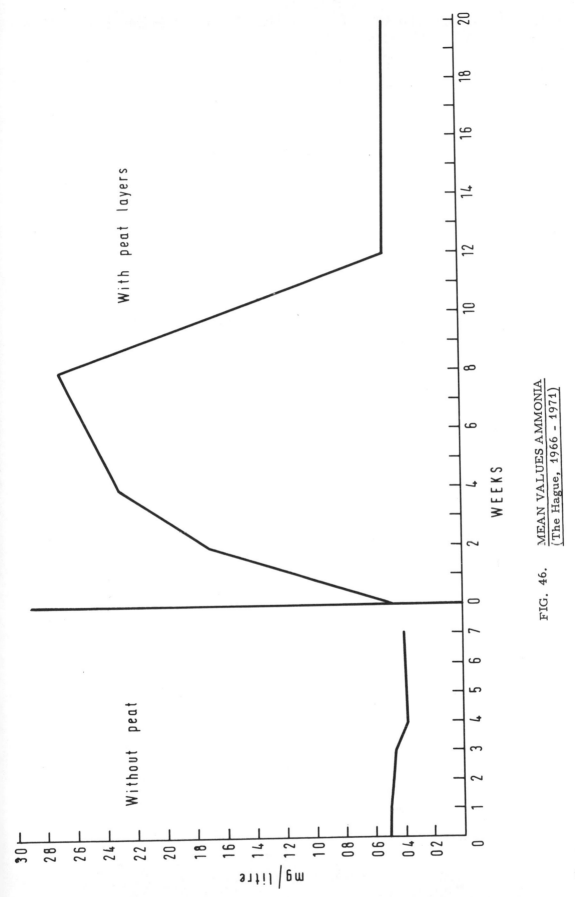

WEEKS

With peat layers

Without peat

mg/litre

FIG. 46.   MEAN VALUES AMMONIA
(The Hague, 1966 – 1971)

147.

THE ROUTINE DISCHARGE OF EFFLUENTS TO THE UNDERGROUND — PRACTICE AND CONTROL

by N.J. Nicolson
Thames Conservancy, Reading, Berks, England

1. INTRODUCTION

My brief is one solely of providing information regarding the practice of discharging effluents to the underground strata and applies largely to the Thames Conservancy area.

Prior to the introduction of main drainage into rural areas, most domestic properties in such locations were drained to septic tanks. The quality of effluents from such tanks was invariably outside Royal Commission Standards (i.e. suspended solids 30 mg/litre, Biochemical Oxygen Demand 20 mg/litre) and therefore not suitable for discharge to water courses within the Thames Conservancy area under the Thames Conservancy Acts, i.e. before the 1951 Rivers (Prevention of Pollution) Act. The only practicable solution to minimize pollution of surface streams was to suggest the disposal of such effluents by means of sub-irrigation drainage systems. Before the Water Resources Act the only legal powers available for the protection of underground resources for public supply were contained in the Water Act (Sections 18 and 21) by which water undertakings could seek, by way of byelaws, to prohibit or regulate actions which were likely to lead to pollution of underground strata.

2. EFFECT OF THE WATER RESOURCES ACT

The Water Resources Act required existing discharges to the underground strata to obtain consent from the appropriate river authority, with the responsibility placed upon those already making such discharges to obtain consent to be allowed to continue to do so. The consent procedure prescribes the location, maximum volume and minimum quality conditions for the discharge. Arising from the provisions of Section 72 of the 1963 Act, a number of toxic discharges were stopped while other existing discharges were permitted subject to appropriate quality conditions. The Conservators' policy, based on discussions with the Ministry of Housing and Local Government, was formulated such that any septic tank discharge taking place via sub-irrigation drains less than four feet deep did not require consent unless the discharge was direct into chalk. The reasoning was that self-purification via the top soil would take place and that no immediate alternative was available anyway. All trade effluent, including cooling water, required consent irrespective of the depth of the discharge. New discharges have been, and continue to be, dealt with according to individual quality considerations. To date, 601 consents to discharge underground have been issued by the Thames Conservancy under Section 72 of the Water Resources Act, of which some 538 are still in force.

Most of these are for domestic sewage effluents from small treatment plants to which one or more dwellings are connected. Discharges of sewage effluent in excess of 5000 g.p.d. include 14 from private and 22 from local authority sources; of these, four are in excess of 100 000 g.p.d. A few sewage effluents have a condition included in the consent that 2 mg/litre free chlorine shall be present in the discharge. This condition was imposed following representations from water undertakings, the idea being that a chlorinated effluent is inherently safer than an unchlorinated one. Because the contact times are nominal, there is no question of such effluents being sterile; bacteriological counts in such effluents, however, confirm that the numbers of bacteria are very much lower than those occurring in unchlorinated effluents. Discharges to underground of trade effluent include 35 in excess of 5000 g.p.d. of which 28 are cooling water; 17 of these are in excess of 100 000 g.p.d. of which 14 are cooling water. All these effluents of larger volume originate from conventional treatment plants, are normally of good quality and differ from conventional effluent discharges only in that they are discharged underground. The majority are sewage effluents and the criteria used in deciding whether to allow any particular discharge include consideration of such quality factors as nature of effluent, particularly any trade component, and such physical factors as the geology of the area, depth of discharge point and particularly the proximity to the zones of influence of any significant abstraction boreholes. Such considerations are often done independently by water undertakings and the river authorities prior to meetings to discuss whether any particular discharge should be allowed, modified or opposed.

Good liaison has been maintained over the years with all abstractors, private and public, and it has been the practice to inform such abstractors regarding proposed new discharges. When serious concern has been expressed and objections to the proposals made, the Conservators have usually supported the objectors.

3. LOCATION OF DISCHARGES

Fig. 47 gives the location of the larger effluent discharges in the Thames Conservancy area and shows clearly the rural location of the discharges and the predominance in the chalk aquifers. From a consideration of recharge volume (i.e. from rainfall) it can be seen that the dilution afforded these effluents is very large. Added to this a degree of self-purification that takes place in the upper part of the strata confirms that the pollution load from such sources is minimal being probably of less concern than the load from agricultural sources. It has never been demonstrated that any routine discharge of sewage effluent underground in the Thames Conservancy area has given rise to a significant deterioration of the water underground. One trade discharge of about 125 000 g.p.d. from large laboratory premises to soakaways has been taking place for many years. Although there is no evidence whatsoever of a quality deterioration, two

149.

water undertakings, together with the Thames Conservancy, have been studying the situation closely for some time. These studies include a consideration of direction of travel, estimates of time of travel and dilution ratios in order to assess the likelihood of contamination from the discharge. These studies continue and particular attention is being focussed on any analytical techniques available that would enable organic compounds in the microgram per litre range to be detected.

4.   SOME MEANS OF PREVENTING POLLUTION

Since the introduction of the Water Resources Act, several alleged pollution incidents have been reported arising from isolated spillage. One, arising from a weed-killer application, has been confirmed analytically and in one or two others the evidence, circumstantially at least, supports the claim. None of these incidents was considered to be sufficiently serious to warrant discontinuing any abstraction, although, potentially, spillages constitute a major pollution hazard. There is little that can be done by river authorities to prevent pollution by accidental spillage. Steps can, and are, taken to establish the cause of such pollution and wherever possible to mitigate its effects. The only further steps that could practicably be taken to minimize this kind of pollution would seem to be the following.

(a)   The formulation of codes of practice on the handling, transport and storage of potentially polluting materials.

(b)   Amendment of the law to provide that failure to conform with codes of practice, or to report the occurrence of accidental pollution, is recognized as equivalent to "causing or knowingly permitting" pollution to occur.

(c)   The taking of positive steps to educate the public at large, including school children, on the need to prevent pollution occurring.

Finally, the greatest risk to underground waters lies not with the disposal of domestic sewage but with the disposal of farm slurries, refuse and industrial wastes (both solid and liquid) to quarries and tips. Control over this type of discharge is excluded under the 1963 Act but it is to be hoped that the Disposal of Poisonous Wastes Act of 1972 will provide the much needed control.

In verbal comment supplementing his paper, N. J. Nicolson presented Table 13 on the quality of crude and treated sewage and Table 14 on the proportions by volume of underground discharges of various effluents within his authority's area.

TABLE 13

PROPORTION OF LOAD REMOVED BY TREATMENT

	Solids	B. O. D.	Ammonia as N
Crude sewage mg/litre	300	360	40 to 50
Royal Commission standard	< 30	< 20	(say 10)
% removal	> 90	> 94	(say 75 to 80)

TABLE 14

PROPORTION OF UNDERGROUND DISCHARGES BY VOLUME

Effluent	Underground m. g. d.	Total m. g. d.	%
Sewage	2. 0	185	1. 1
Trade	1. 2	51	2. 4
Cooling	5. 8	285	2. 1
Total	9. 0	521	1. 7

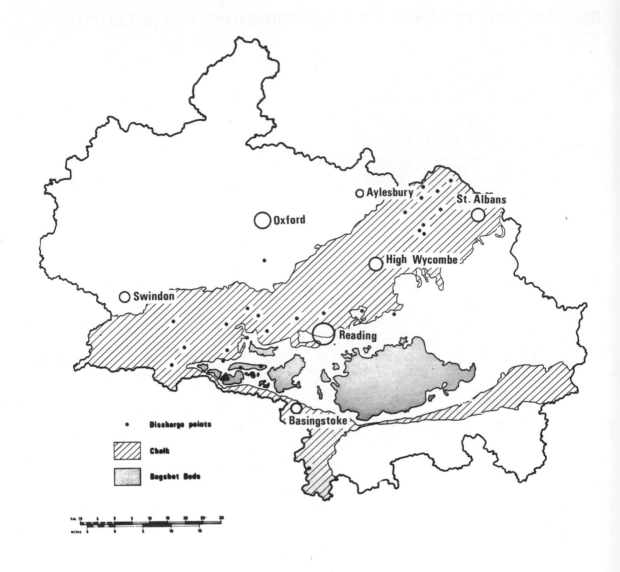

Aylesbury

St. Albans

Oxford

High Wycombe

Swindon

Reading

• Discharge points

▨ Chalk

▨ Bagshot Beds

Basingstoke

km. 10   5   0   10   15   20   25   30
miles 5   0   5   10   15

FIG. 47. THAMES CATCHMENT AREA: MAJOR UNDERGROUND
SEWAGE EFFLUENT DISCHARGES

RESEARCH AT WPRL RELATED TO GROUNDWATER POLLUTION PROBLEMS

by H.A.C. Montgomery and A.B. Wheatland
Water Pollution Research Laboratory, Stevenage, Herts, England

Although it is only very recently that the Water Pollution Research Laboratory (WPRL) has begun to study groundwater pollution as such - in a project referred to at the end of this paper - several investigations carried out during the last thirty years have had a bearing on the problem. These investigations, which are reviewed here and put into context, include studies of groundwater recharge by surface spreading of river water and sewage effluents; characteristics of percolate from tipped refuse; performance of septic tanks; and aerobic biodegradation of cyanide.

1. ARTIFICIAL RECHARGE OF BUNTER SANDSTONE

In the late 1950's the Laboratory took part in experiments to test the feasibility of recharging aquifers in the Bunter sandstone, north of Nottingham, with water from the River Trent (1). The particular points at issue were the extents to which ammonia would be removed and biochemical oxygen demand satisfied, because the Trent water contained a considerable proportion of imperfectly purified sewage effluents and would have been quite unsuitable as a supply of raw water.

Preliminary experiments were, however, made with a well-nitrified sewage effluent of unusually good quality which was already being disposed of by passage into the sandstone. A portion of the effluent, fortified with ammonia (approximately 4 mg/litre as N), was diverted to an area not previously used for disposal. Sampling points were provided at depths of 30 and 60 cm. Continuous application of the effluent at a rate of 90 litre/m^2 per hour resulted in a percolate which still contained an appreciable concentration of ammonia. Intermittent application for 12 hours per day, however resulted in almost complete removal (from 8 to 0.3 mg N/litre) at temperatures in the region of 15°C, and subsequent work showed that almost equally good performance could be obtained at winter temperatures. Detailed analysis of the results showed that the mechanism of removal almost certainly consisted of cation exchange followed by nitrification, presumably by attached nitrifying bacteria, the cation-exchange capacity being thereby regenerated. The added nitrogen was leached out as nitrate and a marked fall in pH value, to between 6 and 7, usually occurred. Most (84 to 87 per cent) of the BOD was satisfied during percolation but the Permanganate Value test (4 hours at 27°C) indicated the presence of residual organic matter.

The ion-exchange capacity of the soil for ammonium ion was found to vary very markedly with the experimental conditions. Thus the exchange capacity of the top 7.5 cm of soil, expressed as mg N per 100 g of soil, was 23.1 for a solution of ammonium chloride in distilled water (231 mg N/litre), 3.87 for a simulated sewage effluent with about 7 to 8 mg N/litre, and only 0.9 for River Trent water (2.7 mg N/litre). The exchange capacity fell as the depth from which the soil was taken increased.

Soakage experiments were then made, on a previously unused plot, with water from the River Trent which was applied at 90 litre/m^2 per hour for 12 hours daily. After a period of maturation, removals of ammonia in excess of 80 per cent, from an initial value around 4 mg N/litre, were observed at a depth of 73 cm, and the average removal of BOD, though inferior to that obtained in the experiments with sewage effluent, was still 67 to 80%, depending on the sampling position. Similar experiments carried out during the winter indicated that removal of ammonia would be affected by temperature and would decrease during prolonged periods when soil and water temperatures were low (about 5°C) but would not be unduly affected by low temperatures for 1 or 2 days only.

Percolation resulted in removal of all of the manganese and most of the zinc and nickel in the river water applied. Some removal of iron, chromium, lead, and copper also occurred. It is not known whether the concentrations of metals would eventually have increased sufficiently to inhibit nitrification in the soil, though it seems likely that the metals would have been bound in highly insoluble forms of relatively low toxicity.

The removal of bacteria during percolation ranged from 81 to 96.5%, depending on the sampling position.

In recent years the subject of recharge of the Bunter sandstone with water from the Trent has been studied by the Water Resources Board in connection with the Trent Economic Model. A supplementary study of the effect of prolonged percolation on the quality of the water has been made at WPRL and a preliminary account of this work has been given (2).

The experiments were designed to study the effect of slow percolation of River Trent water through a column containing an aerobic unsaturated layer consisting of 1.2 m of top-soil followed by 3.3 m of Bunter sand, and then through a saturated zone of 6 m of Bunter sand with an intermediate sampling point at 3 m. The conditions were more representative of those occurring in practice than was the case in the earlier experiments and there was a specific interest in the possibility of the simultaneous removal of organic matter and nitrate by denitrification, which is known to occur in some methods of effluent treatment. Water from the Trent was settled and fortified as required with partially purified sewage and with ammonium sulphate to compensate for the self-purification which occurred during storage. The rate of flow was intended to be equivalent to a minimum period of retention of 25 d in the saturated zone, but ponding caused a reduction in the rate of flow, eventually to only 18% of that used initially. In a repeat experiment the rate of flow was maintained by omitting the layer of top-soil but other experimental difficulties occurred which rendered interpretation of the results difficult.

However, enough results were obtained to show that almost complete nitrification and satisfaction of BOD occurred, although less than half of the applied COD and organic carbon were removed. In the second experiment, the average COD fell from 26 to 17 mg/litre, the average concentration of organic carbon from 7.6 to 4.4 mg/litre, and the pH range from 7.3 to 7.6 to 5.5 to 6.1. An independent experiment showed, however, that a reduction in pH value to 5.4 could be brought about merely by keeping Trent water in contact with Bunter sand for 15 hours, showing that nitrification was not the only cause of the increased acidity observed in the percolation experiments.

The average concentration of oxidized nitrogen rose from 10.3 to 13.1 mg/litre in the second experiment, consistent with nitrification of the ammonia added. The fall in the average concentration of oxidized nitrogen from 11.8 to 5.9 mg/litre, which was observed in the first experiment, could not be attributed to denitrification caused by oxidation of the organic matter in the water, because the COD of the water scarcely changed; denitrification in the layer of top-soil seems a more likely explanation.

The experiments confirmed that nitrification of the ammonia and satisfaction of the BOD of River Trent water can readily be achieved, but the simultaneous removal of dissolved organic matter and nitrate by denitrification, though theoretically probable in the absence of air, was not demonstrated.

2. BIODEGRADABILITY OF PERCOLATE FROM REFUSE TIPS

The Ministry of Housing and Local Government's experiments at Bushey showed that a highly polluting liquid was produced, both by percolation of rain water through 'dry' domestic refuse and by tipping refuse into standing water (3). The Laboratory's part in the experimental work was to carry out long-term respiro-metric tests on the liquids produced.

The first test was made on percolate from refuse which had been tipped dry about two and a half years previously. By this time the biodegradation of the refuse was well advanced and the BOD of the sample was similar to that of a good sewage effluent. However, the sample still contained 269 mg/litre organic carbon, 182 mg/litre ammoniacal nitrogen, and 53 mg/litre organic nitrogen. Nitrification occurred during incubation in respirometers although there was a loss of 20 mg/litre of nitrogen (possibly by decomposition of ammonium nitrite) over the 62-day period of the experiment; 34 mg/litre of organic nitrogen were still present, on average, at the end of the period. More interestingly, only about one third of the organic carbon was removed, as shown by direct chemical analysis, and an even smaller proportion as calculated from the oxygen uptake measured in the respirometers. On the other hand, the organic matter in the percolate from refuse tipped directly into water was shown, by calculations based on the composition of the contents of the respirometer, to be almost completely degraded in 20 days at 25°C (Fig. 48). In this experiment, the percolate was comparatively fresh, the sample being taken about 3 months after the end of the 15-month period of tipping; the initial concentrations of organic carbon and ammoniacal nitrogen in the sample were 34 and 26 mg/litre respectively and its BOD was 73 mg/litre. The curves in Fig. 48 are characteristic for samples containing biodegradable organic matter, ammonia, and a low initial concentration of nitrifying bacteria, and the 'steps' at 10 to 15 d and at 16 to 20 d correspond to the conversion of ammonia to nitrite and of nitrite to nitrate.

The results of these experiments show that domestic refuse can give rise to liquors in which, depending probably on the time which has elapsed since tipping, the organic matter is either easily or very incompletely degradable by aerobic biological action, and also confirm that such liquors do not prevent nitrification.

More recently, rough calculations have been made, using the results of the Bushey experiments, to estimate the order of magnitude of the pollution of well waters which might occur if all the percolate from a refuse tip reached a point of abstraction (4). For the purposes of calculation it was assumed that 5000 m^3/d were abstracted from a circular area of radius 1970 m, that infiltration amounted to 150 mm/year, that 100 tonnes of refuse were tipped on the site each day, and that the polluting load from this was proportional to that leached during a period of 28 months from the single batch of refuse tipped dry at Bushey. It was shown that the oxidizable (i.e. ammoniacal + organic) nitrogen, chloride, sulphate, and 5-day BOD in the well water would increase by about 8, 18, 15, and 44 mg/litre respectively. Increases of this order would be unlikely to occur in practice because some removal of oxidizable nitrogen and BOD would occur in the aquifer and because not all the percolate would reach the point of abstraction.

Dissolution of compounds of iron and manganese, leading to water supply problems, might also be expected to result from the percolation of a strongly reducing liquor through the ground, and it was mentioned (4) that this effect could perhaps be prevented by arranging for nitrification of the leachate in a bed of sand or gravel. The resulting excess of nitrate would be available to heterotrophic bacteria as a source of oxygen and would prevent the development of strongly reducing conditions. A nitrified leachate might also be sprayed on to newly tipped refuse, encouraging aerobic fermentation and also destroying much of the nitrate.

3. RIVER POLLUTION BY 'POT-ALE'

A severe case of river pollution caused by percolation from a soakaway occurred at a distillery site visited by the Laboratory's staff in 1966. The main polluting load was derived from about 250 m^3/week of 'pot-ale' - the residue from the distillation of whisky-having a 5-day BOD of about 25 000 mg/litre. This was mixed with spent lees and wash water and treated only by sedimentation before discharge to a soakaway in a gravel stratum. As a result, a spring 300 m away had become polluted and had an odour of hydrogen sulphide. The stream which received the spring water contained a very heavy growth of filamentous bacteria ('sewage fungus') which is a typical result of the presence of biodegradable organic matter in solution. Masses of the growth were decomposing on the stream bed and the outbreak was regarded as particularly objectionable because of the importance of a salmon fishery in the river which received the polluted stream water.

Alternative methods of disposal of the effluent were recommended.

4. COMPOSITION OF SEPTIC TANK LIQUORS

Two studies of septic tank performance have been made at WPRL. The first, by Pettet and Jones (5), was undertaken to compare the performance of various types of septic tank and to study the treatability of the liquors by biological filtration; the sewage used was purely domestic in origin. As no very clear-cut differences in performance were observed using septic tanks of different design, the results are considered together.

The most obvious characteristic of the septic-tank effluents was their variability: the extreme concentrations of suspended solids, for example, were 15 and 716 mg/litre. The unweighted arithmetic mean BOD values and concentrations of suspended solids in the effluents, discharged during the three periods of operation, are shown in Table 15; the tanks were emptied between the separate periods of operation. Effluents deteriorated towards the end of the first period and one year was considered to be too long an interval between emptying.   Good final effluents (as judged by the BOD values and concentrations of ammonia) were produced by biological filtration of the septic-tank effluents as long as the loading did not exceed that used in conventional sewage-treatment practice.   Determinations of residual organic matter in the final effluents were not made, however.

TABLE 15

MEAN COMPOSITION OF SEPTIC-TANK EFFLUENTS

Period	Suspended solids	5-day BOD
	mg/litre	
First (12 months)	154	277
Second (6 months)	118	319
Third (6 months)	98	357

The second investigation was undertaken to see whether there was any truth in allegations that the performance of septic tanks was impaired by the presence of synthetic detergent materials (6).   Two identical septic tanks were fed with equal volumes of detergent-free effluent from WC's at the Laboratory.   After a period of maturation, the biodegradable surfactant, Dobane JNX, was fed to one of the tanks at a concentration of 20 mg/litre (as Manoxol OT), later increased to 50 mg/litre;   the other tank was kept as a control.   Initially, and also in the presence of 20 mg surfactant/litre, the tanks gave similar performance, but the average BOD of the effluent from the tank receiving 50 mg surfactant/litre was 206 mg/litre whereas that from the control tank was 152 mg/litre.   The effluent quality from both tanks was better than shown in Table 15, possibly because of the different nature of the feed liquor. About 60% of the surfactant was removed at both concentrations.   Most of the contents of the tanks were removed after two and a half years and feeding was resumed as before but with 50 mg surfactant/litre added to the experimental tank from the outset.   The effects of the surfactant during this period were;

(a)   to delay scum formation by 4 weeks,

(b)   to give an effluent of average BOD 234 mg/litre as compared with 187 mg/litre in the control, and

(c)   to give a higher concentration of 'volatile' acids - average 127 mg/litre (as acetic acid) compared with 63 mg/litre in the control.

Effect (c) suggests partial inhibition of anaerobic digestion by the surfactant - a very likely result. The effluent from the experimental tank in the latter period contained on average 94 mg suspended solids/litre, 113 mg ammonia/litre (as N), and 19 mg surfactant/litre, and had an average BOD of 234 mg/litre of which nearly half could probably be attributed to the 'volatile acids' and much of the remainder to the suspended solids and to the surfactant.

The flow of septic tank liquor, which would normally enter the ground, was found in the earlier investigation to be about 100 to 150 litre per person served per day. A septic tank not followed by a biological oxidation plant might therefore be expected to discharge each day of the order of 10 g each of ammoniacal nitrogen, suspended solids, and 'volatile acids' and perhaps 2 g of surfactant material per person served.

5. BIODEGRADATION OF CYANIDE

A soluble cyanide such as sodium cyanide dissolving in rain water would initially give a strongly alkaline solution but it would be expected that such a solution would quickly be buffered in the ground - except perhaps in pure chalk or limestone or inert strata - to a near-neutral condition. The buffering might be assisted by the uptake of carbon dioxide and by dilution with groundwaters containing carbon dioxide or bicarbonate. The cyanide would then be largely in the form of un-ionized hydrogen cyanide, HCN, because HCN is an exceedingly weak acid with a pK value of approximately 9.2. Certain heavy metals and some of their compounds react with cyanide solutions to give soluble complex cyanide ions which exist in equilibrium with cyanide ion and hence with HCN. Complex cyanide ions, in order of increasing stability, are formed by zinc, cadmium, copper, nickel, and cobalt, among other metals. Iron forms the well-known and stable ferri- and ferrocyanide ions which are unusual in that they do not appear to be in a straightforward state of equilibrium with cyanide ion. The chemistry, especially the analytical aspects, of solutions of free and complex cyanides has been discussed extensively in papers by staff of the WPRL (7)(8)(9)(10).

The chemistry of a cyanide solution issuing from an industrial tip, or passing through strata containing compounds of any of the metals mentioned, might therefore be fairly complicated and the exact composition of such a solution would be impossible to predict. Further unknowns would be the rate of volatilization of HCN and its probable very slow rate of chemical hydrolysis to ammonium formate. However, detailed and extensive work at the Laboratory on the treatment by biological oxidation of effluents containing cyanide probably has a bearing on the rates of biodegradation of cyanides to be expected in the ground.

Pettet and Thomas (11) showed, as a by-product of work on the effect of cyanide on sewage treatment, that 30 mg/litre of cyanide in sewage was broken down completely by biological filtration, after 69 days' acclimatization. More recent work, by Pettet and Mills (7), showed that cyanide in sewage was readily degradable in the upper layers of a biological filter, that the main nitrogenous product was ammonia which could be nitrified lower in the filter, and that the removal of cyanide was inhibited partially by nickel, slightly by copper, and hardly at all by zinc or cadmium. It was also shown that cyanide could be destroyed in the absence of sewage. Ferrocyanide was apparently oxidized to ferricyanide but was not otherwise removed to any very great extent. Ware and Painter (12) showed that the organism responsible for removal of cyanide was probably a strictly autotrophic <u>Actinomycete</u> which grew best in a solution containing 40 mg/litre of cyanide.

Investigations were made of the important factors of acclimatization (7)(11)(13), temperature (13)(14), and pH value (13)(15). Acclimatization always took a few weeks and tended to occur most readily and reliably in the presence of organic matter in the form of sewage or peptone (5 mg/litre). Once acclimatized, a filter retained its ability to remove cyanide for at least 5 weeks after the application of cyanide was stopped. Breakdown of cyanide was readily achieved between 10° and 35°C, but not outside this range. Simple cyanide solutions were treatable at pH values as high as 9.6 but previous neutralization with acid tended to be beneficial when complexing metals were present. This suggests that complex cyanide ions are not available to the organisms responsible for the breakdown, because calculations show that a reduction in pH value causes the increased dissociation of complex cyanide ions (10). Consideration of chemical equilibria also suggests that free HCN is probably the chemical species utilized by the micro-organisms.

The implications for groundwater pollution of the Laboratory's work on cyanide removal are that, after a period of acclimatization lasting a few weeks, concentrations of cyanide up to perhaps 100 mg/litre would probably be effectively removed from a solution percolating into the ground from an industrial tip unless;

(a) the ground temperature were below 10°C,

(b) the pH value were outside the range 6 to 9.5 in the absence of zinc, cadmium, copper or nickel, or 6 to 7.5 in the presence of any of these metals,

(c) the solution became anaerobic for any reason,

(d) ferro- or ferricyanide were formed, or

(e) the organisms which can utilize cyanide were absent.

Moreover, if conditions at the site were very unfavourable, a considerable quantity of cyanide might be washed into an aquifer by rain before there was time for a cyanide-removing biological film to develop in the intervening strata (although such a film might eventually develop in the aquifer itself if the conditions were suitable). Clearly there is a need for further studies taking actual local conditions into account before the results of the work described here can be applied with any confidence to the prediction of the fate of cyanide in tips.

6. WATER POLLUTION BY INDUSTRIAL TIPPING

As a result of public concern about tipped waste materials such as cyanide, the Laboratory has just started a new research project in association with staff of the Water Resources Board. The aim is to study the potential pollution hazards to both surface and groundwaters resulting from the tipping of solid and concentrated liquid waste materials. Although the emphasis will be on the more toxic substances it is intended also to include municipal refuse in the investigation.

Chemical and hydrogeological studies will be made, at selected sites, to ascertain the composition and fate of waters issuing from tips, and the subsequent changes in composition. Laboratory investigations of the behaviour of some of the more hazardous substances will be undertaken where necessary in an attempt to predict their fate in aquifers.

Although it will obviously be impossible to predict the pollution hazard of every possible substance tipped at every possible site, it is hoped that the studies will yield background information which will help materially in the development of a safe, economical, and logical policy towards tipping.

ACKNOWLEDGEMENT

Crown copyright. Reproduced by permission of the Controller, H. M. Stationery Office.

REFERENCES

1. WHEATLAND, A.B. and BORNE, B.J.
Wat. Waste Treat.,1961, **8**, p.330.

2. WHEATLAND, A.B.
Contribution to discussion of paper by Frank, W.H.
Proceedings of Artificial Groundwater Recharge Conference, University of Reading, 1970
Medmenham, The Association, 1972.

3. HOUSING AND LOCAL GOVERNMENT Ministry of
Pollution of Water by Tipped Refuse.
London, HMSO, 1961.

4. WHEATLAND, A.B.
Wat. Treat. Exam. ,1969, **18**, p.54.

5. PETTET, A.E.J., and JONES, E.E.
Wat. sanit. Engr. , 1953, **3**, p.446.

6. TRUESDALE, G.A., and MANN, H.T.
Survr munic. Cty Engr. ,1968, **131**, No. 3953, p.28.

7. PETTET, A.E.J., and MILLS, E.V.
J. appl. Chem. , 1954, **4**, p.434

8. MONTGOMERY, H.A.C., GARDINER, D.K., and GREGORY, J.G.G.
Analyst, 1969, **94**, p.284.

9. MONTGOMERY, H.A.C., and STIFF, M.J.
Proceedings of International Symposium on the Identification and Measurement of Environmental Pollutants, Ottawa, June 1971, p.375.

10. MONTGOMERY, H.A.C.
Paper presented at 6th International Conference on Water Pollution Research, Jerusalem, June 1972.

11. PETTET, A.E.J., and THOMAS, H.N.
J. Proc. Inst. Sew. Purif. ,1948, Part 2,p.61.

12. WARE, G.C., and PAINTER, H.A.
Nature,1955, **175**, p.900

13. DEPARTMENT OF SCIENTIFIC AND INDUSTRIAL RESEARCH
Water Pollution Research 1959.
London,HMSO, 1960, p.35.

14. WARE, G.C.
Wat. Waste Treat. 1958, **6**, p.537.

15. DEPARTMENT OF SCIENTIFIC AND INDUSTRIAL RESEARCH
Water Pollution Research 1958.
London, HMSO, 1959, p.44.

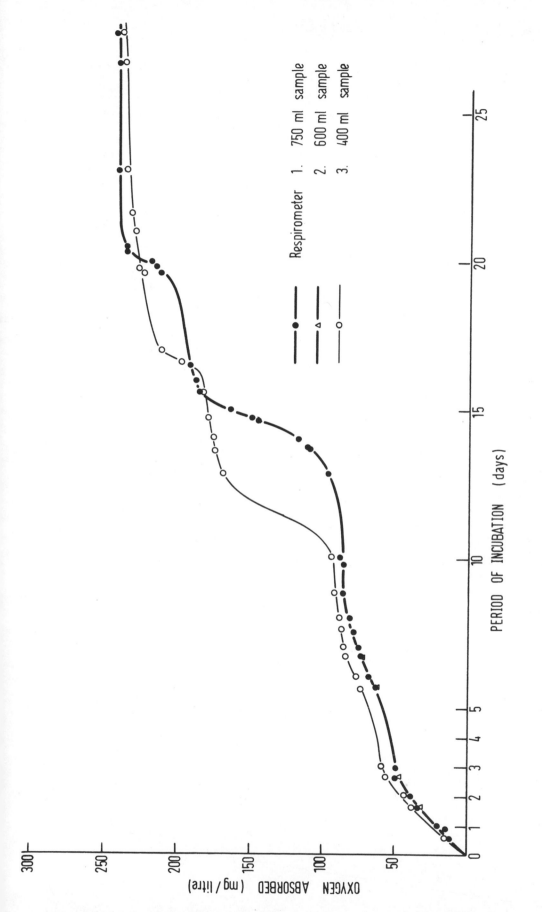

FIG. 48. ABSORPTION OF OXYGEN BY A SAMPLE OF WATER FROM THE WET REFUSE-DISPOSAL PIT AT BUSHEY (Temperature of incubation 25°C; Sample taken 18 March 1957)

DISCUSSION ON CONTAMINATION BY LIQUID WASTES

Inorganics in liquid wastes

1. A case history presented by the Great Ouse River Authority* involved a mixed pollution, containing bromine in simple anionic form, which provided a good reference tracer against which the degradation of phenolic substances could be seen. A parallel was evident with K^+ as the conservative indicator of leachates from refuse tips, whose organics decay.

2. If liquid wastes are discharged to open excavations or ponded on land surfaces, much higher rates of infiltration will occur as compared with the natural unsaturated movement of rainfall. Consequently such wastes have less chance of aerobic degradation above the groundwater zone and their acceptability will depend critically on dilution and dispersion within that.

Other inorganic problems

3. The fate of heavy metals was seen as a continuing concern in the monitoring of dune recharge works. It was vital to ensure that present-day practice, which benefited from the removal of metal ions by adsorption, did not lay in trouble for the future.

4. Sulphides (notably iron pyrites) arose in various strata and considerable experience in managing their oxidation products is available from the mining geology literature (see Bibliography in this volume).

5. G. D. Nicholls supplemented his paper with Fig. 49, p. 166, on the solution equilibria of iron compounds. The presence of iron oxides in sedimentary strata was certainly not always of secondary sulphide origin; many red sandstones in the UK were desert deposits whose iron oxide was deposited ab initio amongst a silica matrix.

6. National Coal Board practice in the compaction of colliery spoil heaps was instanced by T. E. James as a useful measure to counteract percolation (see written contribution, p. 252.)

Silage and sewage effluents

7. The liquors emanating from silage heaps should be regarded as untreatable and highly unsuitable for acceptance by any strata as a direct percolate. By constructing an impervious base for a silage heap, the liquors could be collected and then disposed of by spraying onto grassland.

*See paper on p. 296-298.

8.   Caution was urged by R. F. Packham, who doubted the wisdom of putting a waste discharge, such as a sewage effluent into underground strata without positive evidence as to the waste's degradability.   Existing discharges could provide valuable practical evidence on the fate of organics: such evidence would not be forthcoming if sampling remained so far from the input zone that no contamination could be found.

     In reply, N. J. Nicolson said that total organic carbon analyses on a series of water samples from both "control" and "suspect" boreholes had been carried out.   The results showed interesting variations, but the levels were close enough to the limit of detection of the technique to make interpretation a matter of some conjecture.   As a result, it was hoped to employ more sensitive techniques to isolate organic matter in the μg/litre range - perhaps carbon adsorption or ion exchange resins - in an attempt to establish whether any non-biodegradable organic materials in the discharge appeared in the water at the operating boreholes a few miles away.

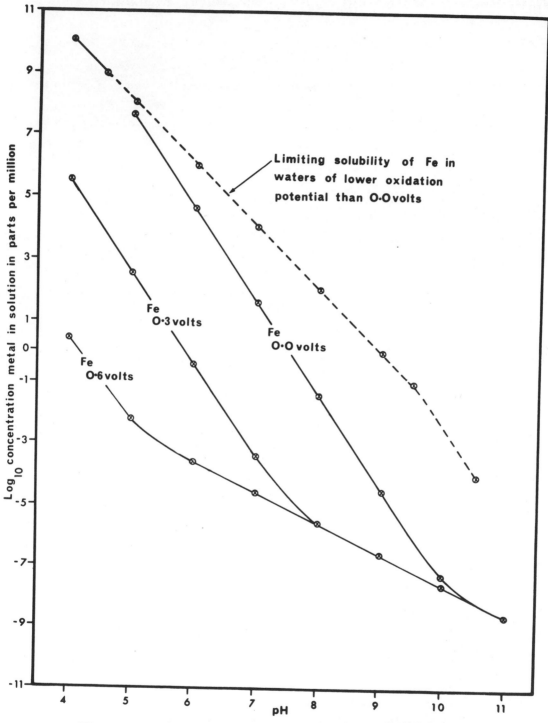

FIG. 49. SOLUBILITY OF Fe IN SULPHUR-FREE WATERS
UNDER VARYING CONDITIONS OF ACIDITY AND
OXIDATION POTENTIAL

MICROBIOLOGICAL ASPECTS OF GROUNDWATER POLLUTION

by M. Hutchinson
The Water Research Association, Medmenham, Bucks, England

1.   INTRODUCTION

Until recently society has been principally concerned with pollution that can be seen, namely the large stretches of deoxygenated and lifeless rivers. The pollution of underground waters has tended to be overlooked possibly with the attitude 'out of sight, out of mind' and it has been left to the diligence of water authorities who appreciate the dangers to protect this vital source of supply.

There exists a dilemma here between the necessity of safeguarding health on the one hand and preventing inconvenience and unreasonable expense on the other. It is desirable to know how close wells and boreholes may be located to sewage and waste disposal sites without fear of contamination. This problem has taken on greater significance by the necessity in certain countries to supplement the available groundwater by methods of artificial recharge and also by the ever increasing spread of urbanization and industrialization with associated problems of waste disposal.

The role of micro—organisms in groundwater pollution may be beneficial or detrimental depending on the types of organism involved and the specific circumstances in which they are present.

The beneficial activities of micro-organisms are manifold, but chiefly concern those which prevent the accumulation of pollutants in the environment. These include:

(i)   Processes of purification in which pollutants, including biological agents, are degraded to innocuous products. This includes the reduction in numbers of pathogenic organisms released into the environment.

(ii)   The cycling of certain essential elements in nature including carbon, nitrogen, phosphorus and sulphur.

(iii) Degradation of man-made pollutants such as pesticides (1), herbicides (2), detergents and other unnatural chemical compounds.

The same processes when taken to extremes by excessive pollution may give rise to certain problems associated with groundwaters. Pronounced microbial activity may result in secondary aspects of pollution; depletion of the dissolved oxygen in the water leading to biological nitrate reduction possibly to ammonia, reduction of sulphate to sulphide with precipitation as ferrous sulphide; the mobilization of iron from the soil under conditions of reduced oxygen tension only to be deposited in other regions of the aquifer through the activities of iron bacteria. Contamination of the water with iron and sulphides may render the water unpotable without treatment. Excessive growths of micro-organisms resulting from the introduction of large amounts of organic pollution in localized areas, growth of iron bacteria together with iron precipitation may result in the blocking of natural infiltration, particularly during groundwater recharge (3). Iron bacteria are frequently encountered in groundwater situations and in particular those subject to a degree of organic pollution. This was evident in the episode described by Taylor (4) where disposal of burnt sugar and biscuit wastes to a flooded gravel pit resulted in profuse growths of iron bacteria in a nearby well.

Brighton (5) reported that the fermentation liquor from silaged pea haulms drained into a swallow hole and polluted a nearby chalk borehole resulting in profuse growths in the borehole of filamentous iron bacteria (Leptothrix and Crenothrix) and filamentous sulphur bacteria (Beggiatoa) growing at the expense of hydrogen sulphide generated in the water by enhanced activities of sulphate-reducing bacteria (Desulfovibrio). The combined activities of these organisms severely restricted the yield of the borehole and rendered the water unpotable as a result of intense taste and odours. This necessitated renovation of the borehole by chlorination and acid treatment.

2. PUBLIC HEALTH ASPECTS OF MICROBIAL POLLUTION

One of the major concerns with drinking water, including groundwater supplies, is the health hazard resulting from its recent contamination by sewage and human excrement. The diseases transmitted by water primarily originate in the alimentary tract from whence pathogenic organisms pass into the environment in the faeces or urine of man, from either a diseased person or a carrier. It should be remembered also that livestock, including wild animals and birds might be infected by similar pathogens and serve as reservoirs of infection (6).

2.1. Agents of enteric diseases of man transmitted by water

2.1.1. Cholera.

This is the most serious disease that may be transmitted by polluted water as the figures for Europe in the nineteenth century would suggest. In 1854, in Britain alone there were over 20 000 deaths from cholera. Although not all the cholera was water-borne, polluted water was an obvious vehicle in its transmission as the famous Broad Street pump outbreak confirmed. The causative agent is <u>Vibrio cholera</u> (<u>V. comma</u>) which may exist in 1.9 to 9.0% of symptomless carriers (7) or 9.5 to 25% in the case of the El Tor biotype (8). These carriers continue to excrete the pathogens intermittently for up to 30 to 40 days (7), though chronic carriers have been known to excrete vibrios for as long as 15 months (9).

The development of sewage disposal facilities and protected community water supplies was largely responsible for its eradication from Europe. However, the Far East still remains a focus from which in recent years new epidemics have arisen for example, in Egypt in 1947 (10), Iran and Iraq in 1964, Odessa in Russia and in Guinea in Africa in 1970 (11). This spread may reflect the international mobility of populations in modern times and stresses the importance of maintaining high standards of protection for water supplies.

2.1.2. Salmonella.

Over 500 serotypes of salmonella are known to be pathogenic to man and although in Britain food poisoning is the most common manifestation of salmonellosis, typhoid and paratyphoid fevers are the most serious and man is essentially the sole reservoir of infection. The causative agents are <u>Salm. typhi</u> and <u>Salm. paratyphi</u> respectively. The number of individuals excreting salmonellae has been reported as less than 1% in England and Wales (12) and United States (13).

Salmonellae are frequently found in healthy cattle at a level of 13% in the United States and 14% in the Netherlands (14). In a recent survey the incidence for pigs in England and Wales was 7% and 3% for Denmark (15). Salmonellae are commonly found in wild animals and birds (16). It is hardly surprising therefore that salmonellae are frequently isolated from the polluted environment including sewage (17), farms and abattoirs (18) and rivers and streams (19)(20). Taylor reported that at the intakes of the Metropolitan Water Board on the Rivers Thames and Lee, 1 to 3 salmonellae/litre could be isolated from approximately one-third of the samples throughout the year.

2.1.3. Shigella.

There are at least 32 serotypes of which <u>Shig. sonnei</u> and <u>Shig.</u> <u>flexneri</u> subgroups account for over 90% of all isolates from humans in the United States; <u>Shig.</u> <u>dysenteriae</u> and <u>Shig. boydii</u> comprise less than 1%. In Britain <u>Shig. sonnei</u> is the most common. Estimates of the average number of individuals excreting Shigella has been put at 0.46% in the United States, (21) and 0.33% in England and Wales (12). Shigellae are rarely found in animals (21).

Most epidemics are food-borne or spread by person-to-person contact, particularly amongst children in schools, though water-borne incidences have been demonstrated. It is probably the absence of an adequate technique that limits the demonstration of this agent in many more situations (22).

2.1.4. Enteropathogenic Escherichia coli.

Some eleven serotypes of E. coli frequently cause a gastroenteritis which can be fatal, especially in new-born infants (23). Enteropathogenic E. coli have been isolated from 2.42% of 24 864 normal children over a period of 5 years in England and Wales (12); 1.2% of 172 normal infants in Houston, Texas (24); 1.8% of hospital staff at Cincinnati but 15.5% of 385 mothers of new-born infants (25) and 6.4% of food handlers in Louisiana (13). They have also been found in farm animals (26).

Enteropathogenic E. coli have been isolated from natural water courses; between 1963 and 1965 in Hamburg, the incidence was 0.4% for well waters and 0.1% surface waters (27). Geldreich (28) reported the presence of enteropathogenic E. coli in streams and lakes polluted by faeces of warm-blooded animals but their occurrence was probably less than 1% of the faecal coliform population.

2.1.5. Viral pathogens.

Over 100 different enteric viruses that infect man have been identified. In a survey, Cook and Smith (29) found that 4.5% of healthy children were excreting enteric viruses.

The principal viral types of concern are the following.
(i)   The agent(s) responsible for infectious hepatitis
(ii)   Polio virus
(iii)   Coxsackie virus
(iv)   ECHO viruses

Paradoxically, according to Moseley (30), infectious hepatitis is the only viral disease for which a water-borne route of transmission has generally been accepted, yet the agent responsible has yet to be confirmed. This makes the task of establishing the source of infection with this disease, very difficult. Although Polio, Coxsackie and ECHO viruses can readily be isolated from sewage and sewage polluted river waters, the role of water as the primary mechanism of dissemination in various described episodes of viral disease is difficult to prove unequivocably. However, these viruses must be considered as potential water-borne pathogens. In a recent epidemic at Montrose, 40% of the population succumbed to gastro-intestinal disorders following the supply of incompletely treated river water. Coxsackie and ECHO viruses were isolated together with <u>Shigella</u> <u>sonnei</u> and various other potential pathogens (31).

Gastroenteritis, possibly of viral origin, is perhaps the commonest water-borne disease though it is rarely associated with any recognizable pathogen. Between 1911 and 1937 in Britain, of 21 outbreaks of disease attributable to public water supplies and sufficiently grave to appear in public health statistics, gastroenteritis accounted for 7439 cases, dysentery 2800 cases and enteric fever, including paratyphoid fever, 1237 cases (32). In the United States, in over 300 outbreaks of water-borne disease involving over 100 000 persons, 198 outbreaks were of gastroenteritis, 35 of dysentery and 99 of enteric fever (33). Not all of these outbreaks were attributable to contamination of the source and, indeed, most of these outbreaks have been the result of gross contamination of the water systems due to obvious defects such as backsiphonage of sewage contaminated water, failure of water treatment facilities etc.

The presence of large numbers of viruses and bacteria not necessarily pathogenic but derived from faeces and foreign to a particular person may in itself be sufficient to give rise to symptoms of disease (34).

2.1.6. Animal parasites.

Ova of various parasitic worms and cysts of <u>Entamoeba</u> <u>histolytica</u> are shed in faeces and pollute the environment. Fortunately in Britain there is little evidence of such diseases but one should be on guard against their introduction for some of these forms are extremely resistant to disinfection as practised at present (35).

A number of cases of Ascaris infection in Hertfordshire in 1923, were traced to infection of a local well supply (36).

2.2. <u>Factors controlling the spread of water-borne disease</u>

In order for water-borne disease to come about, the causative agent must be present in the environment, it must contaminate the water supply, survive and be consumed by another individual. The chances of this person becoming infected depend on:

(i)   The susceptibility of the individual.

(ii) The virulence i.e., the number of organisms necessary to establish themselves within a host.

(iii) The numbers of pathogens in the environment.

(iv) The survival of the pathogens.

2.2.1. The susceptibility of the individual.

Although it is possible to change the susceptibility of the individual to certain diseases by inoculation, this becomes less practicable with others in view of the number of different infective agents for many of which effective vaccines do not exist.

2.2.2. The virulence of the infective agent.

The long history of water-borne typhoid fever compared with paratyphoid fever of water-borne origin has usually been explained in terms of fewer typhoid bacilli being required for clinical infection. Paratyphoid fever, being more a food-borne than water-borne disease, would suggest a larger infective inoculum commensurate with prior multiplication on contaminated food. However, recent experimental work by Hornick and Woodward (37) in which typhoid bacilli were administered to volunteers gave the following results:

(i)   A dose of 10^3 organisms failed to cause symptoms in any of 20 volunteers.

(ii)   A dose of 10^7 organisms led to clinical disease in 16 out of 32 volunteers (50% infection).

(iii) With a dose of 10^9 organisms 40 out of 42 showed clinical symptoms (95% infection).

These findings were similar to earlier work with chimpanzees which showed that a high dose of typhoid organisms were required to produce typhoid fever (38).

Against this evidence must be set incidences of disease in which the level of pollution must have been extremely low. In these instances it has been suggested that low levels of pathogens penetrate the defensive mechanisms of the stomach and on arrival in the intestine they multiply to sufficient numbers to produce a clinical condition.

Plotkin and Katz (39) have reviewed the subject of minimum infective doses and their own data indicates that as little as 1 tissue culture infective dose * of Polio Virus 3 may be infective for infants.

2.2.3. The numbers of pathogens in the environment.

The number of salmonellae in faeces of cases has been put at 10^5 to 10^7/g with symptomless carriers usually excreting smaller numbers (6).

---

* Tissue culture infective dose is that dilution of the virus suspension that produces only a 50% infection of a series of tissue cultures. This technique is similar to the statistical MPN method for counting bacteria.

Sabin (40) found 10^5 tissue culture units of Polio Virus/g of faeces. In experiments with volunteers, Neefe et al (41) found the numbers of infectious hepatitis virus/g of faeces to be 10^4 to 10^5.

Clarke et al (42) using experimental data for the USA found that the ratio of enteric viruses to coliform bacteria in human faeces was 1:65 000. From a mean coliform density of 46×10^6/100 ml for domestic sewage they deduced that the probable number of enteric viruses would be 700 units/100 ml. Similarly, for a mean coliform count of 10^4 to 10^5 for surface water polluted with sewage, the virus concentration would be 1.5 units/litre in cold months to 15 units/litre in warm months.

A summary of the incidence of viruses in water and factors affecting their survival and removal has been given in WRA Technical Inquiry Report, TIR 192 (43).

2.2.4. The survival of pathogens.

The period of time bacterial and viral pathogens may survive the various stages of the hydrological cycle is very important in designing safe water systems. Unfortunately precise times cannot be stated, only wide limits based on experimentation and experience.

Survival is influenced by the following.

(i)   The nature of the suspending medium: faeces, sewage, soil, water etc.

(ii)   The physical environment: temperature or light intensity.

(iii)   The presence of other biological agents including bacteria, phages, protozoa and other predatory organisms.

(iv)   The presence of biocidal agents, either present as natural antagonists from other biological forms or as chemical pollutants.

Table 19, p. 189, sets out typical periods of survival for various pathogenic forms.

2.3. Measures taken to control the spread of pathogens

The success of any pathogen relies on its transmission to another susceptible host and it is upon the interruption of this cycle that public health engineering has concentrated by:

(i)   Isolation and treatment of sewage before release into the environment.
(ii)   Protection of water supplies from sewage pollution.

173.

Unlike surface water supplies which rely on water treatment processes for decontamination, with underground supplies greater attention is placed on natural infiltration through the soil cover together with protection of the supply from pollution by:

(i)   Careful siting of the abstraction with respect to actual or potential sources of pollution.

(ii) Sound construction of the well or borehole and its immediate confines to exclude surface, and therefore recently contaminated water. This should also take into consideration the possibility of flooding.

(iii) Correct operation of the well by avoiding excessive or erratic drawdown of the water table resulting in accelerated ingress of more remote sources of pollution.

Water treatment may be necessary with shallow wells or springs as indicated by the outcome of chemical and bacteriological tests. These may include aeration for a water deficient in oxygen, iron or hydrogen sulphide removal; coagulation and/or filtration; lime softening.

Disinfection should be practised as a final line of defence to safeguard the quality of all supplies but never used as the sole process to make a polluted water potable. It should only be applied when the physical, chemical and biological quality of the water has attained satisfactory limits.

2.4. Natural mechanisms for the removal of pollutants in underground waters

It has long been supposed that the infiltration of water through the soil cover is an effective process of water treatment, so much so that further measures of purification, including disinfection, are sometimes claimed to be unnecessary. This supposes that percolation of surface water containing bacterial, viral, biological and chemical pollution undergoes purification as it penetrates into, and travels with, the groundwater by processes of:

(i)   Filtration
(ii) Adsorption
(iii) Biodegradation
(iv) Death due to excessive competition for nutrients and oxygen, antagonism, predation or simply detention.
(v)   Dilution by dispersion in the groundwater.

2.4.1. Filtration.

This is essentially governed by the factor of particle size both of the filter medium and the pollutant. The process in itself is not very effective for even with a fine grain size of 0.15 mm, the smallest pores are over 20μ in diameter and considering that bacteria generally vary from 0.5 to 5μ in length and viruses are approximately one-hundredth of this size, they are unlikely to be retained by this mechanism alone. Of course, soil containing a significant clay content would tend to aggregate to form a more effective filter.

2.4.2. Adsorption.

This is an interface phenomenon and consequently fine grain minerals are the most effective. Adsorption may be by simple physical attraction (Van de Waals forces) or coulombic effects in that negatively charged minerals adsorb positively charged particles such as iron or aluminium hydroxide flocs, carbonates, as well as cations. This may then give the particle a net positive surface charge which conditions it for adsorption of negatively charged colloidal matter of an organic nature, including bacteria and viruses. Dissolved anions such as nitrates and phosphates may similarly be concentrated from the flowing water phase. All of these processes are essentially reversible so that although movement of pollution may not be completely arrested it may be significantly retarded.

2.4.3. Biodegradation.

The presence of adsorbed organic matter and essential inorganic nutrients such as nitrates and phosphates together with certain saprophytic bacterial types, results in their growth. Using the dissolved oxygen in the water the organic compounds are in part oxidized to provide energy (dissimilation) or partly converted to cell material (assimilation). The oxygen in the water rapidly becomes depleted and this is largely the controlling factor in the degree of purification achieved and also the extent in depth of the biologically active zone. In general, this is confined to the uppermost centimetre (3). Depending on the level of organic pollution this may lead to the formation of a biological filter mat which with maturation becomes an even more efficient strainer.

The net result of this biological activity is the conversion of organic pollutants, including biological agents, to innocuous inorganic compounds such as water, carbon dioxide, nitrates, phosphates and sulphates (mineralization). The efficiency of the biological oxidative process depends on the supply of oxygen, a temperature conducive to active growths of the bacteria and adequate detention time.

At greater depth, products of incomplete dissimilation are utilized by other bacteria, possibly under microaerobic or anaerobic conditions in which case nitrates and sulphates may be utilized as oxidants. These are in turn reduced to ammonia, nitrogen gas or sulphides.

2.4.4. Decline in the numbers of pathogens.

Whilst there are probably no specific mechanisms for the selective removal of pathogens, these essentially parasitic organisms are obviously at a disadvantage in this highly competitive environment with low temperatures and limited availability of suitable nutrients. Predation by protozoa and lower animal species; attack from bacterial viruses or predatory bacteria such as Bdellovibrio species or even antagonism from antibiotic substances produced by other organisms are all responsible for the decline in numbers of biological forms. Although this reduction is not exclusive to pathogens, in view of their relative density to other bacterial types, their removal is virtually complete. However, some may survive for it is known that coliforms pass through slow sand filters and they have been known to penetrate into groundwater supplies.

2.4.5. Dispersion by dilution.

Having entered the saturated zone, the organisms are diluted to a limited extent as they travel in the groundwater proper. It is claimed that turbulence is a factor in this, though certain evidence on the movement of pollutants in groundwater suggests that unless there are major salinity differences the pollution is largely retained in the upper interface of the zone of saturation and only undergoes slight dispersion in a lateral direction. Unlike surface waters, it would be unwise to rely on dilution of pollution as a protective measure in view of the uncertainty of its occurrence and degree. There can be only a small change in view of the relatively slow rates of flow.

2.5. Factors controlling the spread of pollution underground

2.5.1. Phase of infiltration.

As indicated, the travel of pollution may be separated into two phases.

(i)   Vertically through the unsaturated zone.

(ii)   Horizontally, once the percolating water has reached the zone of saturation (groundwater table).

The significance of these two phases of infiltration as far as groundwater pollution is concerned is of immense importance, for it is mainly in the unsaturated zone, an area which is essentially aerobic, that the major part of purification occurs largely through biological agencies. Once pollution enters the groundwater it is carried passively with comparatively little opportunity for further change.

176.

2.5.2. Geological factors.

The geological composition of the aquifer is obviously of great significance, for granular soils with primary porosity act as far better filters than consolidated rocks having secondary porosity in the form of joints, fractures or solution channels. In the latter circumstances it is impossible to achieve any degree of purification, and extreme care must be taken in the selection of suitable strata for abstraction.

A confined aquifer is comparatively safe provided it is effectively sealed for the ingress of surface waters and the outcrop is adequately protected from obvious sources of pollution.

2.5.3. Topography.

Pollution can only spread in the direction of the groundwater flow existing at that time. This is governed largely by the position of sources of replenishment of the groundwater and the relative size and location of sources of abstraction or loss to the aquifer. It is important therefore to consider sources of pollution in the context of direction of groundwater flow and the position of pumping wells.

3. STUDIES TO EVALUATE DISTANCE OF TRAVEL OF POLLUTION

Numerous studies have been made to try to resolve the expected distance of travel of pollution under various hydrogeological conditions in order to give guidance on:

(i)   Siting of sources of pollution in relation to water abstraction facilities.

(ii) Methods of disposal of sewage and refuse.

(iii) Groundwater recharge.

The following is a survey of the findings, supplemented by illustrations taken from the literature of known incidences of pollution.

3.1. Laboratory studies of removal of bacteria and viruses by filtration

(i)   Robeck (44) studied the penetration of coliforms (10 000 to 20 000/ml) in water through saturated sand columns in the absence or presence of detergent (50 mg/litre ABS).

TABLE 16

NUMBER OF DAYS FOR THE PENETRATION OF COLIFORMS IN THE ABSENCE OR PRESENCE OF ALKYL BENZENE SULPHONATE (50 mg/litre ABS)

Sand used:	Chillicothe		Newtown				
Flow rate (cm/day):	9		12		24		0
Presence of ABS:	+	-	+	-	+	-	+
15	-	-	-	-	9	9	0
30	6	11	16	15	41	41	0
60	68	62	63	70	60	60	
120 Distance penetrated (cm)	69	69	114	114	82	82	
180			196	231	119	119	
240			417	417			
300							

Characteristics of sand used:	Mean size (mm)	Coeff. uniformity	Porosity
Chillicothe	0.38	1-37	52%
Newtown	0.18	1-8	43%

These indicate that bacteria did not penetrate to a depth of 3 m even with prolonged infiltration. Furthermore, the degree of penetration was not significantly affected by the presence of 50 mg/litre of detergent.

In intermittent surface spreading experiments in a 1-m diameter column packed to a depth of 1.2 m with Newtown sand and receiving 19×10^{-3} m^3 of septic tank effluent each day, 1 to 10 coliforms/ml were observed to penetrate in the eighth month of infiltration.

In virus studies with Polio type 1, since 10% of the 10 to 30 000 virus units/ml die off in 24 hours a flow-rate of 0.9 to 1.2 m/day and a column depth of only 0.6 m was used to ensure a detention of less than 24 hours. In over 50 hours there was little penetration of the virus under saturated conditions with both Chillicothe and Newtown sands. Counts of 1 to 12 virus units/ml appeared in the effluent after 42 days for Chillicothe sand and 105 days for Newtown sand and the situation did not change over 6 months of continuous infiltration.

The amount of virus removal by the 0.6 m bed of sand varied with flow-rate. At 4.4×10^{-10} m^3/sec/m^2 of sand there was virtually complete removal whereas at 1.158×10^{-7} m^3/sec/m^2 of sand about 80 to 90% penetration

occurred for both sands under percolation or saturated conditions with upward
or downward flows (45).

(ii) Gilcreas and Kelly (46) compared the survival of Coxsackie and Thieler's viruses
with E. coli during percolation through 0.9 m of garden soil. There was only a
50% reduction in virus compared with a 75% reduction in E. coli.

(iii) In studies using radioactive labelled bacterial viruses, Drewry and Eliassen (47)
found adsorption to be the process of removal and this was significantly affected
by the pH of the saturated soil-water system. At pH values around neutrality,
virus adsorption was optimum but at higher pH values it was significantly less
due to an increasing negative charge on the soil particles and/or viral surface
resulting from a proportionate increase in ionization of carboxyl groups of the
proteins beyond the isoelectric point. Similar results were found by Cookson (48)
for virus adsorption on activated carbon.

Virus adsorption by certain soils was enhanced by increasing the cation
concentration in that the cation served to neutralize repulsive negative charges
on virus and/or soil particles. Although, in general, virus removal correlates
with an increase in clay and silt content, ion exchange, and glycerol retention
capacity, certain soils with low values for one or all of these parameters could
still rank highest in terms of virus adsorption. It would seem therefore that
the various tests which are used to characterize a soil, may have little value
in predicting virus removal. All soil columns tested removed over 90 to 95% of
virus within 45 to 50 cm depth of soil. In general there was an increase in viral
removal with decreasing particle size. A sand of particle size 0.12 mm had a
removal efficiency of 99.999% for 60 cm of penetration (47) (49)).

(iv) In studies of rapid gravity filtration at Harvard University using bacterial viruses
it was shown that columns 34 cm deep containing 400 g of fresh sand removed
200 000 virus units/g before reaching a saturation point. Backwashing
of this sand regenerated only approximately 10% of the adsorptive capacity. After
treatment with albumin solution, sand removed very little virus, showing that the
adsorptive sites were fully saturated (50).

(v) Other workers have studied the adsorption of albumin as representing a virus-
sized particle (0.015 μ) and found up to 96% retention by infiltration through less
than 1 m of sandy soil of mean size 0.3 mm that had previously been heated to
760°C to destroy organic matter and biological activity. With a high flow-rate

through a clay-less white sand (mean size 0.21 mm) a 58% removal was observed (51).

3.2. Field studies of removal of microbiological pollutants during infiltration

3.2.1. Travel of pollution in the non-saturated zone.

(a) Latrine studies

    (i)   With a water table 3.5 to 4 m below the bottom of a pit latrine, Caldwell (52) found that coliforms did not penetrate 30 cm below or 30 cm laterally from the latrine. Rain carried coliforms down to 1.2 m and the addition of 378×10^{-3} m^3 water daily distributed coliforms to 2 m below the pit with a lateral spread of 0.6 m.

    (ii)   Baars (53) studied the effect of a 1 m deep latrine in 2 m of soil cover (size 0.17 m) overlaying the groundwater at 3.5 m. E. coli could not be detected at depths exceeding 1.5m

(b) Surface spreading (54)(55)(56)

    (i)   At Whittier, the dosage of a secondary sewage effluent containing coliforms of at least 110 000/100 ml gave rise to a concentration of 40 000/100 ml at a depth of 1 m in 12 days. None appeared at 2 m depth even after prolonged spreading upon coarse medium.

    (ii)   At Azusa, spreading a secondary sewage effluent containing 120 000 coliforms/100 ml gave a percolate at 75 cm and 2 m of 6000 coliforms/100 ml. When primary sewage was applied however the 2 m coliform count rose to 1.2×10^6/100 ml.

    (iii)   In experiments with beach sand percolation through 3.5 m reduced 70 000 coliforms/100 ml to 700/100 ml.

    (iv)   In the infiltration of tertiary treated sewage into dune sand on a scale large enough for a city, Baars (53) found E.coli penetration down to 3 m but rarely 6 m. In virgin dune sand most bacteria were removed in the first 30 cm and no E. coli could be detected below 50 cm.

    (v)   At Lodi, California (57) depending on the individual basin, pollution penetrated to 0.6 to 4 m with primary and final sewage effluents. Not until between the third and eighth months was there appreciable penetration down to a depth of 1.2 m. There was no evidence of bacterial breakthrough on changing from final to primary sewage or from fresh water to final sewage (Table 17).

TABLE 17

MPN OF COLIFORMS IN HANFORD FINE SANDY LOAM AT LODI, CALIFORNIA

Depth (m)		Average MPN of coliforms/100 ml.						
		Surface	0.3	0.6	1.2	2.1	3.0	4.0
Basin	Effluent Spread							
A	Primary	4.14×10^6	1.6	32*	0.6	0	0	
	Final	1.79×10^5	1.2	285*	2.1	0	0	
B	Primary	5.7×10^6	20	0	0	0	0	
	Final	1.88×10^5	482	5.6	0.5	0.2	0.1	0
C	–							
	Final	1.88×10^5	148	305	2	0.2	0.1	0.3
D	Fresh water	2.32×10^2	1.6	–	–	0	–	0
	Final	1.64×10^5	0.2	–	–	0	–	0

* Channelling in sand from surface to 0.6 m depth.

(vi) Lysimeter studies of several soil types in the Lodi area revealed a pronounced removal of bacteria in the top 0.5 to 1.0 cm of soil with a secondary zone at a depth of 50 to 60 cm. Comparative counts for five soil types at a depth of 1 m are given in Table 18 (58).

(vii) In recharge pits at Peoria, Illinois, Suter (59) reported the reduction of total bacterial counts in the range 12 000 to 66 000/ml to 5 to 20 ml during passage through 140 m of gravel. About 1% of the samples were positive for coliforms.

TABLE 18

SOIL TEXTURE AND COLIFORM COUNT AT A DEPTH OF
1 m FOR 5 CALIFORNIAN SOILS (58)

	Effective size (mm)	Sewage infiltration rate (cm/day)	Coliform count (MPN/100 ml)
Hanford sandy loam	0.0056	9	4.5×10^3
Hesperia sandy loam	0.002	6	4.5×10^3
Columbia sandy loam	0.0034	9	4.5×10^3
Yolo sandy loam	0.0155	9	2.4×10^6
Oakley sand	0.015	3	2.4×10^5

Initial coliform count was 1.9×10^8/100 ml.

3.2.2. Travel of pollution in the saturated zone.

(a)   Latrine studies

      (i)   In studies of pollution resulting from a latrine 5 m deep and penetrating the groundwater table to a depth of 1.5 m, Caldwell and Parr (60) found that with a groundwater flow of 2 m/day after 3 days a few coliforms were detected 4.5 m away. After 35 days E. coli were detected at a distance of 3 m and after 60 days in 90% of samples at 8 m with an occasional positive sample at 11 m. After 200 days the pollution regressed almost to the latrine indicating that self-purification processes were fully developed.

      (ii)   In another study faecal matter was added to a latrine 2.5 m deep with the groundwater level at only 1.5 to 2 m from the surface and having a flow of approximately 4 m/day (61). Pollution extended to the limit of observation (26 m) but after 16 months this pollution had regressed to 6.5 m from the pit and only an occasional E. coli was recovered.

      (iii)   Stiles and Crohurst (62) studied the pollution of groundwater by a latrine in soil of size 0.13 mm. The pollution travelled as a thin sheet at the surface of the zone of saturation in the direction of the groundwater flow. This applied to both coliforms and fluorescent tracers. Coliforms were observed to travel 19.5 m in 187 days. There was no evidence to suggest lateral movement in the unsaturated zone but when the groundwater level fell,

pollution remained stranded in the capillary fringe, only to travel
further on return of the original groundwater level.

In other experiments the survival of organisms in this situation was
sought by burying faeces in pits in a region of high groundwater. After
three years, E. coli could still be recovered from 3 out of 5 samples
and ova of the parasitic worm Ascaris lumbricoides were found to be
present, but all 57 ova recovered were non-viable.

(b) Percolation into a confined aquifer

(i) In the Santee Project (63) (64) tertiary treated sewage effluent was
pumped into a confined shallow stratum of sand and gravel. An
arrester trench was situated at 450 m with sampling wells at 60 m and
120 m. The flow-rate was 60 m in 2 days.

Most of the bacteria were removed within the first 60 m with little
further removal in the remaining 390 m of travel through this coarse
medium.

Twelve litres of concentrated Polio vaccine were added to the
percolation bed over 3 hours. All samples collected at each of the
sampling locations were negative for virus suggesting that virus did
not penetrate as far as 60 m. Furthermore, over a period of 3 years
of infiltration with sewage effluent, viruses were never isolated from
the bed although Adenovirus, Reovirus, Polio, ECHO and
Coxsackie viruses were at times identified in the oxidation pond
effluent.

(c) Well injection

(i) Ditthorn and Luerssen (65) introduced Bacillus prodigiosus into an
aquifer of 32.8% porosity at a point 21 m from a sampling well.
The bacteria appeared in the well on the ninth day for 10 consecutive
days and after injection ceased pollution could be detected for as long
as 30 days.

(ii) Fournelle et al (66) used an injection well extending 1 to 1.2 m into a
shallow groundwater table, 1.5 to 2 m below the ground surface.
Bacterial pollution, as measured by enterococci, extended only 15 m
in a narrow path of travel 0.45 to 1.2 m in width.

(iii) In the Richmond, California study (67) a confined aquifer was injected with water containing 6, 10, 20, and 27% primary sewage at rates of 19 to 233×10^{-5} m^3/sec. The groundwater flow was from south and east and velocities ranged from 0.457×10^{-4} m/sec to 7.12×10^{-3} m/sec. At an injection rate of 233×10^{-5} m^3/sec of 27% sewage representing a coliform count of 4.7×10^6/100 ml, over significant periods of time, peak counts of 23/100 ml were recovered at the maximum distance of 30 m in the direction of normal groundwater flow. Similar maxima were obtained in other directions at 15 m and 19 m. Increasing the rate of injection from 85 to 403×10^{-5} m^3/sec or the concentration of organisms in the recharged water did not result in travel of pollution to a greater extent. This only shortened the permissible period of recharge before well redevelopment to remove clogging. The introduction of fresh water did not bring about the spread of pollution.

There was no difference in distance travelled between coliform bacteria, enterococci or other bacteria determined by plate counts. However, the rate of travel of bacteria was only approximately one-half that of fluorescein. A period of 33 hours was necessary for travel to 30 m south, but none travelled as far as 68 m south or 30 m east in 41 days although chemical evidence indicated that pollution travelled these distances.

Pollution was also observed to travel against the natural groundwater flow due to recharge effects. After 3 days of recharge at 233×10^{-5} m^3/sec regression of bacterial numbers occurred at all observation wells.

When the logarithm of the number of bacteria remaining are plotted against distance a straight line relationship was found which may be expressed as:

$$\text{Log } N_2 = \text{Log } N_1 - F (r_2 - r_1)$$

where N_1 = Number of organisms at sampling point r_1

N_2 = Number of organisms at sampling point r_2

and $r_2 - r_1$ = Distance between the contamination and sampling points.

F is defined as the 'Filterability' of the system, a physical characteristic of that particular aquifer and is not a function of pressure gradient and therefore not related to groundwater velocity. For example, it was found that there was a 24% decrease in coliform organisms in 0.3 m over a distance of 4 to 19 m at an injection rate of 233×10^{-5} m^3/sec whereas at 107×10^{-5} m^3/sec over the same distance the corresponding decrease per 0.3 m was 26%. It may be concluded that within limits removal depends on distance only and not upon the rate of pollution recharge. Also an approximate tenfold change in groundwater velocity has but a negligible effect on the rate of decrease in bacterial numbers.

McGauhey and Krone (67) conclude that the extent of biological pollution is so limited that reclamation of sewage effluents by recharge would not be limited by concern over public health aspects of microbial contamination.

3.3. Reported distances of travel of bacterial contaminants under operational conditions

Many instances are on record of pollution entering underground supplies from sources of contamination of different origins and severity, located at various distances from the abstraction. Of course, in the majority of these situations the nature and uniformity of the geological strata are unknown but obviously with very rapid or far distant travel, fissuring of the strata is likely to be the major contributing factor. Typical instances of groundwater pollution experienced are given in Table 20.

3.4. Summary of evidence on travel of microbiological pollution

Despite many anomalies in the data observed and in particular that between experimental and actual incidences of pollution which must obviously be the result of geological and/or well operation defects, certain general principles are evident in much of this work.

(i) Although smaller than bacteria, viruses are removed both in the saturated and unsaturated zones with comparable efficiency and in view of their relative numbers compared with bacterial indicators of faecal pollution in natural environments they should not represent a special problem.

(ii) Biological removal is best achieved in aquifers having a uniform, fine or very fine sand with a high clay content, namely one with a maximum surface

185.

area/volume ratio conducive to adsorption.

(iii) Organic pollutants in water may compete with bacteria and viruses for sites of adsorption. Cations aid removal by neutralizing repulsive negatively-charged sites on both biological and soil particles. High levels of organic pollution resulting in an efficient biological filter mat may assist in the removal of pathogens.

(iv) Competition and antagonism is perhaps the most important factor in microbial reduction, particularly in the unsaturated zone. Longevity is of minor importance in penetration bearing in mind the published findings on survival of pathogens in natural situations compared with groundwater retention in shallow aquifers. This survival is greatest in relatively unpolluted waters under conditions of low temperature.

(v) Bacteria and viruses travel passively with the flow of water and may during times of recharge or abstraction, move in a direction contrary to that of the normal groundwater.

Within limits, the removal of bacteria and viruses varies logarithmically with the distance travelled. The rate of removal with distance is a function of the aquifer termed "Filterability." For any degree of filterability removal depends on distance only and not upon the rate of pollutant recharge.

(vi) Two distinct phases of infiltration exist, namely that in the essentially aerobic, unsaturated zone where biological processes of degradation proceed and the other in the saturated zone which may be depleted of oxygen. Removal of pollution in the unsaturated zone is therefore far greater, with the result that the extent of biological travel appears to be limited to approximately 3 m in depth. For a uniform saturated system biological penetration is not limited until 15 to 30 m of travel in the groundwater whereas, with fissured strata the extent of travel is unlimited and the degree of purification virtually non-existent.

4. RECOMMENDED SAFETY ZONES

There is undoubtedly a potential health risk in the siting of water abstractions in relation to sources of groundwater pollution such as:

(i)   Contaminated surface water

(ii)   Ponds for sewage treatment, cesspools etc.

(iii)   Agricultural wastes

(iv)   Stormwater discharge pits

(v)   Drains and sewerage pipelines

(vi)   Refuse tips and spoil heaps

Because of the many variables of geology, chemistry of pollutants, hydrology and biology in addition to climatic factors and topography, it is obviously impossible to lay down precise safe distances for any particular well and potential source of pollution. However, certain guidelines are given in the literature as being at least a reasonable working guide. These are set out in Table 21.

Romero (68) has rationalized these somewhat for areas underlain by unconsolidated strata by taking into consideration the important distinction between

(i)   The zone of saturation lying in excess of 3 m below the source of
        contamination, and

(ii)   Systems in which the zone of saturation is in contact with, or very near the
        source of contamination.

These recommended safe distances are set out in Table 22.

Areas underlain by consolidated rock subject to fracture or solution channels should be examined very critically hydrogeologically, topographically and by chemical and bacteriological analyses because of the potential of extensive travel of pollution. In this connection tracer studies and an investigation of bacterial types present may afford some evidence of the origin or destination of such waters (4).

5.   DISCUSSION

To date reliance has been placed essentially on experience and the ability to construct sound groundwater facilities. However, with the ever-increasing spread of urbanization, and a need to utilize water resources often to their limits, it is possibly asking a great deal of subsurface filtration to maintain safety in every instance .

Regulations for the protection of groundwater supplies as written in byelaws resulting from the 1945 Water Act possibly err on the side of safety in many instances in that such measures may be interpreted as being too restrictive for a society with an ever-increasing housing problem. However, it can be argued that comparable installations with less substantial zones of protection are taking unnecessary risks and that the cost of purchasing land to afford the necessary protection is less expensive than undertaking the necessary hydrological investigations to prove the safety of a particular site. Obviously these questions cannot be resolved satisfactorily which is why the course followed is assessment of the safety of any particular situation on the quality of groundwater obtained in terms of bacteriological and chemical analyses of the raw water. It is important therefore that chlorine should not be applied in the borehole itself since on several occasions the first indication of contamination in a supply has been detected by routine bacteriological tests of the water before disinfection (4)(69). Frequent and regular water analyses are important and should always be applied during floods, drought and rain following drought.

The greatest safeguard of all lies in the use of adequate disinfection for all supplies. This should be operated under constant supervision or by automatic means of control with the necessary fail safe and alarm systems. An adequate period of contact after disinfection should be guaranteed before the water reaches the first consumer.

TABLE 19

SURVIVAL OF PATHOGENS IN NATURAL SITUATIONS

Infective agent	Species	Suspending medium	Treatment	Temp. (°C)	Original numbers	% Reduction	Time	Ref. No.
Cholera		Stored R. Thames water		10 to 18	7×10^4 to 13×10^6	99.9	7 days 21 days	70
		Natural water	Autoclaving				1 hr to 13 days	71
							Increased survival by several days	
		Synthetic water	Common saprophytes added			99	6 hrs	72
Salmonella	typhi	Faeces		Cold weather Hot weather			4 months 2 weeks	73
	paratyphi B	Faeces		37			4 days	74
	paratyphi B	Faeces		15			10 days	75
	typhi	Faeces					<12 days	76
	paratyphi B	Trickling filter	Depending on retention			84 to 99	4 days	17
	typhi	Anaerobic digests				25 to 92.4		77
	typhi	Activated sludge		20 to 25 10 to 15		85	4 hrs 8 to 14 days	
	typhi	Sewage	In light	Room temp.			83 days >38 days	78
	typhi	Sewage	In dark	10 to 15			14 days	79
	typhi	Cesspool					165 days	
		Subsoil drain					158 days	
	paratyphi	Septic tank liquor					61 days	80
		Well water					54 days	
	paratyphi	In sewer	Using sewer swabs				70 days 72 days	81

continued......

TABLE 19 (continued)

Infective agent	Species	Suspending medium	Treatment	Temp. (°C)	Original numbers	% Reduction	Time	Ref. No.
Salmonella	paratyphi	Farm pond water		21 to 29			>14 to 16 days	82
		Farm effluent in irrig. water	In dark	Summer Winter			20 days 60 days	83
	typhimurium	Storm water		20 10		99 95	10 days >14 days	84
	typhi	Tap water	In light	Room temp.	5 to 10 x 10^8/ ml		211 days	85
		Distilled water Normal saline Well artificially contaminated					443 days 153 days Several weeks	86
	typhi	River water	In dark	37 18 10 0	10^5/ml		2 weeks 4 weeks 5 weeks 9 weeks	87
	typhi	River water		10 to 21	10^6/ml	99.9	7 days	
Shigella		Well artificially contaminated					22 days	86
	flexneri	Farm pond water	Mixed bacteria	20 37			12 days 12 to 15 hrs	82 88
		in vitro expts. Faeces Soil Water		Temps down to -45			145 days 135 days 47 days	89

continued........

TABLE 19 (continued)

Infective agent	Species	Suspending medium	Treatment	Temp. (°C)	Original numbers	% Reduction	Time	Ref. No.
Viruses	Coxsackie Polio ECHO	Farm pond	–	4 – 20	1		up to 9 weeks up to 12 weeks	90
	Coxsackie	Loamy soil	Addition of 1 M MgCl	3 – 10			170 days	91
	Polio & ECHO	Distilled water Spring or well water	Addition of faecal bacteria				Survival reduced by bacteria	92
	Polio	River water		4			188 days	93
	Coxsackie	Sewage	Dark bottles	8 20			50 days <20 days	94
		Distilled water		8 20			>272 days 41 to 135 days	
		R. Ohio	Autoclaved	8			12 to 16 days	
		R. Ohio		8			150 to 171 days	
		R. Ohio	Autoclaved	20			Approx. 6 days	
		R. Ohio		20			Approx. 102 days	
	Coxsackie & Theilers	Spring water Sewage		8 to 10 8 to 10			} >200 days	46
		Spring water Sewage		20 to 30 20 to 30			} >140 days	
	Polio virus 3	R. Lee		5 to 6 15 to 16 22	20 000 PFU/ 100 ml	99.9 95 99.95	>63 days 7 days 11 days	95
		Soil		3 to 10			150 to 170 days survived longer in moist soil and at pH 7.5	96

TABLE 20

REPORTED INCIDENCES OF GROUNDWATER POLLUTION

Source of pollution	Substrate	Indication of pollution	Travel Distance	Travel Period	Remedial treatment	Period for subsidence	Disease resulting	Ref. no.
1. Fractured sewer			75 m		Repaired sewer		270 cases of typhoid	97
2. Sewer blockage	Chalk	2500 $E.coli$/100 ml rising to 90 000/100 ml. Nitrates and chlorine demand	50 m laterally to a depth of 38 m		Repair using cast iron			4
3. Flooded gravel pit into which burned sugar and biscuits had been added	Chalk	Progressive increase in $E.coli$ count. Finally 8/8 samples positive with mean count of 57/100 ml.	30 m laterally to a depth of 24 m	Continuous over 8 months	Filled in pit with inert material			98
4. Sewer blockage	Chalk	110 $E.coli$/100 ml	50 m to a depth of 38 m		Repair			4
5. Flooding caused sewers to surcharge	Chalk	25 000 $E.coli$/100 ml	to a depth of 24 m	3 - 4 days	Close station pump to waste	12 days		98
6. Foul drains in precinct of pumping station	Chalk	2 - 25 $E.coli$/100 ml	9 m laterally to a depth of 40 m		Repair drains	3 months		69
7. Fractured earthenware drain	Chalk	18 $E.coli$/100 ml	13.5 m laterally to a depth of 42 m		Repair drains; increase chlorine	2 days		69
8. Surcharge of sewer	Chalk	180 $E.coli$/100 ml	36 m laterally to a depth of 33 m			2 weeks for $E.coli$ 6 weeks for coliforms		69
9. Sewer fracture	Chalk	3 $E.coli$/100 ml rising to 16 $E.coli$/100 ml	1200 m laterally to a depth of 53 m	1 day of pumping	Closed down for 6 days	3 days for $E.coli$ 8 weeks for coliforms		69
10. Surface drainage entering swallow holes at N. and S.Mimms	Chalk	Bacteria and dyes	15 km					4
11. River	Limestone	$E.coli$ and salt	approx. 0.5 km	24 hr for salt				6
12. River	Limestone	Coliforms and Cl.welchii	approx. 1 km	79 - 90 hr	Concreted river bed	Very short time		6
13. Manured market garden polluted spring	Soil cover	Coliforms and nitrates	100 m		Abandoned			6
14. Ditch receiving manure		$E.coli$	nearby		Removed pollution			6
15. Well polluted from earth stable yard	Chalk	Faecal bacteria	100 m deep		Concreted yard	1 month		6
16. River entering abandoned well and penetrating to used well		Pollution proved with salt	240 m	17 hr for salt			1100 cases of typhoid and dysentry	99
17. Well suddenly polluted following dredging of river bed		Bacterial	54 m lateral at 27 m deep					100
18. Sewage spreading basin	Sand & Gravel 6 to 60 m deep	$E.coli$	75 m laterally to a depth of 48 m	2 months				101
19. Septic tanks 2 and 3 m from well with defective cover at time of flood	Limestone	coliforms were present in 47% of samples in village	Spreading groundwater from polluted well at periphery to other wells towards the center of the cone of depression.				89 cases of infectious hepatitis	102
20. Silage fermentation liquor entering a swallow hole and contaminating well water	Chalk	Coliforms. Later filamentous iron and sulphur bacteria	0.5 km 12 m deep well	2 days for pH to rise after addition of lime to swallow hole.	Filled with quicklime			5

TABLE 21

COMPARISON OF RECOMMENDED SAFE DISTANCES IN METRES BETWEEN WELLS AND POLLUTION SOURCES
(AFTER ROMERO (68))

Source of pollution	California	Federal Housing Authority	US PHS	Suffolk County, N.Y., Hlth Dept. Aug. 1958 (103)	Colorado	Britain (Memo 221; 1939, 1948) (104)
Septic tank	15	15	15	a. 30 m separation between well and sewage disposal facilities.	a. 30 m from point of juncture between well casings and aquifer.	"A well or borehole should be made watertight to such a depth as will prevent any surface pollution from entering... The water undertakers should acquire a sufficient area of land surrounding the site of well or borehole to enable them to protect the immediate surroundings."
Sewer line with watertight joints	3	3	3			
Other sewer lines	15	15	15			
Percolation field	30*	30*	30*	b. Minimum well depth of 15 m with at least 12 m below water table,	b. Minimum horizontal distance between well casings and potential source of contamination is 7.5 m.	"Where the well or borehole is on or near the outcrop of the strata from which the water is drawn it should be the routine duty of the water undertakers' staff to make regular and frequent inspections of the area within at least 3.2 km of the site with a view to detecting possible causes of pollution. Particular attention should be paid to any cesspools and soak-aways... a map showing details of any sewerage system within this area should be kept".
Adsorption bed	30*	30*	–			
Seepage pit	30*	30*	30	c. When 30 m is unattainable this distance can be reduced to a minimum of 20 m provided that for every 1.5 m horizontal decrease, the depth below the water table is increased by 0.6 m.		
Abandoned well	15	15	15			
Cesspool	–	–	45			
Other sources of pollution	as recommended	as recommended	–			

*Horizontal distance may be reduced by 15 m if point of well casing and aquifer are separated by a well-defined, continuous impervious strata.

TABLE 22

RECOMMENDED SAFE DISTANCES IN METRES BETWEEN WELLS AND VARIOUS SOURCES OF POLLUTION

(AFTER ROMERO (68))

Source of pollution	UNDERLAIN BY UNCONSOLIDATED ROCKS		UNDERLAIN BY IGNEOUS METAMORPHIC OR CONSOLIDATED ROCKS
	Areas in which zone of saturation is greater than 3 m from the source of contamination	Areas in which zone of saturation is less than 3 m from the source of contamination	
Septic tank with absorption trench or seepage bed	7.5 - 30 m *•+	22.5 - 30 m *•	An extensive survey is necessary to determine the likelihood of extensive pollution travel via fissures. In lieu of an extensive survey, the minimum distance between source of pollution and well should be 30 m.
Septic tank with seepage pit (if possible pit should terminate at least 1.2 m above water level).	15 - 30 m *•	22.5 - 30 m *•	
Septic tanks with no absorption system.	15 - 30 m *•	22.5 - 30 m *•	
Sewer lines with water-tight joints	3 m•	7.5 m•+	
Abandoned well.	7.5 m•+	as recommended	
Cesspool (not recommended).	15 m•	as recommended	
Other sources of pollution.	as recommended	as recommended	

*Measured as the shortest distance from the well to either the septic tank outlet or edge of adsorption system.

•Distance may be reduced or extended depending on the individual situation.

+For wells in the vicinity of 7.5 m from source of pollution, proper sealing to a depth of not less than 9 m.

REFERENCES

1. WRIGHT, S. J. L.
    Degradation of herbicides by soil micro-organisms.
    "Microbial Aspects of Pollution"; ed by
    G. Sykes and F. A. Skinner;
    London, Academic Press, 1971, pp 233-254

2. CRIPPS, R. E.
    The microbial breakdown of pesticides In
    "Microbial Aspects of Pollution"; ed by
    G. Sykes and F. A. Skinner;
    London, Academic Press, 1971, pp 255-266

3. McGAUHEY, P. H. and
    KRONE, R. B.
    Soil mantle as a wastewater treatment system
    Berkeley, California University, Sanitary
    Engineering Research Lab. Rep No. 67-11,
    1967; 201 p

4. TAYLOR, E. W.
    The pollution of surface and underground
    waters.
    J. Brit Wat. Wks Ass. , 1960, 42, pp 582-603

5. BRIGHTON, W. D.
    Pollution of chalk boreholes by filamentous
    organisms.
    Proc. Soc. Wat. Treat. Exam. , 1958, 7, pp 144-156

6. HOLDEN, W. S. ed.
    Water Treatment and Examination,
    London, Churchill, 1970

7. POLLITZER, R.
    Cholera, WHO Monograph Series No. 43;
    Geneva, World Health Organization, 1959

8. YEN, C. H.
    Cholera in Asia
    Proceedings of Cholera Res. Symposium, PHS
    Publ. No. 1328, 1965, p 346.

9. DIZZON, J. J.
    Carriers of El Tor in the Philippines
    Proceedings of Cholera Res Symposium,
    PHS Publ. No. 1328, 1965, p 322

10. POLLITZER, R.
    Cholera advances in historical perspective
    Proceedings of Cholera Res Symposium,
    PHS Publ. No. 1328, 1965, p 380

11. NATIONAL COMMUNICABLE
    DISEASE CENTER
    Morbidity and Mortality Weekly Rept. 1970,
    19, 5 Sept, p 349

12. PUBLIC HEALTH LABORATORY
    SERVICE AND SOC. MED. OFFICERS
    OF HEALTH
    Monthly Bull. 1965, 24, p 376

13. HALL, H. E. and
    HAUSER, G. H.
    Examination of faeces from food handlers for
    Salmonella Shigella Enteropathogenic
    Escherichia coli, and Clostridium perfringens
    Appl. Microbiol; 1966, 14, pp 928-933

14.	ROTHENBACKER, H.J.	J. Am. Vet. Med. Ass. ,1965, <u>147</u>, p 1211

14. ROTHENBACKER, H.J. J. Am. Vet. Med. Ass. ,1965, <u>147</u>, p 1211

15. PUBLIC HEALTH LABORATORY SERVICE AND SKOVGAARD, N. and NIELSEN, B.B. Salmonellas in pigs and animal feeding stuffs in England and Wales and in Denmark. J. Hyg. 1972, <u>70</u>, pp 127-141

16. WATER POLLUTION CONTROL FEDERATION Annual literature review, 1969. Wat. Poll. Control Fed. , 1970, <u>42</u>, p 1059

17. McCOY, J.H. The presence and importance of Salmonellae in sewage. Proc. Soc. Wat. Treat. Exam. 1957, <u>6</u>, pp 81-89

18. LEE, J.A. <u>et al</u> Salmonellae on pig farms and in abattoirs. J. Hyg. ,1972, <u>70</u>, pp 141-151

19. METROPOLITAN WATER BOARD Rept. Res. bact. chem. biol. Exam. London Water. , 1963-1964, <u>41</u>, p 31

20. METROPOLITAN WATER BOARD Rept. Res. bact. chem. biol. Exam. London Water. , 1964-1965, <u>42</u>, p 18

21. RELLER, L.B. <u>and others</u> Shigellosis in the United States; 5 - year review of nationwide surveillance. Am. J. Epidemiol. ,1970, <u>91</u>, pp 161 - 169

22. BURMAN, N.P. Discussion. <u>In</u> Bacteria in sewage treatment processes. J. Proc. Inst. Sew. Purif. ,1966, <u>6</u>, p 5

23. AMERICAN PUBLIC HEALTH ASSOCIATION Control of communicable diseases in man 11th Ed. , New York, Am. Publ. Hlth Ass. ,1970

24. YOW, M.D. <u>and others</u> Association of viruses and bacteria with infantile diarrhoea. Am. J. Epidemiol. , 1970, <u>92</u>, pp 33 - 39

25. COOPER, M.L. <u>et al</u> J. Diseases Children. 1959, <u>97</u>, pp 255

26. GLANTZ, P.J. <u>and</u> KRADEL, D.C. <u>Escherichia coli</u> serogroup 115 isolated from animals: Isolation from natural cases of disease. Am. J. Vet. Res. ,1967, <u>28</u>, pp 1891-1895

27. MULLER, G. Zentr Bakteriol Parasitenk. Abt I Orig. 1967, <u>203</u> p 464

28. GELDREICH, E.E. Waterborne Pathogens <u>In</u> "Water Pollution Microbiology" ed by R. Mitchell New York, Wiley, 1972, pp 207 - 241

29. COOK, G.T. <u>and</u> SMITH, A.J. Enterovirus excretors in residential nurseries Monthly Bull. Min. Hlth & Publ. Health Lab Services, 1962, <u>21</u>, p 47

30. MOSELEY, J. W.

Transmission of viral diseases in drinking water
In "Transmission of viruses by the water route"
Ed. G. Berg.
New York, Interscience, 1967, pp 5-23

31. GREEN, D. M. et al

Waterborne outbreak of viral gastroenteritis
and Sonne dysentery.
J. Hyg., 1968, 66, p 383

32. HEALTH, Ministry of

Report of Chief Medical Officer
London, HMSO, 1937, p 159

33. ELIASSEN, R. and
CUMMINGS, R. H.

Analysis of waterborne outbreaks, 1938 - 1945.
J. Am. Wat. Wks Ass., 1948, 40, p 509

34. MOORE, B.

The health hazards of pollution. 'Microbial
Aspects of Pollution' ed by G. Sykes and
F. A. Skinner
London, Academic Press, 1971, pp 11-32

35. MacLEAN, R. D.

Imported intestinal parasites
Medical Officer, 1958, 99, p 61

36. HEALTH, Ministry of

Report no. 31, 1925
London, HMSO 1925

37. HORNICK, R. B. and
WOODWARD, T. E.

Appraisal of typhoid vaccine in experimentally
infected human subjects.
Trans. Am. Clin. & Clim Ass., 1967, 78, pp 70

38. EDSALL, G. et al

Studies on infection and immunity in
experimental typhoid fever. I Typhoid fever
in chimpanzees orally infected with
Salmonella typhosa
J. Exp. Med. 1960, 112, p 143

39. PLOTKIN, S. A. and
KATZ, M.

Minimal infective doses of viruses for man
by the oral route.
'Transmission of viruses by the water route'
New York, Interscience, 1967, pp 151-164

40. SABIN, A. B.

Am. J. Med. Sci., 1955, 230, p 1

41. NEEFE, J. R. et al

An epidemic of infectious hepatitis apparently
due to a waterborne agent
J. Am. Med. Ass., 1945, 128, p 1063 - 1075

42. CLARKE, N. A. et al

Human enteric viruses in water.
Source survival and removability.
Proceeding of First International Conference
on Water Pollution Research.
Pergamon Press, 1964 pp 523 - 542

43. WATER RESEARCH ASSOCIATION
Technical Inquiry Report
TIR 192

Notes on viruses in water:
Factors relating to their survival and removal.
Medmenham, The Association, 1969, 9 p

44. ROBECK, G.G. Groundwater contamination studies at the sanitary engineering center.
Cincinnati, R.A. Taft Sanitary Engineering Center, Tech Rept W61-5, pp 193 - 197

45. ROBECK, G.G. et al Effectiveness of water treatment processes in virus removal.
J. Am. Wat. Wks Ass., 1962, 54, 1275-1292

46. GILCREAS, F.W. and KELLY, S.A. Relation of coliform organism test to enteric virus pollution.
J. Am. Wat. Wks Ass., 1955, 47, p 683

47. DREWRY, W.A. and ELIASSEN, R. Virus movement in groundwater
J. Wat. Poll. Control Fed., 1967, 40, (4) pp R257-271

48. COOKSON, J.T. Mechanism of virus adsorption on activated carbon.
J. Am. Wat. Wks Ass., 1969, 61, pp 52 56

49. ELIASSEN, R. et al Studies on the movement of viruses in groundwater.
Ann. Repts., Water Quality Control Res. Lab., Stanford Univ., Stanford, Calif., 1964-1967

50. CHANG, S.L. The present status of knowledge on destruction and removal of enteric pathogens in the water environment.
U.S. Publ. Hlth Service; Bureau of Water Hygiene and Federal Water Pollution Control Administration, Water Res Lab
Cincinnati, 1969., pp 134, 29, 5

51. FILMER, R.W. and COREY, A.T. Transport of virus sized particles in porous media.
Colorado, State University
Sanit Engng Papers. No. 1., June, 1966

52. CALDWELL, E.L. Pollution flow from a pit latrine when permeable soils of considerable depth exist below the pit.
J. Infect. Disease, 1938, 62, p 125

53. BAARS, J.K. Travel of pollution and purification en route in sandy soil.
Bull. Wld Health Org., 1957, 16, pp 727-747

54. ARNOLD, C.E. et al Report upon reclamation of water from sewage and industrial wastes in L.A. County, Calif.
Planograph, 1949
Abstracted in Public Health Engng Abstracts, 1950, xxx, 5, 23

55. STONE, R. and GARBER, W.F. Sewage reclamation by spreading basin infiltration
Proc. Am. Soc Civ Engrs., 77, No 87 Sept, 1951

56. STONE, R.   Land disposal of sewage and industrial wastes.
                                        Sewage ind. Wastes, 1953, _25_, pp 406 - 418

57. GOTAS, H. B.   Field investigation of waste water reclamation
                                        in relation to groundwater pollution.
                                        California State Water Pollution Control
                                        Board Publ. No. 6, 1953

58. UNIVERSITY OF CALIFORNIA   Studies in water reclamation.
                                        Sanit Engng Laboratory, Univ Calif.,
                                        Tech Bull No 13, 1955, p 43

59. SUTER, M.   The Preoria recharge pit: Its development
                                        and results.
                                        Proc. Am. Soc. Civ. Engrs., J. Irrig. Drain.
                                        Div., 1956, _82_, (IR3), 17 pp

60. CALDWELL, E. L. and   Groundwater pollution and the bored hole
    PARR   latrine
                                        J. Infect. Disease, 1937, _61_, p 148

61. CALDWELL, E. L.   Pollution from a pit latrine when an
                                        impervious stratum closeby underlies the flow.
                                        J. Infect. Disease, 1937, _61_, p 270

62. STILES, C. W. and   The principles underlying the movement of
    CROHURST, H. R.   B. coli in groundwater with resultant pollution
                                        of wells.
                                        Publ. Hlth Rept., 1923, _38_, p 1350

63. CALIFORNIA DEPARTMENT   A study of sewage effluent purification by
    OF PUBLIC HEALTH   filtration through natural sands and gravels
                                        at Sycamore Canyon at Santee.
                                        California, Bureau of Sanit Engng., Dept Publ
                                        Health, 1965.

64. MERRELL, J. C. et al   The Santee recreation project, Santee,
                                        California. Summary Rept., 1962-1964
                                        Public Hlth Service, Publ. No. 999 WP-27,
                                        1965, 69 pp.

65. DITTHORN, F. and   Experiments on the passage of bacteria
    LUERSSEN, A.   through soil
                                        Engng. Record. 1909, _60_, (23), p 642

66. FOURNELLE, H. J. et al   Experimental groundwater pollution at
                                        Anchorage, Alaska. Publ. Health Rept.,
                                        1957, _72_, p 203

67. McGAUHEY, P. H. and   Report of investigation of travel of pollution.
    KRONE, R. B.   California State Water Pollution Board,
                                        Publ. No. 11, 1954, 218 p.

68. ROMERO, J. C.   The movement of bacteria and viruses through
                                        porous media. Ground Wat. 1970, _8_, (2),
                                        pp 37-48

69. SHEPHERD, J. M.   Pollution of groundwater supplies by sewage.
                              Proc. Soc. Wat. Treat. Exam. , 1962, 11,
                              pp 11-16

70. HOUSTON, A. C.   Studies in Water Supply
                              London, Macmillan, 1913

71. PESIGAN, T. P.   Studies on the viability of El Tor Vibrio in
                              contaminated foodstuffs, fomites and water.
                              Proceeding of Cholera Res. Symposium,
                              Public Health Service PHS Publ. No. 1328,
                              1965, p 317

72. PILLAI, S. C. et al   Influence of activated sludge on certain
                              pathogenic bacteria.
                              Indian Med. Gaz. , 1952, 87, pp 117-119

73. HARVEY, D.   J. Roy. Med Cps. , 1915, 24, p 491 In Water
                              Treatment and Examination,
                              London, Churchill, 1970

74. FLETCHER, W.   J. Roy. Med. Cps. , 1917, 29, p 545
                              In Water Treatment and Examination
                              London, Churchill, 1970

75. FLETCHER, W.   J. Roy. Med. Cps. , 1918, 30, p679
                              In Water Treatment and Examination
                              London, Churchill 1970

76. DOLD, H. and   Z. Hyg. Infektkrankh. , 1944, 125, p 444
     KETTERER, M.   In Water Treatment and Examination,
                              London, Churchill, 1970

77. RUCHHOFT, C. C.   Studies on the longevity of Bacillus typhosus
                              (Eberthella Typhi) in sewage sludge.
                              Sewage Wks J. , 1934, 6, pp. 1054-1067

78. WILSON, W. J. and
     BLAIR, M. E. M.   J. Hyg. , 1931, 31, p. 138

79. BAIRD, T. T.   Correspondence,
                              Lancet, 1955, Aug 13, p 348

80. METROPOLITAN WATER BOARD Rept Res bact. chem. biol. Exam.
                              London Water. , 1955, 35 p 39

81. HARVEY, R. W. S. and   Lancet, 1955, 16 July, p 137
     PHILLIPS, W. P.

82. ANDRE, D. A. et al   Survival of bacterial enteric pathogens
                              in farm pond water.
                              J. Am. Wat. Wks Ass. , 1967, 59 pp 503 - 508

83. BRAGA, A.   Research on the survival of Salmonellae in
                              farmyard effluent used in a fertilizing
                              irrigation plant.
                              Water Pollution Abstracts 1967, 40, 845

84. GELDREICH, E.E. et al   The bacteriological aspects of storm water
                                    pollution
                                    J. Wat. Poll. Control Fed. ,1968, 40 pp 1861-1872

85. SHREWSBURY, J.F.D. and   A note on the absolute viability in water of
     BARSON, G.J.   S. typhi and the dysentery bacilli.
                                    Brit. Med. J. 1952, 1. p 954

86. BARTOS, D. et al   Die Lebensdauer der enteralen Krankheit
                                    Krankheitserregenden Bakterien im
                                    Brunnenwasser.
                                    Z. Hyg. Infektionskrankh. 1947, 127, pp 247-254

87. METROPOLITAN WATER BOARD   Rept. Res. bact. chem. biol Exam
                                    London Water. , 1908 to 1915, 1 - 11

88. HENTGES, D.J.   J. Bact. ,1968, 97, p 513

89. MIRZOEV, G.G.   Gigiena i Sanit (USSR) 1968, 33, p 437

90. JOYCE, G. and   Survival of enteroviruses and bacteriophage
     WEISER, H.H.   in farm pond waters
                                    J. Am. Wat. Wks Ass. ,1967, 59, pp 491-501

91. CARLSON, G.F. et al   Virus inactivation on clay particles in natural
                                    waters.
                                    J. Wat. Poll. Control Fed. ,1968, 40, R89-R106.

92. SQUERI, L. et al   Boll Ist Sieroterap, Milan, 1968, 47, p 595

93. RHODES, A.J. et al   Prolonged survival of human poliomyelitis
                                    virus in experimentally infected river water.
                                    Canad. J. Publ. Hlth. ,1950, 41, p 146

94. CLARKE, N.A. et al   Survival of Coxsackie virus in water and sewage.
                                    J. Am. Wat. Wks Ass. ,1956, 48, pp 677

95. POYNTER, S.F.B.   The problem of viruses in water
                                    Proc. Soc. Wat. Treat. Exam. ,
                                    1968, 17, pp187-198

96. BAGDASARYAN, G.A.   Survival of viruses of the enterovirus group
                                    (Poliomyelitis, Echo, Coxsackie) in soil and
                                    on vegetables.
                                    J. Hyg. Epidemiol. ,Microbiol, and Immuniro. ,
                                    Prague, 1964, 8, (4), pp 497-505.
                                    Cited in Bull. Hyg. ,1965, 40, (7), p 780

97. HEALTH Ministry of   Report on the outbreak of enteric fever in the
                                    Malton Urban district. Report on Public
                                    Health and medical subjects, no. 69
                                    London, HMSO; 1933

98. MacLEAN,R.D.   The effect of tipped domestic refuse on
                                    groundwater quality: A survey in North Kent
                                    Proc. Soc. Wat. Treat. Exam. ,1969, 18,
                                    p18-34

99. WARRICK, L. F. and   Pollution of abandoned well causes Fond du
     TULLY, E. J.   Lac epidemic.
                                 Eng. N. R. 104-410-1 March 6, 1930

100. GROUNDWATER   Newsletter
                                 Ground Wat. , 1971, 9, (3), p 42

101. BOGAN, R. H.   Problems arising from ground water
                                 contamination by sewage lagoons at Tieton,
                                 Washington.
                                 In U. S. Dept. Hlth. Ed. & Welfare;
                                 Cincinnati R. A. Taft Sanit. Engng. Center,
                                 Tech Rept W61-5, 1961, pp 83 - 87.

102. VOGT, J. E.   Infectious hepatitis outbreak in Posen,
                                 Michigan.
                                 Cincinnati, R. A. Taft Sanitary Engineering
                                 Center. , Tech Rept W61-5, 1961, pp 87 - 91

103. FLYNN, J. M.   Impact of suburban growth on ground water
                                 quality in Suffolk County. N. Y.
                                 Cincinnati R. A. Taft Sanitary Engineering
                                 Center, Tech Rept W61-5, 1961, pp 71 - 82

104. HEALTH, Ministry of   Safeguards to be adopted in the day to day
                                 administration of water undertakings.
                                 London, HMSO; 1939, Memorandum 221,
                                 Revised 1948.

INCIDENTS OF ALGAL AND BACTERIAL POLLUTION NEAR BRIGHTON
by S.C. Warren
Water Department, Brighton, Sussex, England

1. BACTERIAL POLLUTION OF UNDERGROUND CHALK WATER

Marked evidence of bacterial pollution of water at Burpham Pumping Station in 1964 on increasing abstraction from 1 to 4 m.g.d.

1.2. Potential sources

(a) Infiltration from River Arun.

(b) Leaking cesspools in Burpham, situated on outcrop of Upper Chalk. Need to discover source causing pollution.

Earlier work showed presence of two chalk waters supplying boreholes, the harder water having two escape routes east and west of village controlled by tidal influence of river (Figs. 50, 51, 52 and 53). Daily samples of each borehole water were examined bacteriologically and chemically with chemical examination of river water during years 1964 to 1968.

Results were graphed. Chemistry of borehole waters plotted superimposed on one sheet showing parameter changes simultaneously (Table 23). Graphs showed that softer chalk water supplied BH2, 3 and 4 from January to May. During May, upward divergence occurred of plotted values for BH3 and 4 waters, producing wave form in correlation with tidal influence on river. The bacterial contents of the three waters varied in similar manner (Figs. 54 and 55), but featured increasing numbers of coliforms with increasing tide height with drastic decrease before tide had reached maximum in monthly cycle.

The following observations were possible

(a) Bacterial pollution greatest in BH4 and least in BH2.

(b) Bacterial pollution greater in second than first half of any year.

(c) BH2 water chemically stable throughout year.

(d) BH3 and 4 waters changed with tide cycle, being harder at high than at low tide.

(e) Changes in BH3 and 4 waters in last half of any year caused by admixture of harder underground water and were coincident with increased bacterial pollution.

(f) Amount of admixture governed by

   (i)   abstraction rate

   (ii)   relative levels of hard/soft water tables

   (iii) tide height in River Arun

(g) If river infiltration was involved, observed crests of wave form would have been troughs because river is softer than either chalk water.

Observations demonstrate that river not involved in underground water changes and bacterial pollution not derived from river.

Position of harder water escape route i.e. holes in river bank west of village at low tide or spring east of village at high tide. Increasing tide height stops escape into river and directs water under village, eastwards towards boreholes. If chalk under village contaminated by wastes from leaking cesspits, water contaminated also. Under low tide conditions less hard water directed to Burpham P.S., liquid waste accumulates in chalk. Successive higher tides direct more hard water toward boreholes, removing sewage and resulting in less pollution at highest tides as little sewage remains in chalk.

Only such a mechanism could cause type of pollution observed and is the only one fitting facts produced during investigation.

To combat this pollution village has been sewered.

TABLE 23

	River Arun	River Arun (Hole in East Bank)	Burpham P.S. BH4 (Low Tide)	Burpham P.S. BH4 (High Tide)	Burpham Spring
pH	7.60	7.40	7.30	7.35	7.40
Alkalinity	68	241	197	229	262
Chloride	27	36.4	17.1	21.5	29.0
Temporary Hardness	68	241	197	229	262
Permanent Hardness	42	39	25	13	20
Total Hardness	110	280	222	242	282

2. POLLUTION OF UNDERGROUND CHALK WATER WITH ALGAE

Arundel Pumping Station situated on outcrop of Upper Chalk (Fig. 56). Six in. borehole drilled in 1956, 71 ft deep, lined to 30 ft with 6 in. solid tubes. Pumping rate 43 000 g. p. h. Engine Room Floor Level 19. 25 ft AOD. Rest level of borehole water 8 ft 6 in. AOD.

Twenty-four in. borehole drilled and acidized 1969, 200 ft deep lined to 30 ft with solid tubes, remainder lined with slotted tubes except for bottom 40 ft which again is solid tube. Top of lining tube 18 ft AOD. Rest level of water 9. 9 ft AOD. Situated 47 ft SE of 6 in. BH and 120 ft from south end of lake. Twenty-four in. BH test pumped for 7 days in October 1969 at between 70 000 and 94 000 g. p. h. Thereafter water supplied from 6 in. BH. Area of Swanbourne Lake 13 acres. Water level 11 ft 3 in. AOD. Fed by chalk springs in lake bed. Also on site an old dug well 18 ft deep by 12 in. diameter with heading in NE direction 36 ft x 10 ft 5 in. This discharges water into mill stream.

First complaint by consumer of "hairs" in domestic water supply 4 August 1970. Microscopic examination showed these to be zygnema, a filamentous alga. During August, 119 samples of raw and treated water from reservoirs and distribution mains examined. All contained zygnema and other organic debris as also did lake water. Arundel PS shut down end of August.

Further 108 samples examined during September but with district being supplied from another Station; algae not found in many samples, although still present in borehole waters when these pumped to waste.

Considering source of pollution it was thought possible that acidization and test pumping of 24 in. BH at high rate had cleared blocked fissure connecting with lake and this was source of biological pollution. Circumstantial evidence very strong but not proven.

Possible courses of action

    (a)   Continue use of water and risk complaints.

    (b)   Find and seal fissure.

    (c)   Remove organisms by treatment.

    (d)   Abandon source.

Action to be taken consists of developing alternative supply from springs feeding the lake (at R on Fig.56) and the cress beds (at G and H). Chemical character of these springs virtually identical (see Table 24). A new supply based on tapping the spring flow could obviate pollution either by grouting the aquifer or other engineering work isolating the springs from hydraulic connection to the lake or by leaving a residual springflow into it .

TABLE 24

ABBREVIATED ANALYSES OF WATERS IN BOREHOLES AND ADJACENT ESCAPE POINTS

Reference point	6 in. Borehole	24 in. Borehole	(R) Spring in lake	(L) Weir	(G) 15 in. Pipe at cress beds	(H) 18 in. Pipe at cress beds	(D) Well in heading system	(F) 10 in. Pipe into millstream	(P) Outlet from (D) into millstream	(Q) Outlet of (L) into millstream
Date taken	23.7.70	17.10.69	23.3.71	16.3.71	16.3.71	16.3.71	16.3.71	16.3.71	16.3.71	16.3.71
pH	7.30	7.55	7.30	7.40	7.30	7.30	7.20	7.20	7.20	7.20
Alkalinity	200	201	194	195	194	194	206	205	206	201
Chloride	20.8	22.7	21.2	22.4	22.4	22.4	23.0	23.4	23.4	22.2
Temporary hardness	200	201	194	195	194	194	206	205	206	201
Permanent hardness	24	22	34	33	35	35	29	33	32	30
Total hardness	224	223	228	228	229	229	235	238	238	231
Nitrate (NO_3)	2.7	2.7	2.5	2.2	2.3	2.2	3.1	3.3	3.2	3.0
Sodium (Na)	12.0	11.6	12.0	12.0	11.6	11.6	12.4	12.4	12.9	12.4
Potassium (K)	0.8	0.9	0.9	0.9	0.8	0.8	1.6	1.6	1.0	1.0

Letter symbols refer to Fig. 56.

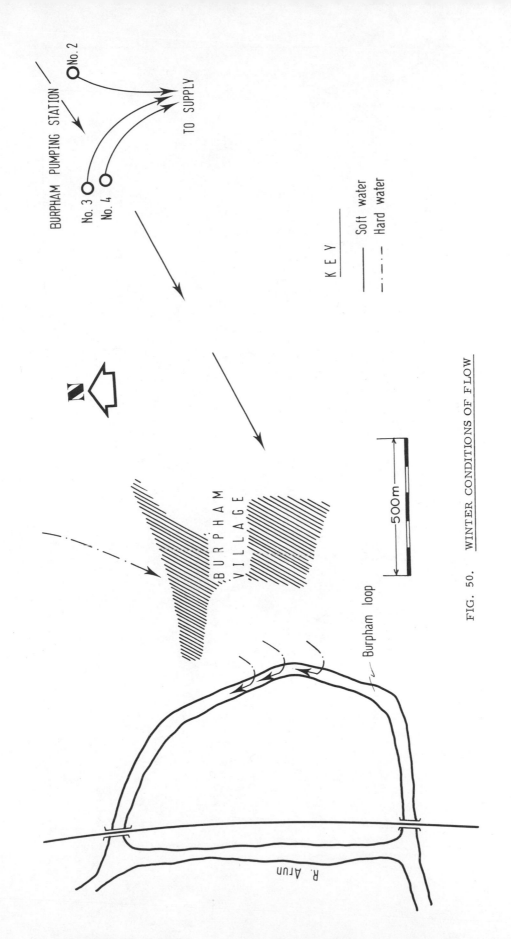

BURPHAM PUMPING STATION

No. 2

No. 3
No. 4

TO SUPPLY

K E Y

——— Soft water
—·—·— Hard water

BURPHAM VILLAGE

500m

Burpham loop

R. Arun

FIG. 50.   WINTER CONDITIONS OF FLOW

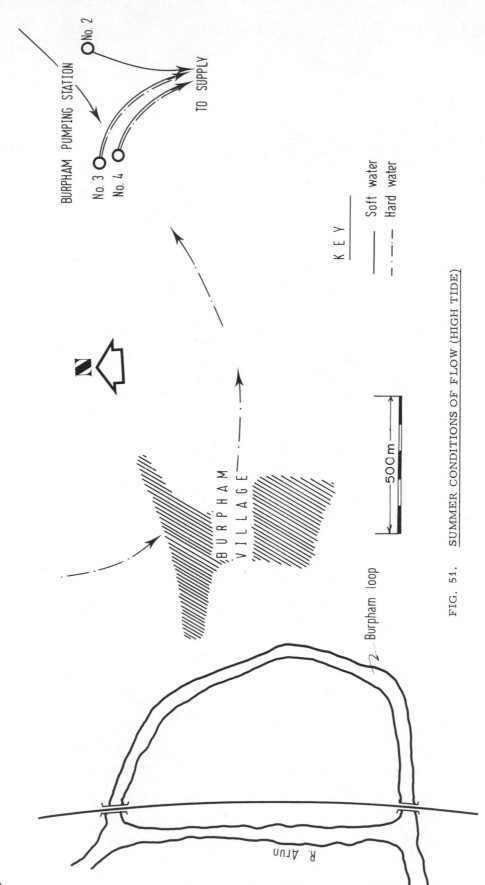

FIG. 51. SUMMER CONDITIONS OF FLOW (HIGH TIDE)

KEY
———— Soft water
—·—·— Hard water

BURPHAM PUMPING STATION

No. 2

No. 3
No. 4

TO SUPPLY

N

BURPHAM VILLAGE

500 m

Burpham loop

R. Arun

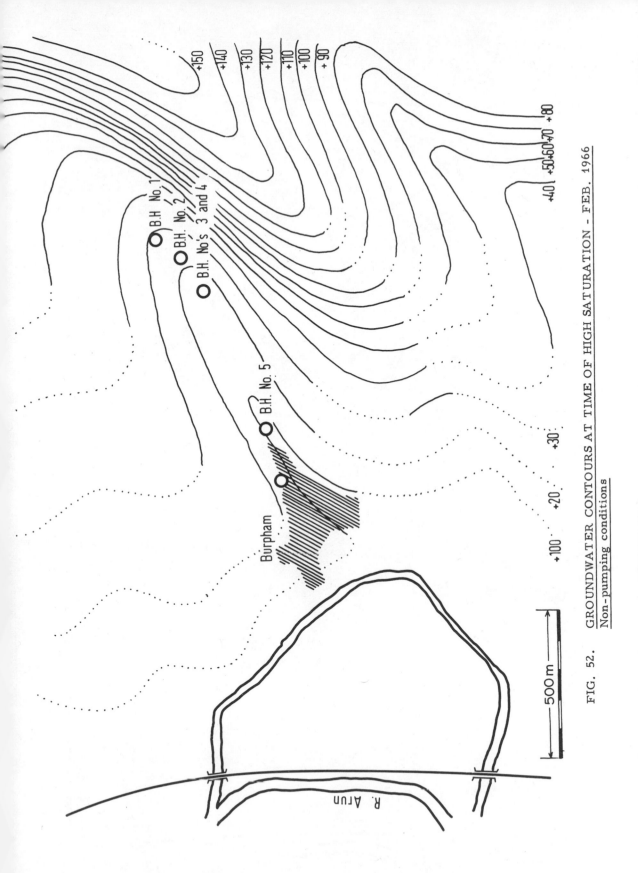

FIG. 52. GROUNDWATER CONTOURS AT TIME OF HIGH SATURATION – FEB. 1966

Non-pumping conditions

209.

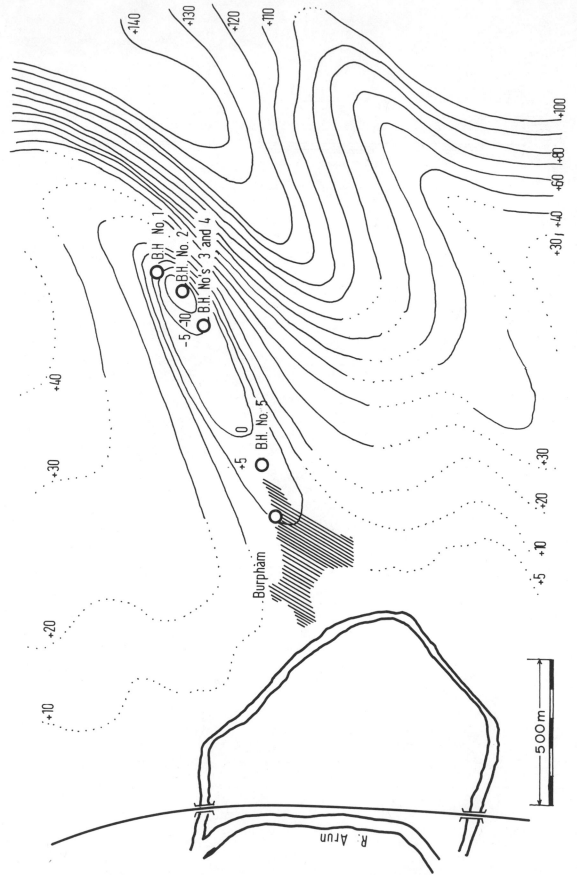

FIG. 53. GROUNDWATER CONTOURS AT TIME OF LOW SATURATION – SEPT. 1967

Pumping conditions

210.

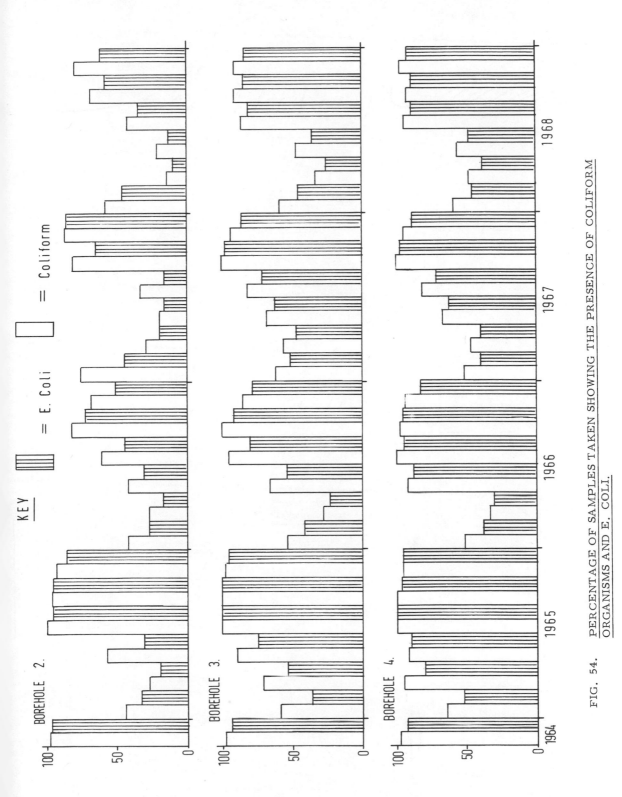

FIG. 54.   PERCENTAGE OF SAMPLES TAKEN SHOWING THE PRESENCE OF COLIFORM
ORGANISMS AND E. COLI.

211.

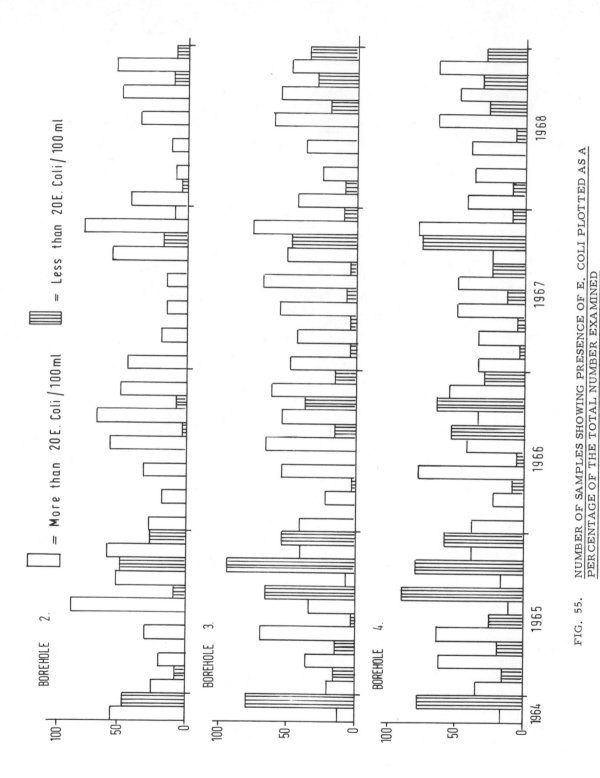

= More than 20 E. Coli / 100 ml

= Less than 20 E. Coli / 100 ml

BOREHOLE 2.

BOREHOLE 3.

BOREHOLE 4.

1964   1965   1966   1967   1968

FIG. 55.   NUMBER OF SAMPLES SHOWING PRESENCE OF E. COLI PLOTTED AS A
PERCENTAGE OF THE TOTAL NUMBER EXAMINED

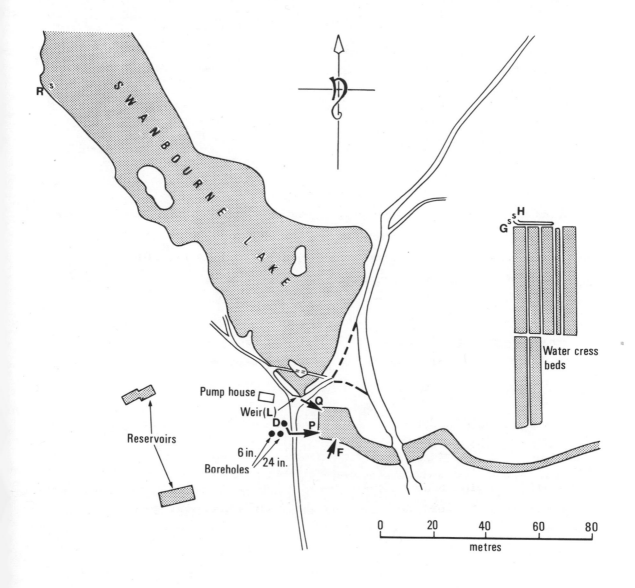

FIG. 56.   LOCATION OF ARUNDEL PUMPING STATION

DISCUSSION ON MICROBIOLOGICAL POLLUTION

Technical points

1.   The risks of bacterial pollution have to be seen against the general level of public health.   The infamous Broad Street incident in the last century when 600 people died of cholera was set against a backdrop of a quarter of a million deaths due to cholera in the UK that same year.

2.   In the surveillance of waters for public water supply, there is good reason to adhere to the bacteriological standard because there is ordinarily a ratio of coliform bacteria to virus in the order of 10^5:1.

3.   An instance was cited of a borehole water being suddenly polluted, as detected by daily coliform counts, by a fractured sewer in the waterworks enclosure.   Some years later a waterlogged ditch, dug to contain a safety sleeve for a fuel oil pipe, gave similar trouble.   Fissuring between the surface and the water table was suspected therefore, perhaps an imperfect grout seal at the well head.   In any situation where a borehole is at risk from pollution near to itself, one would recommend frequent monitoring of the water rather than chlorinating down the borehole, which obscures the evidence.   In the case mentioned, the contaminated water was subjected to superchlorination and dechlorination, no coliform organisms being detected in the finished water.

4.   The oxygen and microbiological regimes below an urban waste tip are closely linked, as shown by Exler's Fig. 70, p. 240.

5.   Warning was given lest a too literal interpretation of the bacterial movement quoted in Tables 21 and 22 of Hutchinson's paper be uncritically transferred to UK conditions.

Research needs

6.   In planning the research on the fate of cyanide waste tips, the WPRL experience on biodegradation of cyanide in percolating filters was relevant.   These only achieve full efficiency after an acclimatization phase in aerobic conditions.   What is not yet known is whether a similar biodegradation would occur under anaerobic conditions.

7.   Studies on bacterial movement underground, in relation to pore and fissure geometry.

Administrative comment

8.   In the opinion of M. Hutchinson it was unforgiveable not to disinfect groundwaters. Hence microbiological pollution and viral transmission are issues of vital importance only where primitive water supply conditions occur or where the disinfection process is unreliable

DEFINING THE SPREAD OF GROUNDWATER CONTAMINATION BELOW A WASTE TIP

by H.J. Exler
Bavarian Geological Service, Munich, Germany

SUMMARY

Results of an investigation of groundwater pollution by tipped refuse have shown that contaminated groundwater travelled at least 3000m. This concerned a relatively narrow zone which became progressively sharper with distance. The investigation involved a general survey of groundwater temperature and electrical conductivity as well as supporting evidence of pollution by analysis for chloride, permanganate value and nitrate. The measurements from a network of observation bases gave only a momentary indication of the state of pollution which was capable of being misinterpreted. The frequency of observations of such a network over a long period must be chosen such that all variables, as well as overall trends, are known.

1. INTRODUCTION

Since 1967 the Bavarian Geological Office has been investigating the extent and range of groundwater contaminants derived from a domestic and industrial refuse tip located in Southern Bavaria and operated since 1954. The tip lies on fine to coarse sandy gravel, the top soil layer having been previously removed. Direct contact between the refuse and groundwater does not occur. So far, about 4 million m^3 of refuse and covering has been deposited over an area of about 153 000 m^2.

At the start of the investigation six sampling points and a spring at some distance were investigated (Fig. 57). Chemical and bacteriological analyses (total plate count and E. coli) were made. However the sampling points were so placed that some gave only slight and variable indication of possible contamination and others none at all. Only the observation point 8, a well, indicated by its smell and taste that the groundwater was contaminated. This water, after being impounded, eventually found its way into a public supply system.

2. HYDROGEOLOGICAL ASPECTS

The refuse tip lies partly on fluvio-glacial gravel and partly on redeposited valley gravel. The terrain includes a heavily eroded terrace edge having a maximum height of between 1.5 to 2.0m. The gravel consists mostly of carbonate rocks and is poorly classified. In a single sample practically all particle sizes from 10 or 20 cm to fine sand occurred. The proportion of a single size fraction can vary rapidly both horizontally

This paper originally appeared as "Ausbreitung und Reichweite von Grundwasserverunreinigungen im Unterstrom einer Mülldeponic" in GWF-Wasser/Abwasser, 1972 (3), 101-112. Grateful acknowledgement is made to that journal for permission to republish in translated form.

and vertically.

The thickness of the Quaternary in which the aquifer lies varied between 3.05 and 5.70m. This was less on account of the moderately articulated and proportionately flat terrain surface than because of the extreme articulation in the profile of the groundwater bed.

The groundwater bed was formed from fresh water sediment of the late Tertiary. It consisted chiefly of marly clay, clayey marl and marl, in which fine to medium sand was included. Fine and medium shingle and thin banks of pyrites are likely to occur, but were not observed in the borings, which generally went only 2m at most into the groundwater bed.

The general lie of the terrain surface and of the Tertiary surface was north-north-east, which applied also to the groundwater. The gradient was between 0.0025 and 0.004 and the mean hydraulic conductivity was of the order of 5×10^{-3} m/sec. The flow velocity was between 5 to 10 m/day, the principal direction being north-north-east. The thickness of the groundwater varied between 0.42 and 3.0m at low water level and between 0.94 and 3.97m at high level. The groundwater temperature varied during the period of observation between 6.5 and 12.5°C. This was without the effect of warming from exothermic reactions in the tip.

3. GROUNDWATER INVESTIGATIONS AT THE TIP

In order to determine whether and to what extent the groundwater was polluted by the refuse, well 9 (Fig.57) was bored into the underflow directly at the foot of the tip and fitted with screen and lining. To establish a test blank, Bore 10 was installed in the upper side of the tip with 4 in. PVC screen filters. A water-level recorder was also installed.

After completion of both bores, a pumping test was carried out on Bore 9 for 94 hours during the period 18 to 21 January 1967, and samples were taken at certain intervals and analyzed chemically. The manner and intensity of chemical pollution of the groundwater were thereby determined and a possible method for its reduction by pumping observed. Blank observations on Bore 10 were also made. Altogether during the pumping test, 408m^3 of heavily polluted water were taken from Bore 9. The groundwater temperature was uniformly 19°C.

The chemical analysis of the water showed that the chemistry of the pumped water had not changed between when 8 or 400m had been pumped out. In contrast, comparison between the analyses of the upstream and downstream water showed clearly the marked changes suffered by the groundwater in passing through the gravel under the tip:

(a) Colourless groundwater acquired a yellow-brown to yellow-green colour reminiscent of liquid manure.

(b) The water acquired a pungent odour typical of the tip itself.

(c) The water temperature was raised from between 7° and 10°C to 19°C. (The highest temperature was 24°C, measured in Bore 9 in June 1968.)

(d) The chloride concentration increased from 34 mg/litre to 453 mg/litre, i.e. 13-fold.

(e) The ammonium content rose from 0.9 mg/litre to 275 mg/litre, corresponding to a factor of 300.

(f) The permanganate value rose from 25 mg/litre to 590 mg/litre, i.e. by a factor of 23.

(g) Because of the oxygen-demand in the water, sulphate diminished from 116.6 mg/litre to 31.5 mg/litre, i.e. about one-quarter.

These results interpret the tabulated data on p. 241 and left no doubt about the extraordinarily strong pollution of the groundwater by the tip. It was then necessary to investigate the spreading and range of the contaminated groundwater in the near vicinity of the tip.

4. INVESTIGATION OF GROUNDWATER IN THE VICINITY OF TIP

The investigations described below ranged from 800m upstream (south) of the tip to about 1500m downstream (north)(Fig. 57). After establishing the direction of flow of the groundwater by measurements at the existing test sites, additional bores were made as follows:

5 bores downstream from the north-eastern corner of the tip at a very acute angle with the general direction of groundwater flow (Bores 14-18).

Bores 5a and 13 on the northern edge of the tip, in order to localize the exit of the polluted water.

Bores 11 and 12 in the upstream of the tip.

All nine of the new bores penetrated the entire depth of the groundwater stream and also 1 to 2m into the groundwater floor. Finally they were fitted with 4 in. PVC screens and bore linings for the uppermost 0.5m and for the standpipe, ordinary galvanized iron pipe was used. The bores were capped with 4 in. safety caps which could not be opened by unauthorized persons. Moreover, the standpipe was protected by a 50 cm high concrete pedestal.

4.1. Experimental methods

After the network of observation points had been installed, various methods were tested to see if an indication of the relative pollution gradient in the groundwater could be obtained without having to sample and analyze water from all of the bores. What was required was a method for surveying the whole network of test points rapidly. Therefore, for standard measurements a combination was chosen of:

(i)   Electrical conductivity of the groundwater (corrected to 20°C.)

(ii) Groundwater temperature.

For the measurement of electrical conductivity of groundwater in the bores a special probe was developed which consisted of a cell (cell constant $\simeq$ 1.0), a temperature compensation resistance and a thermocouple (Fig. 58). A measurement was thereby obtained of the total dissolved ions in the groundwater. Because of the temperature differences in the groundwater, it was expedient to use an apparatus which compensated for the effect upon electrical conductivity.

As was shown at the outset of the investigation, the chloride component increased sharply in groundwater passing under the tip. Since this was the predominant ion present in the groundwater, the electrical conductivity was affected accordingly. Chloride in the underflow cannot be removed and its concentration is reduced only by dilution. On this basis, the chloride concentration, as indicated by electrical conductivity, allowed the path of the pollution to be traced throughout the zone in which the organic components were decaying or already decayed.

Near to the tip, the increase in groundwater temperature was a guide to the total pollution of the water. Groundwater temperature could only be used as an indicator near to the tip because the heat, which originated from exothermic reactions in the tip, was lost fairly rapidly. The porous medium through which the groundwater flowed favoured on the one hand a heat exchange between polluted and unpolluted water but hindered their mixing. For this reason, substances which could be reduced in concentration only by mixing were transported further than those which decayed.

Another possible method for indicating the intensity and range of polluted ground-water without chemical analysis was the measurement of fluorescence under UV radiation. This indicates, above all, the organic substances which have reached the groundwater. For more exact investigations examination of the fluorescent spectrum is advisable. However, it was sufficient in this situation merely to measure the total spectrum. Data were thereby obtained which, like the increase of groundwater temperature, were proportioned to electrical conductivity. However, this measurement was less convenient than the measurement of either temperature or electrical conductivity in that water had to be sampled and taken to the laboratory.

Chemical analyses were performed by the laboratory of the sewage works and bacteriological examination by the Department of Health only for the greater distances.

4.2. Results

Fig. 59 represents an example of the twice-monthly results for electrical conductivity and temperature for 27 March 1968. It can be seen that the highest value of electrical conductivity and the highest temperature do not coincide. The highest temperature was $18.5^{\circ}C$ in Bore 14 and the highest conductivity on the north edge of the tip was 4 milliSiemens (mS) in Bore 13. The still higher value of 4.5mS was observed about 350m from the tip in Bore 15. These values show that electrical conductivity, as a measure principally of inorganic pollution, does not always reflect the rise in temperature, which (with attenuation) is due to the exothermic processes associated with the organic pollution.

Fig. 60 shows results taken on 30 April 1968, in which the temperature rise directly on the north edge of the tip was clearly observable. Bore 13 in fact diminished in temperature from 12° to $11^{\circ}C$, but in Bore 9 an increase from 12.5° to $22^{\circ}C$ was indicated. Bore 14 however, showed a tendency to diminish from 18° to $11^{\circ}C$. Moreover the upstream side of the tip to the south in Bore 12 increased from 11.5° to $13.5^{\circ}C$. The highest temperature on this occasion ($22^{\circ}C$) coincided with the highest electrical conductivity (5mS) and occurred in Bore 9. Exceptions are Bores 13 and 14, both with temperatures of $11^{\circ}C$ but conductivity values of 3.4 and 0.7mS respectively. Inclusion of the chloride concentrations and permanganate values (Fig. 61)gives the following comparisons:

TABLE 25

Bore No.	Temperature ($^{\circ}C$)	Permanganate Value (mg/litre)	Electrical conductivity (mS/cm)	Chloride concentration (mg/litre)
14	11	110.6	0.7	172
13	11	158	3.4	350
9	22	790	5.0	650
5a	8.5	11.4	0.68	84

These results showed a relationship between temperature and permanganate value as well as one between electrical conductivity and chloride concentration. Bore 9 showed a great deal of inorganic as well as organic constituents in the groundwater. In Bore 13 the organic components were present in only moderate concentration compared with the inorganic. In contrast, Bore 14 in the north-eastern corner of the tip carried only moderate concentrations of both organic and inorganic constituents. In Bore 5a values were measured which could have been caused by only very small pollution of the groundwater by organic and inorganic substances. Here on the north-western corner of the tip, relatively low values of pollution were measured consistently.

Comparison of permanganate values and temperature taken further in a northerly direction (Fig. 61) leads to the conclusion that no further temperature increase occurs, although according to the permanganate value, an organic loading is still present. A greater spreading of the temperature increase in the groundwater first takes place (Fig. 62) if the temperature of the tip and its surroundings increases. This applies even upstream of the tip and shows that it is the tip itself which is the source of the heat. Only in this way can the upstream warming be explained, and that it occurs during spring and early summer coinciding with an increasing activity in the tip.

During the course of further investigations, increased temperature was nearly always observed at Bores 3 and 16, about 750 m from the tip (Fig. 60). Temperature increase could only be observed further than this when the groundwater temperature at the tip was sharply increasing. Increased temperatures were noted in Bores 18 and particularly 8 at a distance of about 1300m. However, detection of increased temperatures at the greater distances was usually uncertain.

According to the measurement of electrical conductivity, inorganic contaminants extended in the direction respectively of Bores 9, 17 and 18, i.e. north-north-east. In contrast, the spread of increasing temperature occurred more towards the north-east towards Bores 9, 16, and 3 and northward again to Bore 8. The reason for this inconsistency cannot yet be explained.

Increased electrical conductivity could always be followed along the line to Bores 18 and 8, where in most cases 2 mS or more could be measured. The base value of conductivity varied upstream of the tip in Bore 11 between 0.29 and 0.75 mS. This means that a value of 2mS or more at about 1300 m from the tip is a significant increase, as shown clearly also by the chloride concentration (Fig. 61):

TABLE 26

DATA FOR 30 APRIL 1968

Bore Number	Chloride Concentration (mg/litre)	Electrical Conductivity (mS/cm)
8	160	1.1
18	230	1.9

The observations referred to up to now have been for bores in which uninterrupted increase in conductivity and in part, an increased temperature were measured, as for example was the case with Bore 9 and the series 15, 16, 17 and 18. These bores mark the main path of the polluted groundwater. Other bores, as for example, 3 and 8, stand on the edge of the polluted stream and show therefore variations in conductivity which may be great (Bore 8), slight (Bore 3) or none at all. How the individual bores behaved in respect of their electrical conductivity is shown in Fig. 63. The hope of being able to follow by this representation certain maxima from bore to bore, and hence to obtain an indication of the flow velocity was not fulfilled.

From Fig. 63 the following points may be observed:

(a) The descending tendency in some results (for the Bores 2, 3, 8 and 12) from the end of March to the end of May 1968, were observed only in the upstream and around the edge of the tip. This may be attributed to the precipitation during the period which effected a slow dilution of the normal dissolved content of the groundwater.

(b) The conductivity measurements from 14 June 1968 showed a sudden drop in all values, which was due to a very short-lived 'dilution effect'. During the months of June and July 1968, the conductivity meter at Bore 12 was fitted with a recorder. It was evident from the recorder trace that within 2.5 days (from 10 to 12 June) the conductivity fell sharply and returned during 14 to 15 June to its former value. Such heavy rainfall that produced a large dilution of the upstream could well have affected also the water downstream from the tip, but this was not observed. Both instrument and electrode were functioning correctly. A possible explanation was that there was a dilution by rainwater in the observation bore which did not extend to the aquifer. If precipitation percolates into groundwater, it first fills capillary voids above the water table and slowly spreads downwards into the groundwater stream. These processes could be frequently observed on the groundwater level recorders. The recorder curve shows a sudden strong rise and a slow fall. During the sudden rise in the trace a lot of fresh percolate rapidly enters the bore screen and mixes with the groundwater in the bore. This gives a local dilution which could not happen so suddenly in pores carrying flowing groundwater. Since the bore screen is continuously swept out by flowing groundwater, the diluted groundwater is displaced at a rate depending upon the flow velocity, and the original concentration of dissolved substances is thereby regained.

A meter was installed on Bore 9 during April and May 1968. The trace for this is also included in Fig. 63 which shows clearly how differentiated was its run. If the long-term traces for Bores 9 and 12 are compared with the plots of the individual values of conductivity it is seen that the measurement on a particular day is neither an absolute indication of the behaviour of a particular bore nor representative of the degree of pollution within the whole sampling area.

The investigation established that the contamination could certainly be traced over a distance of 1350m. The spreading did not follow a wide front, but rather a tongue which followed the stream direction of the groundwater out from the tip and became progressively narrower with distance. Results of measurements and chemical analyses for 30 April 1968 are collected in Table 27.

TABLE 27

	Bore no.	Elect. cond. $(\mu S)/cm$	Chloride conc. (mg/litre)	Permanganate Value (mg/litre)	Temperature ($^{\circ}C$)
Upstream	6	400	30	12.6	8.5
	10	490	24	5.7	10.0
	11	410	40	7.9	8.0
	12	420	52	7.3	13.5
North side of the tip	5a	680	84	11.4	8.5
	13	3400	350	158	11.0
	9	5000	650	790	22.0
	14	700	172	110.6	11.0
Downstream (about 1350 m)	8	1100	160	107.5	8.5
	18	1900	230	94.8	8.0

Compared with the upstream results, values obtained for Bore 18 show conductivity, chloride and permanganate value which are 4,7 to 10 and 9 to 16 times higher respectively. If results for Bore 8 are compared with the upstream the increase in conductivity is not so great as with Bore 18, but the reverse is true of the permanganate values. No increase in temperature was detectable at either Bores 8 or 18.

5. INVESTIGATIONS FURTHER FROM THE TIP

After the investigations in the near vicinity of the tip had shown that the pollution in groundwater would be detected at a distance of about 1350m, it was necessary to follow its course further. For this purpose the borehole network was extended and in certain places intensified.

In the near vicinity of the tip three further bores, 19,20 and 21 were sunk, while seven more (22 to 28) were put down at a distance of about 3000m from the north edge of the tip along an east-west line. Measurements were carried out as already described, as well as further chemical and bacteriological analyses for the greater distances.

Measurement of electrical conductivity for 1 June 1970, provides a good example of the spreading of polluted groundwater (Fig. 65) because chemical analyses were also performed again by the sewage works laboratory. A comparison of the direction of groundwater flow (Fig. 64) with electrical conductivity (Fig. 65) shows very good agreement. Furthermore, the distributions of chloride (Fig. 67) and of permanganate value (Fig. 68) show practically the same picture. All the results indicated that in Bore 24, 3000m from the tip, significantly high values would continue to be observed.

However, no stationary state prevailed, as shown by a comparison of conductivity on 1 June 1970 (Fig. 65) and 15 September 1969 (Fig. 66). Although the conductivity at the north edge of the tip was essentially higher on 1 June 1970 than on 15 September 1969, the range of the 2-mS conductivity line was shorter on the latter occasion.

That not only inorganic but also organic pollution travelled 3000m was indicated by the distribution of permanganate value (Fig. 68). The permanganate value in Bores 23 to 26 was certainly higher than in the upstream.

The distribution of nitrate content of the groundwater is set out in Fig. 69 in which it can be seen throughout that where the organic loading was high, the nitrate was completely destroyed (Bores 9,13,14,15,16 and 8). Nitrate was found in Bores 3 (26 mg/litre) and 18 (31 mg/litre). With regard to organic loading, both of these bores were somewhat marginal. That the organic loading begins to diminish north of Bore 8 is shown possibly by the presence of nitrate in Bores 20 (14 mg/litre) and 21 (17 mg/litre). Certainly this is not due to the whole region north of the tip being used for agriculture. For this reason it cannot be deduced that the groundwater overall contains the same nitrate as at the outset. The opposite may have been the case, since there is the possibility that various quantities of dissolved nitrate may reach the groundwater locally in percolate. That means, amongst other things, that in a sample of water in which no nitrate can be found, there has not necessarily been more nitrate reduction than in water, for example, in which 17 mg/litre of nitrate was present. The reverse might have been the case, since the quantities of nitrate liberated from the tip from time to time were not known.

An observation which is certainly not without significance is the fact that for more than a year the tipping of household refuse has been progressively declining in favour of ash. Organic substance is largely absent and that which reaches the groundwater is destroyed in the reduction zone. Moreover, the ash contains, as well as chloride, a lot of calcium oxide, which by its conversion to calcium carbonate, cements and can become impermeable. Had the investigation continued, it would have been shown whether the tipping of ash brings about, in the long run, a reduction in groundwater pollution.

6. COMPARISON WITH ANOTHER TIP

Comparison of the results of the present investigation with some from South Hessen shows both agreement and inconsistencies (2) (3) (4) (5) (6). There is agreement in the existence of so called reduction and transition zones, terminating in the oxidation zone (4)(5).

Fig. 70 shows the typical occurrence of such zones in relation to groundwater below the tip. The occurrence of micro-organisms shows a related pattern. In the South Hessen study the various zones were indicated simply by the suspension of asbestos tape in the test bores. After a few weeks in the reduction zone the asbestos was stained black from iron sulphide, while in the transition zone the stain was from brown ferric hydroxide. The asbestos remaining colourless indicated that the water contained dissolved oxygen.

In the present investigation the reduction zone included Bores 9, 12, 13 and 14 and, more recently, Bore 15 also. Bores 16 to 18 could be classed as the transition zone. Whether this stretched further north would have had to be investigated further. In this connection, it was unfortunately not possible to distinguish in the region where a temperature increase was measured, how much heat originated in the reduction zone, since the tip itself gave out a great deal of heat.

The reduction zone is defined by the consumption of all of the available oxygen through the oxidation of organic substances which have reached the groundwater (2) (3) (4) (5) (6) (7) (8). In anaerobic régimes this leads to the reduction of nitrate to nitrite and the formation of ammonium or free nitrogen gas. Under more extremely reducing conditions, sulphate can be reduced, leading to the precipitation of black iron sulphide. For a while in Bore 9, so much methane was given off that it could be set alight.

The transition zone is distinguishable, above all, by the occasional availability of dissolved oxygen and by the precipitation of ferric hydroxide. That this zone is not stationary, but capable of being displaced, was indicated in Bore 9 in which after the abatement of vigorous methane formation, a period followed during which large flocs of ferric hydroxide formed. Then, after a while, the formation of iron sulphide occurred. This showed that Bore 9 was not permanently in the reduction zone, but fell for a period in the transition zone. Bore 15 appeared to behave similarly for its brown asbestos tape subsequently turned deep black.

There was no agreement with the previous work to be observed in the pattern of spreading and the range of the polluted groundwater. Certainly the quantity and area of the tip affects this, but above all there were striking differences in the hydrogeological properties to be considered, in particular in respect of the three following factors:

(a)   Height (thickness) of groundwater in the aquifer.

(b)   Permeability of the aquifer.

(c)   Gradient of the groundwater.

The velocity of the groundwater is affected by (b) and (c). The thickness of the groundwater was less than in the Hessen investigation but the permeability and above all, the velocity of the groundwater (10 m/d) was greater in the present situation. A comparison of the two investigations shows that the range and pattern of spreading of polluted groundwater depends extensively upon the local hydrogeological conditions. An important consideration is also the time intervals in the tipping of refuse and its composition, in that these affect the dissolved constituents of the groundwater. In the event of similar investigations being undertaken in the future, it should be realized that variations in hydrogeology can affect the results (7). For this reason it is essential before the opening up of a new tip that the hydrogeological conditions are investigated at an appropriate time to estimate what would happen if, in spite of all precautions, water soluble contaminants reached the groundwater. The use of tracers would enable the direction and velocity of groundwater to be established and possibly also the range of various concentrations of contaminants.

REFERENCES

1. FARKASDI, G. and others

    Microbiological and sanitary investigations
    of groundwater pollutants in the underflow of
    waste tips. In German.
    Städtehygiene, 1969, 20, pp 25-31.

2. GOLWER, A., and
    MATTHESS, G.

    Research on groundwater contaminated by
    deposits of solid waste.
    Int. Ass. Sci. Hydrol., 1968, 78, pp 129-133.

3. GOLWER, A., and
    MATTHESS, G.

    Quality deterioration in groundwater caused
    by waste. In German.
    Proceedings of Deutsch. Gewässerk.
    Meeting 1968.
    Deutsche Gewässerk. Mitt. 1969, pp 51-55.

4. NÖRING, F. and others

    The decomposition processes of groundwater
    pollutants in the underflow of waste tips.
    GWF-wasser/abwasser, 1968, 109, (6), pp
    137-142.

5. NÖRING, F. and others

    The effects of industrial and domestic waste
    on groundwater. In German.
    Mem. Congr. Int. Ass. Hydrogeol., 1967, 7,
    pp 165-171

6. GOLWER, K. and others

    Self-purification processes in aerobic and
    anaerobic groundwater. In German.
    Wasser, 1970, xxxvi, pp 64-92.

7. GOLWER, A. and others

    The influence of waste tips on groundwater.
    Der Städtetag, 1971, (2), pp 1-5.

8. GERB, L.

    'Reduced' water. In German.
    GWF-wasser/abwasser, 1953, 94, (4),
    pp 31-36.

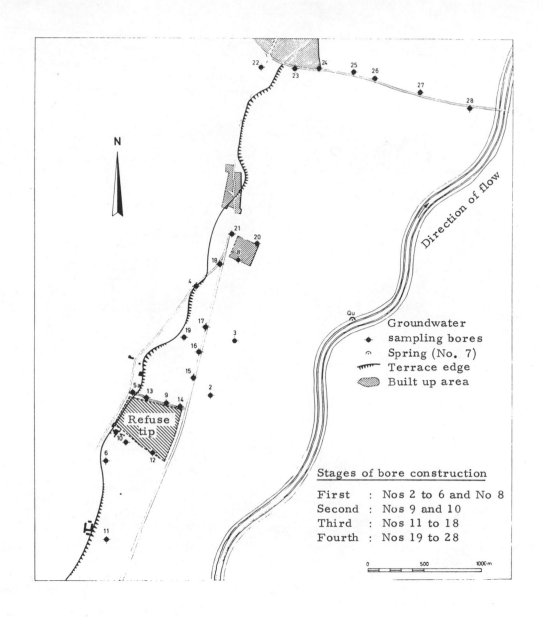

FIG. 57.   DISTRIBUTION OF SAMPLING BORES IN THE INVESTIGATION AREA

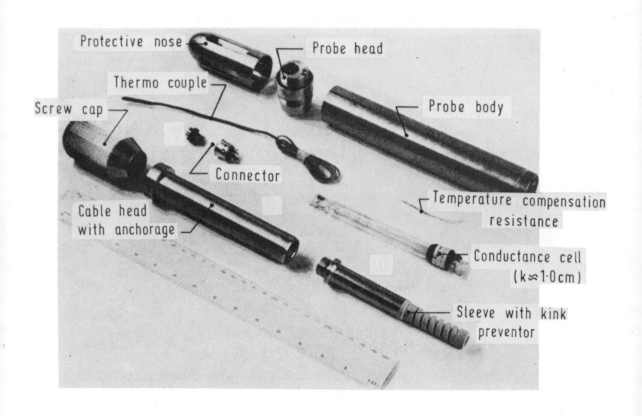

FIG. 58.   BORE MEASUREMENT PROBE

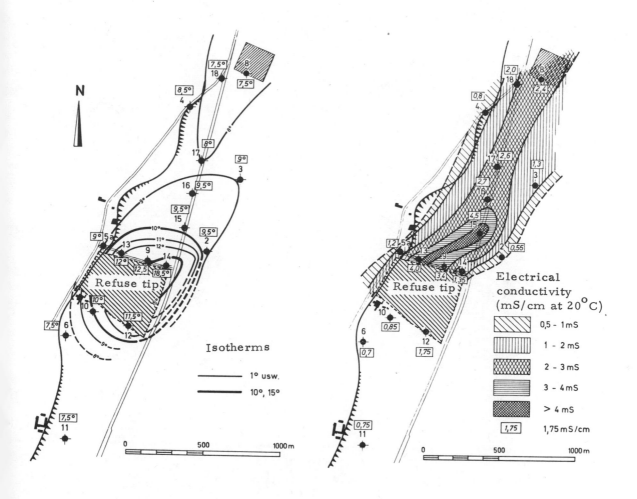

FIG. 59.   ELECTRICAL CONDUCTIVITY AND TEMPERATURE
OF THE GROUNDWATER ON 27 MARCH 1968

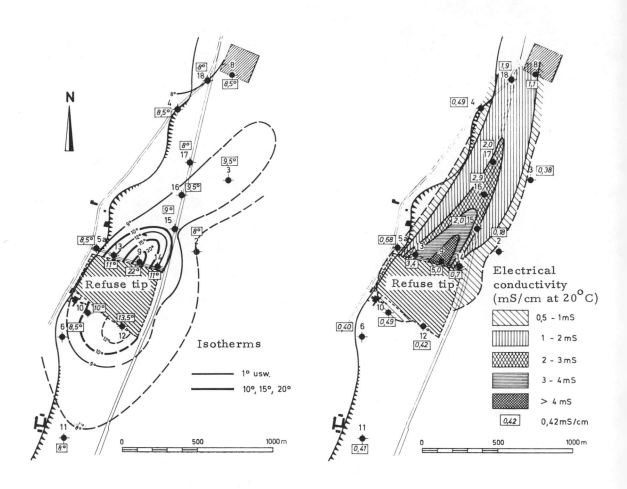

FIG. 60. ELECTRICAL CONDUCTIVITY AND TEMPERATURE
OF THE GROUNDWATER ON 30 APRIL 1968

230.

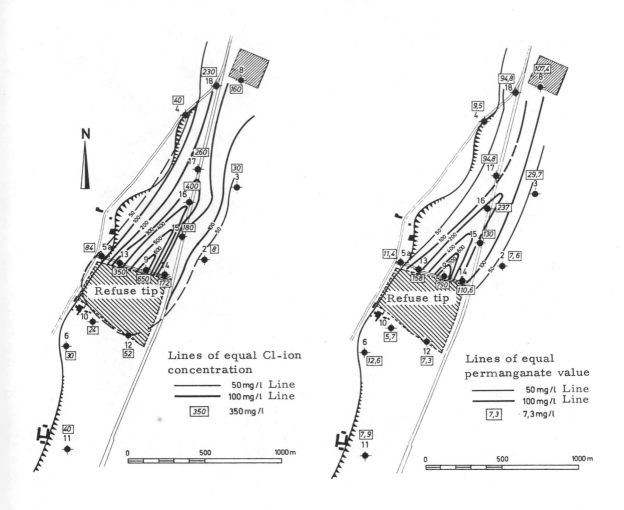

FIG. 61.   SPREAD OF CHLORIDE AND PERMANGANATE VALUE
DURING THE PERIOD 22 TO 29 APRIL 1968

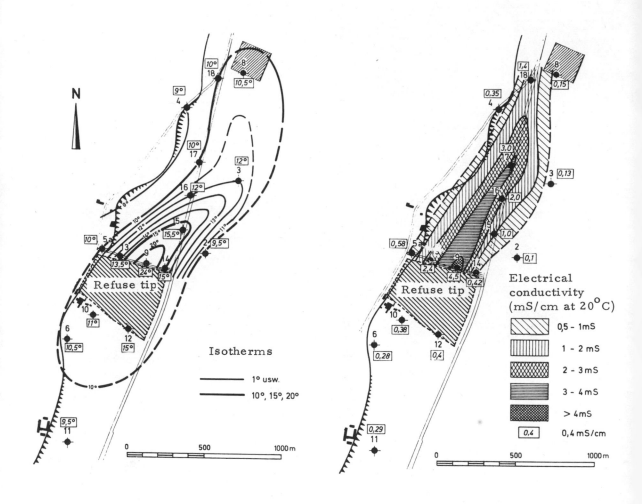

FIG. 62. ELECTRICAL CONDUCTIVITY AND TEMPERATURE OF THE GROUNDWATER ON 14 JUNE 1968

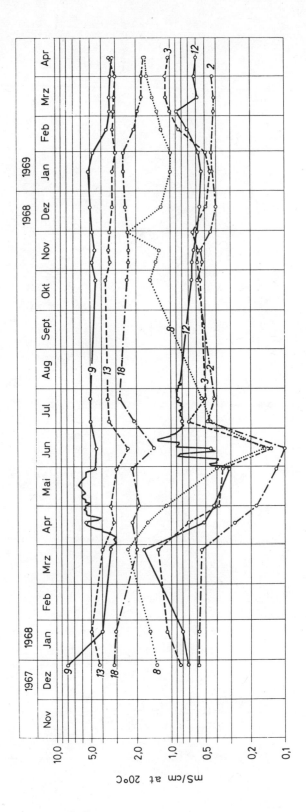

FIG. 63.   RECORD OF ELECTRICAL CONDUCTIVITY

233.

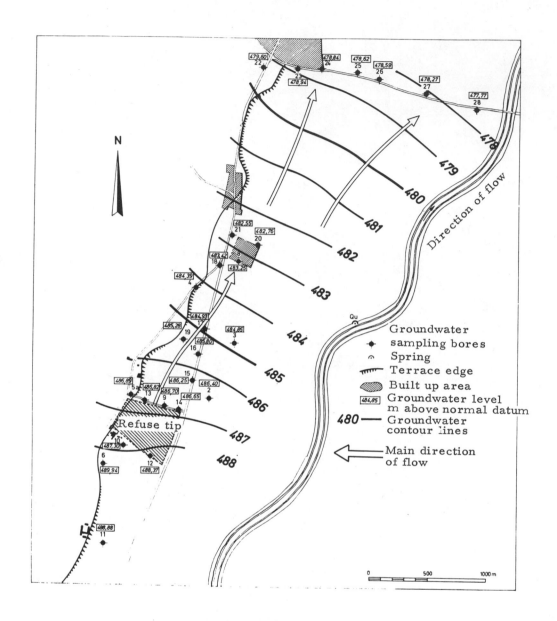

Legend:

- Groundwater sampling bores
- Spring
- Terrace edge
- Built up area
- 484,85 — Groundwater level m above normal datum
- 480 — Groundwater contour lines
- Main direction of flow

FIG. 64.   GROUNDWATER LEVEL AND DIRECTION OF
FLOW OF GROUNDWATER ON 1 JUNE 1970

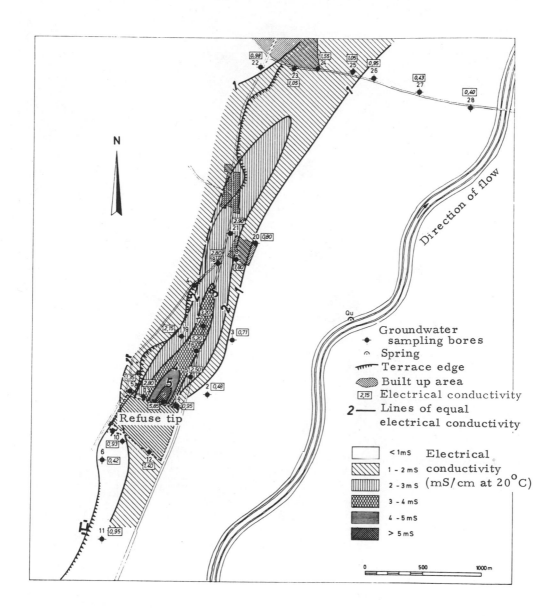

FIG. 65.   ELECTRICAL CONDUCTIVITY OF GROUNDWATER ON 1 JUNE 1970

235.

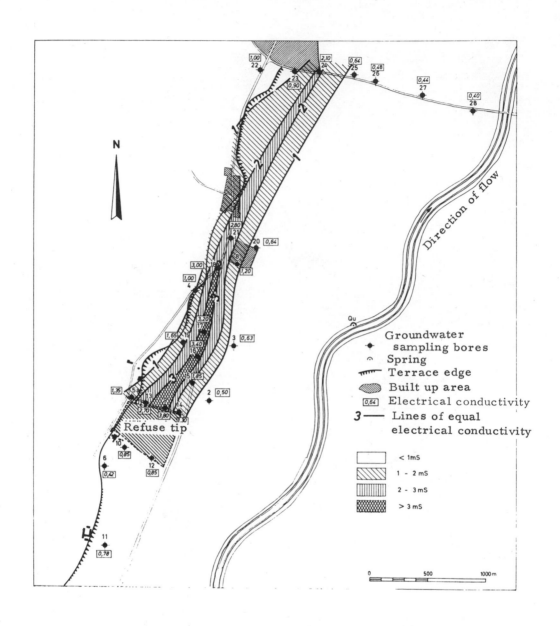

FIG. 66.   LINES OF EQUAL ELECTRICAL CONDUCTIVITY
OF GROUNDWATER ON 15 SEPTEMBER 1969

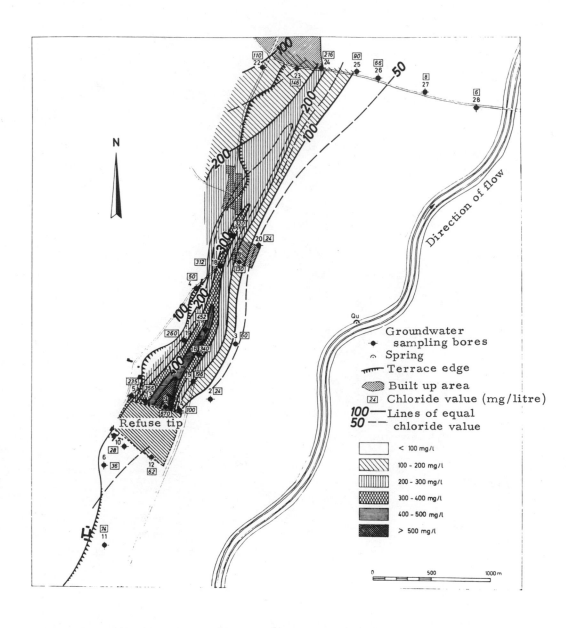

FIG. 67.   DISTRIBUTION OF CHLORIDE IN GROUNDWATER DURING
THE PERIOD 4 TO 7 JUNE 1970

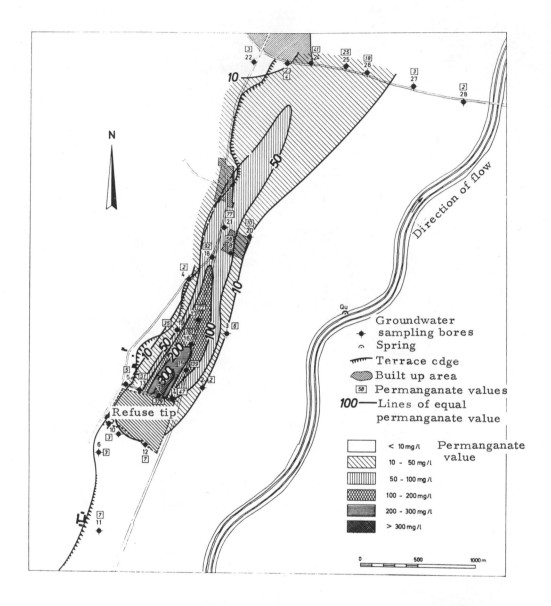

FIG. 68.   DISTRIBUTION OF PERMANGANATE VALUE IN GROUNDWATER
DURING THE PERIOD 4 TO 8 JUNE 1970

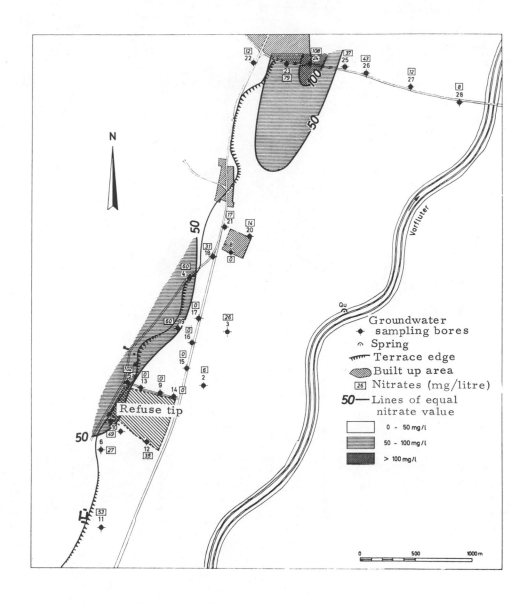

FIG. 69.   DISTRIBUTION OF NITRATE IN GROUNDWATER DURING
THE PERIOD 4 TO 8 JUNE 1970

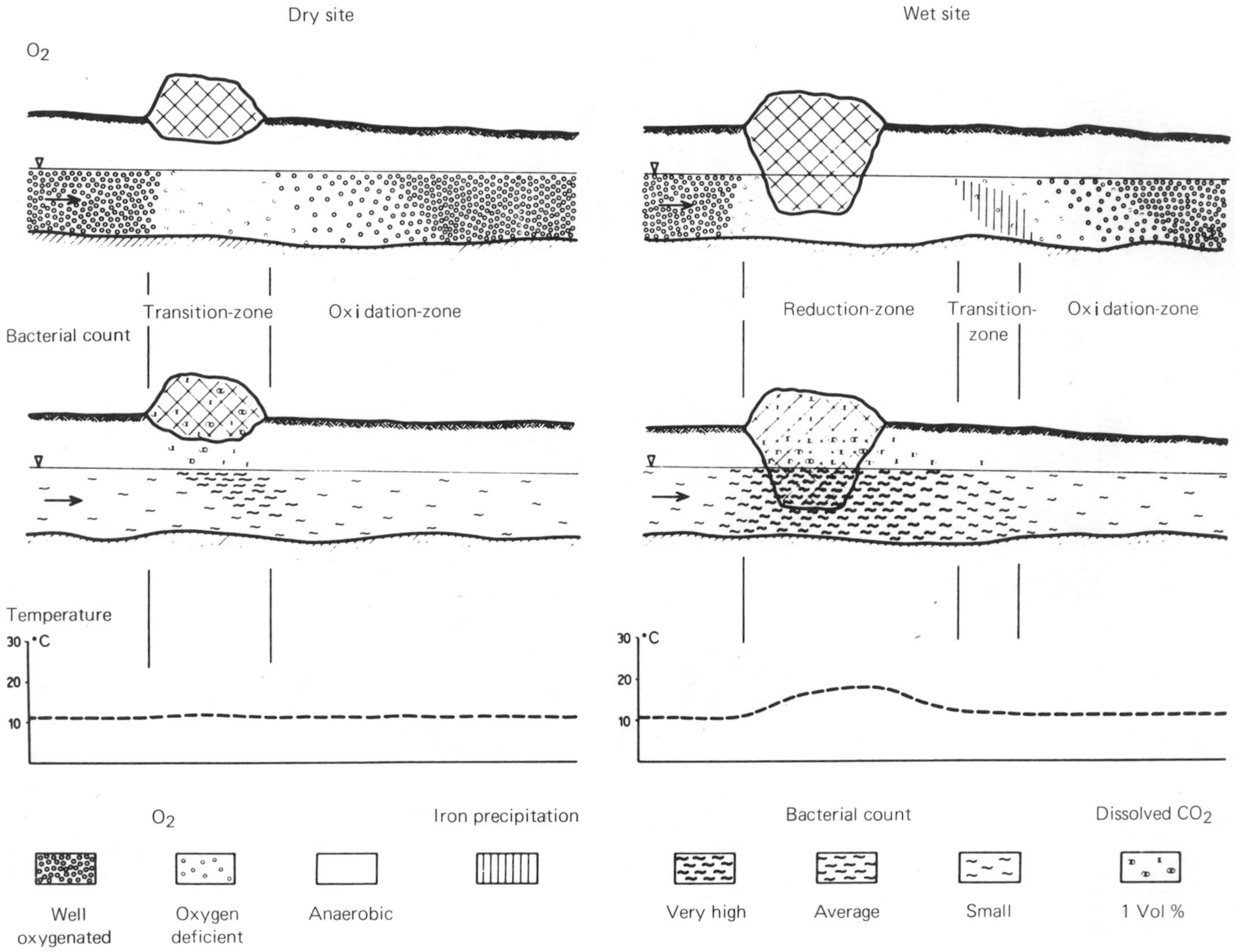

FIG. 70. ZONES OF VARIOUS OXYGEN CONTENTS, BACTERIAL COUNTS AND TEMPERATURES IN GROUNDWATERS FLOWING BENEATH WASTE DISPOSAL SITES

TABLE 28

WATER QUALITY OBSERVED IN BORES ADJACENT TO REFUSE TIP

	Bore No. 10 20/1/67	Bore No. 9	
		18/1/67 (8 m^3 pumped)	21/1/67 (400 m^3 pumped)
Odour	odourless	offensive	offensive
Colour	colourless	yellow-brown	yellow
Chloride (mg/litre)	34	453	433
Sulphate (mg/litre)	116.6	38.0	31.5
Ammonium (mg/litre)	0.9	270	275
Permanganate value	25	590	587

POLLUTION PREVENTION AT TWO REFUSE TIPS

by G.M. Swales and A.S. Davidson
East Surrey Water Company, Redhill, Surrey, England

1. INTRODUCTION

The Company supplies a population of about 326 000 in an area of almost 300 square miles with an average water demand of 20 m.g.d. This area is of considerable scenic beauty, including part of the North Downs and many well-known commons, with high-class residential areas and a few large towns or industrial centres. As a result, tipping of domestic refuse was not encouraged in the area, and other methods of refuse disposal had been employed by most of the local councils prior to about 1956. Since that time, there have been several proposals to tip refuse, mostly from the London area, in disused chalkpits and sandpits which are in fairly close proximity to our boreholes. Following consultation with the Company, controlled tipping has been allowed in several pits with safeguards which include precautions to try to prevent percolate gaining access to the underground water supplies.

2. PRECAUTIONARY MEASURES

2.1. Merstham tip

Following a public inquiry, Croydon Corporation were granted powers for controlled tipping of household refuse at the rate of 70 000 tons per annum on an area of about 21 acres in an old chalkpit on the escarpment in Surrey. The site is in the same valley as the Company's boreholes at Smitham and Purley, which are situated at distances of about $3\frac{1}{2}$ and $4\frac{1}{2}$ miles on lower ground. The total quantity to be tipped was estimated at about one million tons.

The method of protection was to grade the base of the pit to falls and progressively to puddle the bottom with marly chalk, which was already on the site and is known to behave like puddled clay. On this prepared bottom was laid a system of land drains designed to intercept any percolate from the refuse. These drains discharge into a 9 in. cast iron drain leading to a 30 000 gal concrete tank, where the liquid is measured and may be sampled before it is removed by gully emptier to a local sewer.

Tipping started in April 1963 and was carried out in 6 ft layers over the whole site, and tipping finished in October 1971, when over 800 000 tons had been tipped. The quantity of percolate has varied, not only due to rainfall and the degree of percolation (about 50% at this tip), but at times due to areas of uncovered puddled chalk acting as a direct catchment to the tank.

Samples of percolate have been tested in our laboratories since 1963, and the results confirmed that the liquor is highly polluted both chemically and bacteriologically, indicating the potential risk to underground water supplies if suitable precautions are not taken.

2.2. Buckland sandpit

This sandpit is in the lower greensand at its junction with the gault clay, and tipping of household refuse began in the excavated section in 1960 at a rate of about 40 000 tons a year. Tipping has proceeded in several sections of the sandpit, (the rate increased to 50 000 tons a year), and work continues. It is estimated that about 50 acres have now been filled with about 650 000 tons. Here the method of control adopted made use of the property of refuse to absorb moisture, by tipping in 6 ft layers, (to a total depth of about 18 ft) in such a way that no layer is left uncovered by the next layer for more than 4 months, and finally sealing the top with a layer of clay (3 ft). The clay is rolled and graded to falls so that rainfall runs off the tip into the adjacent ditch.

By this method, it was hoped there would be no percolate, and none has been detected. This experiment makes use of the fact that gault clay is available on the site for sealing the top.

The Company have boreholes at The Clears and Buckland, both in the Folkestone Beds of the Lower Greensand, and within $\frac{1}{2}$ mile of the tip. Pumping from these boreholes is limited to short periods in summer, when it may be intermittent. Hence, these boreholes are standing idle for a large part of each year, and the first few samples after start-up show abnormal chemical results.

3. QUALITY OF OUR BOREHOLE WATERS

There is no evidence of any deterioration in bacterial or organic quality due to tipping of household refuse. However, several changes in certain ionic constituents have been noted in the two greensand waters, causing increased hardness, chloride, sulphate and nitrate. We have no definite evidence that the observed changes are due to effluents from refuse tips, but all the constituents which have increased are normally high in percolates from tips of household refuse (including that from Merstham tip). Potassium is considered by us to be the most useful indicator for tracing pollution from refuse tips.

4. CONCLUSION
1. The effluents from dry refuse tips are highly polluted liquors.
2. It is essential that the sites of refuse tips are carefully and correctly selected.
3. The quantity of refuse tipped per year on any particular gathering ground is most important.

4. Potassium is a useful indicator for tracing tip pollution, but there may be other useful indicators for both inorganic and organic constituents arising from present-day refuse (which differs in many respects from the refuse tested by Dr. Key for the Ministry of Housing and Local Government Report published in 1961).

5. One characteristic of anaerobic tip effluents is their ability to induce the growth of fungus or iron-bacteria which will not normally grow in well-oxygenated waters, but may cause musty taste and odour in some underground water supplies. This may be the main difficulty for a water undertaking if pollution by refuse tip effluent occurs.

6. As precautions have been taken at the large tips on our gathering grounds, no serious change in water quality would be expected as a result of tipping household refuse, and none has been detected to date.

DISCUSSION ON OPERATION OF WASTE TIPS AND SANITARY LANDFILLS

1.   Typical oxidation/reduction zones and microbiological regimes below waste tips were described, as a function of the tip being above or within the saturated zone (see Fig. 70, p. 240, supplementing Exler's paper).

2.   Artificial drainage of a tip was illustrated by reference to an Illinois State Geological Survey report*, showing how the percolate from a refuse tip may be intercepted by tile drains or by a pumped well, as in Figs. 71 and 72.

3.   Not all organics decay to simple compounds; the organic matter of tipped refuse will form environmentally undegradable humic substances (see right side of Fig. 73), when oxygen supply is restricted. This is an identical mechanism to the creation of humus or peaty layers in soils.

The humic substances referred to in Hall's paper (p. 102) are the same as "hard" organics in Drost's terminology (p. 136). They are potential blocking agents in artificial recharge operations.

4.   More investigations near sites of known local groundwater contamination to discover the relative importance of degradation and dilution of organics.

5.   Development of analytical techniques for "hard" organics adsorbed onto aquifer material, and employment of total organic carbon determinations.

6.   Drainage of waste tips and spoil heaps as an engineering design problem.

7.   Codes of practice should be set up for the siting, design, operation and inspection of urban waste tips.   Some design objectives are indicated on p. 246.

8.   The criteria for site selection were needed particularly urgently as industry and waste disposal contractors wish to avoid submitting planning proposals that lack a prima facie chance of success.   Such criteria do not pre-empt the water authorities' right to object, but overcome the problems of varying standards and of oversights in major protection matters.

---

*HUGHES, G. M., LANDON, R. A. and FARVOLDEN, R. N. Hydrogeology of solid waste disposal sites in Northeastern Illinois, US Environmental Protection Agency, Rept. SW-12D, 154 pp., 1971.

POSSIBLE OBJECTIVES OF REFUSE TIP DESIGN

by J.A. Cole
The Water Research Association, Medmenham, Bucks, England

The following headings summarize the design objectives put forward by Hughes et al (see footnote, p. 245).

1. Eliminate percolate
   - cater for gas production (cf. Fig. 73)
   - water table may rise after tip capped

2. Migration of percolate under acceptable conditions

   impermeable case
   - some attention in clays because of ion exchange
   - surface seeps to be catered for

   permeable case
   - dilution with regional flow of groundwater
   - movement towards a useless stratum
   - dilution in surface/water

3. Migration and recovery of percolate
   - exploit springs
   - pump wells or install drains (cf. Fig. 71 and 72)

4. Retention and recovery of percolate

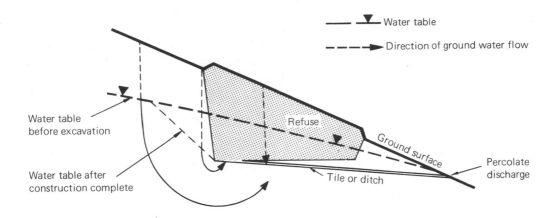

FIG. 71. DRAINAGE OF REFUSE TIPS
GRAVITY DRAINAGE ON SLOPING GROUND

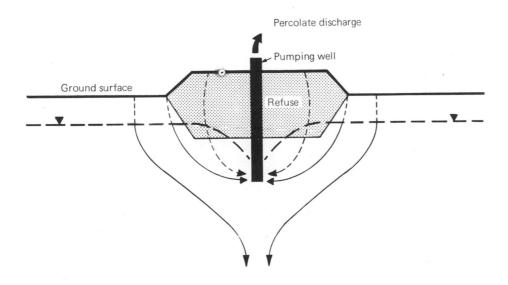

FIG. 72. DRAINAGE OF REFUSE TIPS
DRAINAGE INDUCED BY A PUMPED WELL

247.

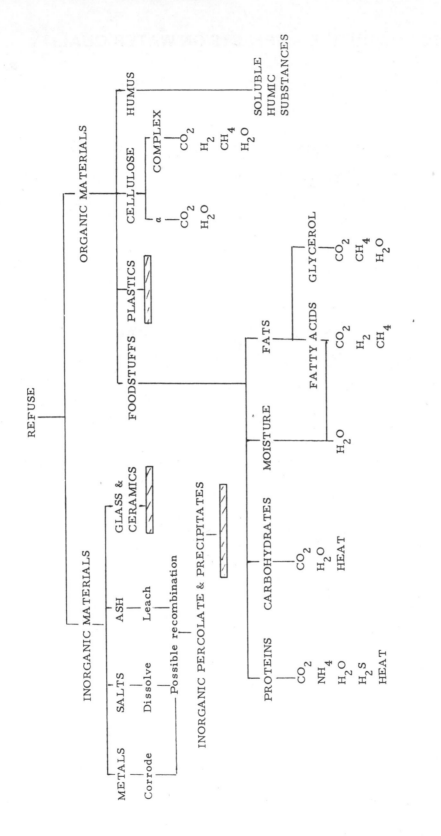

FIG. 73.

THE DIVERSE FATE OF REFUSE

This diagram is a simplified version of that
appearing in Hughes et al (see footnote on p. 245).

▨ denotes undegradable solid

DREDGE MINING FOR RUTILE — EFFECTS ON WATER QUALITY

by R. Herbert
Engineering and Resources Consultants Ltd., Bracknell, Berks, England

1. POLLUTING MECHANISMS

Areas of the Tomago Aquifer (Fig. 74) which supplies Newcastle NSW, Australia, are being mined by dredging for rutile and zircon. Engineering and Resources Consultants Limited (Ercon) were appointed consultants to monitor and advise on effects of mining.

A dredge pond is created 400 ft x 400 ft to a depth of 20 ft in aquifer of 50 ft thick sands. The sand is classified for rutile and zircon then redeposited as tailings including a slimes layer 1 ft thick at pond bottom. Subsequent deterioration in ground-water quality.

2. THE SOILS AND GEOLOGY

The aquifer consists of 50 ft thick aeolian and fluviatile sands - deposited on a coastal plain. They contain a mineralized zone - the Wooloomooloo - and a series of podsol profiles with localized cemented patches. When mined the topsoil, 6 in. is cleared, then redeposited. Slimes layer is concentration of fine-grained material and is high in iron and organics.

3. GROUNDWATER CHEMISTRY

Water quality determined from sampling boreholes.

In unmined areas groundwater iron content 0 to 1 mg/litre total dissolved solids 100 mg/litre colour 20 mg/litre, pH 5, acidity 50 mg/litre.

In mined-over areas iron content rose, after a delay, to up to 60 mg/litre for a period of at least 3 years. Total dissolved solids 200 mg/litre, colour 400 mg/litre, pH 5, acidity 200 mg/litre.

4. BACTERIOLOGY

The major water quality changes occurred near the slimes. Field and laboratory tests were run to investigate the possible biological processes. Two possible causes of pollution have been identified.

a) There is a general bacterial attack on organic matter of slimes.
This releases carbon dioxide and causes reduction of oxidized
iron compounds. When all available oxygen used up, iron goes
into solution in ferrous form as iron bicarbonate. This reaction
associated with high plate count, high CO_2, high acidity but no
drop in pH. All above phenomenon observed.

b) Also, the introduction at depth of aerated mining-pond water
could encourage growth of sulphur oxidizing bacteria, which die
when all oxygen used up. This cellular material is an alternative
source of organic material which would then be subject to the same
reaction described above resulting in high iron concentration.

The above reactions, separately or jointly, have caused deterioration in water
quality. Ercon has specified that the mining companies should dispose of the mining
slimes outside the aquifer boundaries thus eliminating pollution by reaction (a) above.
Monitoring continues at a new mining area to assess the importance of reaction (b)
and any other pollution sources.

5. DISCOVERY OF POLLUTION

Ercon mounted general environmental investigation/protection study which included
installation of 44 boreholes to monitor groundwater quality.

6. COMBATING POLLUTION

Mining company instructed to remove dredge slimes beyond developed aquifer
boundaries. They are providing thickener to concentrate slimes, with disposal pipeline.
Monitoring programme set up to investigate effectiveness of thickener and residual
slimes in mining pond. Soil and water chemical and bacteriological investigation
programmes continuing.

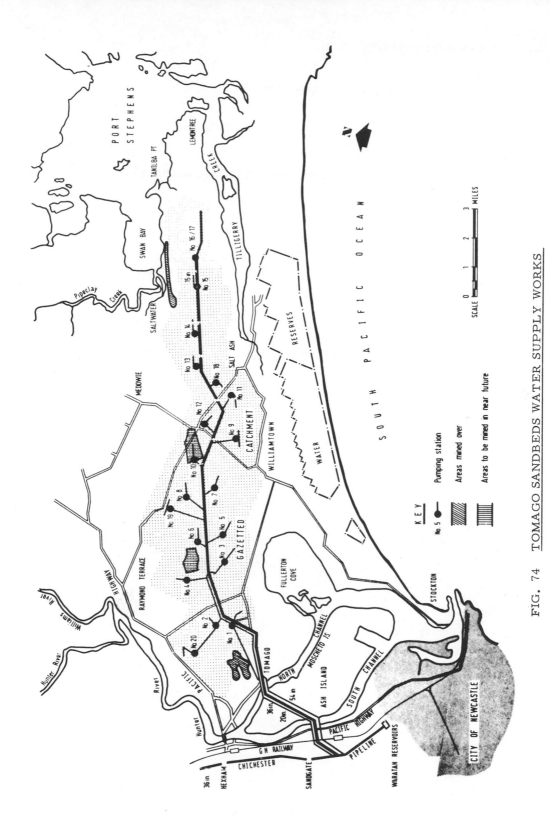

FIG. 74   TOMAGO SANDBEDS WATER SUPPLY WORKS

COLLIERY SPOIL HEAPS

by T.E. James
National Coal Board, Doncaster, Yorks, England

In case it is assumed that groundwater pollution from colliery spoil heaps is excessive, I wish to point out that a detailed investigation of a nearby colliery spoil heap has been carried out and reported*.

The investigated heap consisted of spoil from the same seam, the Barnsley Seam, and of the same age as that mined at Markham Main. The spoil, which was clay-like in nature, was uncompacted when tipped but, nevertheless, appeared to have a low permeability, while weathering had merely affected the surface.

Present day methods of stocking spoil and of utilizing old spoil for new works such as motorway embankments, utilize mechanical compaction and stabilization. These processes produce low permeabilities when measured in the laboratory; typical results obtained from Hatfield Colliery Spoil Heap, also from the Barnsley Seam are appended.

Test	Dry density lb/ft^3 Std. compaction[+] = 108 lb/ft^3	Permeability cm/sec
No. 1.	106	3.5×10^{-5}
No. 2.	100	3 to 7×10^{-4}
No. 3.	92	4 to 7×10^{-3}

[+]As defined in BSS. 1377;1967

Referring again to the Markham Main situation, it should be noted that substantial areas of ground consist of sands and gravels overlying the Bunter Sandstone. Water from Sandall Beat, mine water and water from the colliery boreholes after use enters Fores Drain and as Nicholls points out, seepage from the drain into the aquifer is possible.

---

* "A mineralogical investigation of a spoil heap at Yorkshire Main Colliery"; D.A. Spears, R.K. Taylor and R. Till, Q. Jl. Engng Geol. Vol. 3, 1971, p. 239-252.

DISCUSSION ON CHEMICAL AND HYDROGEOLOGICAL ASPECTS OF FLY ASH DISPOSAL

Chemical and hydrogeological aspects of fly ash disposal

1.   The fate of inorganic constituents from fly ash tips was under study by the Central Electricity Generating Board.   In commenting on p. 96 of Hall's paper, J. Brown reported that studies of fly ash had shown that physical as well as chemical factors could limit the equilibrium concentrations of soluble species attained.   Fly ash contained about 2% of soluble material, the bulk of which arose from Na^+, K^+, Ca^{++} and SO_4^{--} ions, and laboratory and field trials with packed columns of fly ash had shown that:

(i)   the chemical composition of fly ash percolate was, for a given ash type, sensibly independent of bed depth,

(ii)   the time required for percolate to traverse a fly ash bed was a function of bed depth, and deep beds took longer to clear themselves of soluble constituents producing a longer tail to the concentration/ time graph of TDS.

A model of fly ash as an inert substrate supporting very limited amounts of soluble material could account for the observed effects.

Because fly ashes had various soluble constituents, depending on the source of coal of which they form the residue, it was not easy to generalise on their percolate quality. In practical situations where water supplies might have been affected by ash tipping, CEGB had supplied information on the relevant ashes and had taken expert hydrogeological advice.   In no dumping situation to date had any problem emerged with respect to potable waters.

2.   As fly ash tips can infill excavations in pervious strata, their drainage requires to be provided in such a way as to encourage the flow of percolate to surface waters with minimal entry to any deeper strata used for potable supplies.   A schematic diagram of a typical case was given by G. P. Jones (Fig. 75, p. 254).

3.   When investigating the water quality of ash tips special attention has to be given to samples taken near the capillary fringe, which fluctuates seasonally due to varied percolation amounts, and thus is prone to strong leaching action. Fig. 76 on p. 255 shows the type of piezometer head used by the Water Resources Board[*] in sampling from below recharge basins, which present similar problems.

[*] R. H. L. Satchell and K. J. Edworthy The Trent Research Programme Vol . 7
                            Artificial Recharge: Bunter Sandstone
                            (Water Resources Board, Reading, 1972, 76 p.)

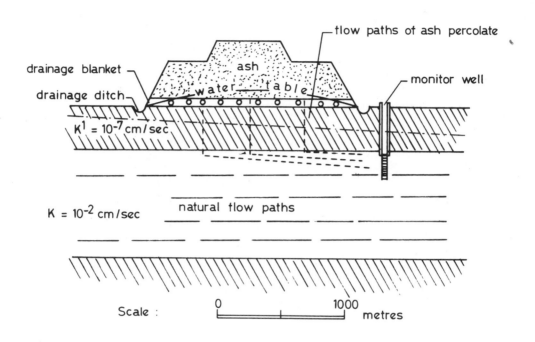

FIG. 75.   SCHEMATIC SECTION OF FLY ASH TIP

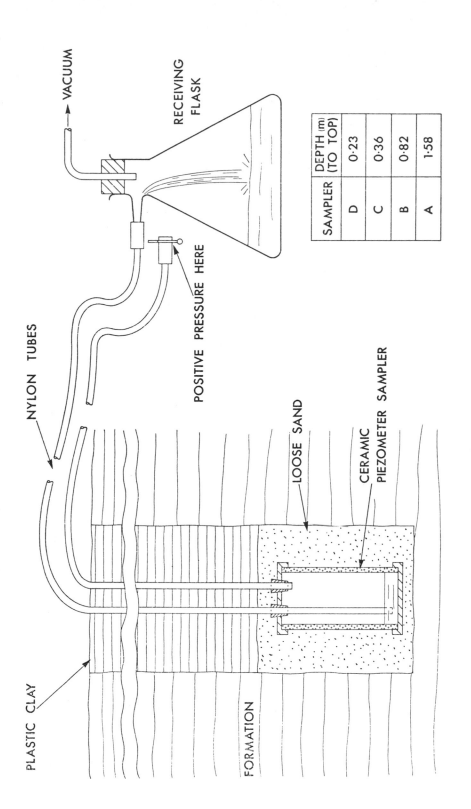

SAMPLER	DEPTH (m) (TO TOP)
D	0·23
C	0·36
B	0·82
A	1·58

VACUUM

RECEIVING FLASK

NYLON TUBES

POSITIVE PRESSURE HERE

PLASTIC CLAY

LOOSE SAND

CERAMIC PIEZOMETER SAMPLER

FORMATION

FIG. 76 WATER SAMPLING DEVICE FOR UNSATURATED ZONE

SALINE POLLUTION OF UNDERGROUND SUPPLY AT LEICESTER LANE BOREHOLE

South Warwickshire Water Board, Leamington, Warwickshire, England

NATURE OF POLLUTANT

Pollution discovered in groundwater.

The pollution is thought to be caused by leaching of salt from a nearby highway depot stockpile of rock salt for road work.

LOCATION

Borehole located at GR SP 329690 (Fig. 77)

North east of Leamington Spa.

DETAILS

Ground level	249.51 OD
Depth	120 ft
Lined to 67 ft with plain tubes	
67-84 perforated tubes	
84-98 plain tubes	
Principal water bearing strata, Keuper sandstone 79 ft to 81 ft	
Rest level	188 OD
Pumping level	173 OD
Normal yield	100 000 g.p.d.

CHEMISTRY

	Before 1963	During (now)	
pH	7.1	7.3	
E.C.	550	1650	μmhos
T. Hardness	330	540	mg/litre
T. Alkalinity	155	165	
Chlorides	30	375	(Fig. 78)
Nitrate	40	50	
Oxy. Abs.	0.1	0.2	

BACTERIOLOGICAL ANALYSES

Plate counts	0	0
Coliforms	0	0
E. Coli	0	0

COMBAT MEASURES

1. Cover salt pile (polythene sheeting).
2. Construct concrete holding bay.
3. Abandon source.

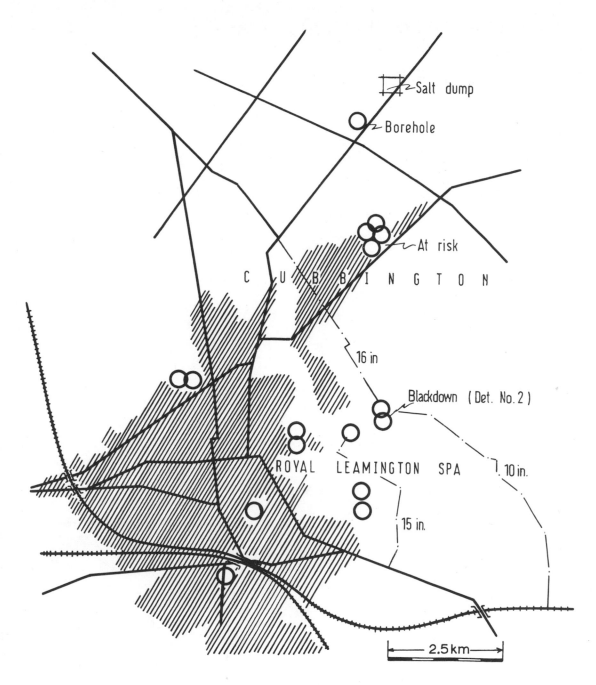

FIG. 77.   LOCATION OF BOREHOLE

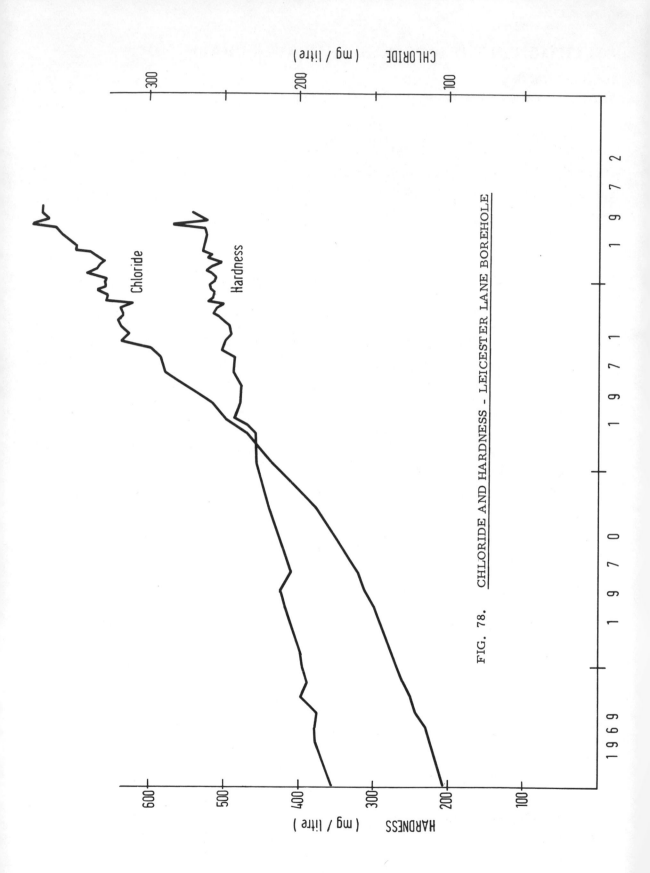

CHLORIDE

(mg / litre)

300

200

100

1 9 7 2

1 9 7 1

Chloride

Hardness

FIG. 78.   CHLORIDE AND HARDNESS - LEICESTER LANE BOREHOLE

1 9 7 0

1 9 6 9

600

500

400

300

200

100

HARDNESS   (mg / litre)

POLLUTION IN THE BRIGHTON SEAWATER CHALK

by S.C. Warren
Water Department, Brighton, Sussex, England

This note concerns seawater intrusion in the Upper Chalk, on the coast of Sussex, affecting the quality of water pumped from a linked system of wells and boreholes at Balsdean. Fig. 79 shows the general layout, together with a picture of how tunnelled headings form horizontal connections between the boreholes.

Winter chloride value 30 mg/litre. Summer values following winters of low rainfall 300+ mg/litre. In summer 1949, Saltdean Pumping Station discontinued owing to chloride of 3000+ mg/litre.

Fig. 80 gives chloride determination (during twice-a-day pumping shifts) plotted against tides in 1949. Some detailed correlation with tidal cycles may be seen, plus a systematic rise in Cl$^-$ level with increase of tidal amplitude.

At this time, factors influencing appeared to be:

    (a)   winter rainfall

    (b)   abstraction rate at Balsdean during summer

    (c)   abstraction rate at other stations to north and

    (d)   state of tide in monthly cycle

During winter 1957, Balsdean recorded conductivity showed pattern similar to flattened version of summer "V" obtained when seawater involved. Chloride was normal showing maximum variation of 3 mg/litre whereas hardness varied by 12 mg/litre. It had been assumed that "V" form in Fig. 80 was function of seawater ingress but this was not the whole explanation for reasons given below.

For seawater, ratio of chloride to hardness is 3:1. In Balsdean water, ratio was 3:12 in winter 1957. This indicated two chalk waters being pumped simultaneously in varying proportions. Chloride and hardness decreased during first three hours of pumping shift and increased thereafter. Time of appearance of harder water in second half of shift in winter equal to time of appearance of high chloride in summer. High chlorides derive from seawater infiltration south of Balsdean. Possible that hard water also from south. If so, same flow conditions underground winter and summer. Conception contrary to previous assumption that all water flows from north to south.

In July 1961 decided to pump one 18 hour shift instead of two shorter shifts. Fig. 81 shows chlorides' recorded "V" form at commencement of pumping, thereafter, a wave form having good correlation with tide though not in phase with it. Uncertain whether infiltration was by horizontal flow in fissures or vertical movement of seawater wedge in chalk. If latter, interface traverse would be finite distance and constant time lag expected. If former, unknown time intervals superimposed on tidal pattern.

To resolve the above uncertainty, Saltdean Borehole was investigated. Conductivity survey indicated fissures at -38 ft and -89 ft OD and that horizontal flow was actuated by Balsdean pumps. Depth sampling also confirmed saline gradient, i.e. surface chloride 83 mg/litre and bottom 2000 mg/litre. Sought proof of underground flow between Saltdean and Balsdean using $(NH_4)_2SO_4$ as tracer. $(NH_4)_2SO_4$ introduced at Saltdean found 12.5 hours later at Balsdean, substantiating in 1961 deduction of 1957 that one of two chalk waters at Balsdean came from south.

Later work showed seawater being transported to Balsdean by other unknown fissure between Saltdean and Balsdean because conductivity recorder at Saltdean showed higher salinity in water flowing southwards on Balsdean shutdown that in that passing northwards during pumping.

Salinity now controlled by programming pumping to take place through low tide period allowing for time lag between tide state and exhibited chloride value, topping up from inland stations which are rested in winter months by general plan of controlled pumping in whole area.

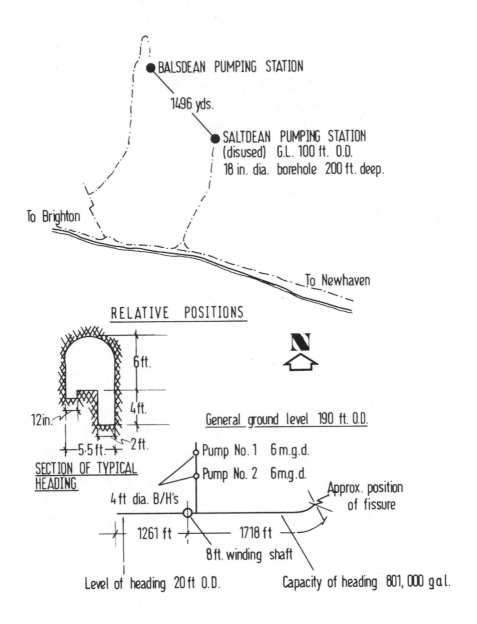

BALSDEAN PUMPING STATION

1496 yds.

SALTDEAN PUMPING STATION
(disused) G.L. 100 ft. O.D.
18 in. dia. borehole 200 ft. deep.

To Brighton

To Newhaven

RELATIVE POSITIONS

6 ft.

4 ft.

12 in.

5·5 ft. — 2 ft.

SECTION OF TYPICAL
HEADING

N

General ground level 190 ft. O.D.

Pump No. 1   6 m.g.d.
Pump No. 2   6 m.g.d.

Approx. position
of fissure

4 ft dia. B/H's

1261 ft — 1718 ft

8 ft. winding shaft

Level of heading 20 ft O.D.   Capacity of heading 801,000 gal.

FIG. 79.   POSITION OF BALSDEAN PUMPING STATION

261.

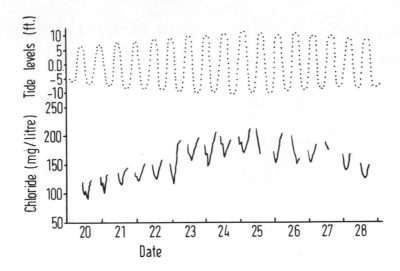

FIG. 80.   BALSDEAN PUMPING STATION: CHLORIDE AND TIDES, SEPTEMBER 1949

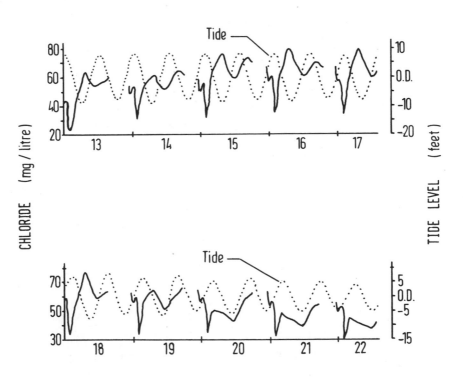

FIG. 81.   BALSDEAN PUMPING STATION - CHLORIDE AND TIDE, JULY, 1961

SALINE POLLUTION OF TWO BOREHOLES AT CANVEY ISLAND
The Essex Water Company, Romford, Essex, England

Nature of pollutant

The pollutant is a saline intrusion. It occurs continuously and originates from the tidal river Thames 500 yds away.

Location

Fig.82 shows the position of the two boreholes in relation to the Thames estuary, the rest water levels for October 1968 and the base of the London Clay. The geological succession at the boreholes is 92 to 94 ft of drift material overlying 94 ft of London Clay which in turn rests on Eocene Sands and Cretaceous Chalk.

Borehole No. 1 is lined to 349 ft, the base of the Eocene Sands and Borehole No. 2 is lined to 296 ft. The main aquifer is the Chalk and as seen from Fig.82 artesian conditions prevail. Groundwater is also present in the drift material and here the water level is dependent on the sea level.

Pollution is thought to enter No. 2 borehole through a damaged casing at the top of the London Clay at -85 ft O.D. It then obtains access to No. 1 borehole through the water - bearing strata below the London Clay.

Chemistry

A typical analysis of the potable water from No. 1 borehole is given by the average results for 1971:

Elect. conductivity	1500	μS/cm
pH	8.4	
Chloride Cl	380	mg/litre
Total hardness	44	
Free alkalinity N	240	
Free ammonia N	0.52	
Fluoride F	2.3	
Opt. Dens.	0.03	275 nm/40 mm

No. 2 borehole shows chlorides well in excess of 1000 g/litre but a detailed analysis is not available.

Bacteriological analysis

No. 1 borehole   Coliform organisms - absent in 100 ml.

                         E. Coli   - absent in 100 ml.

The pollution first became evident about 1956, and showed as high chlorides from the bores. From experiments carried out it was discovered that what had been regarded as a naturally saline source, was in fact an intrusion of salt water into No. 2 borehole.

Combat measures

In 1957, to combat the saline intrusion, efforts were made to grout No. 2 borehole but this failed. Grouting was difficult as the bore lining tubes were installed one inside the other. The present arrangements are to pump No. 2 borehole continuously to waste while No. 1 borehole is pumped into supply. In this way the saline intrusion is pumped away and the potable supply is maintained unadulterated.

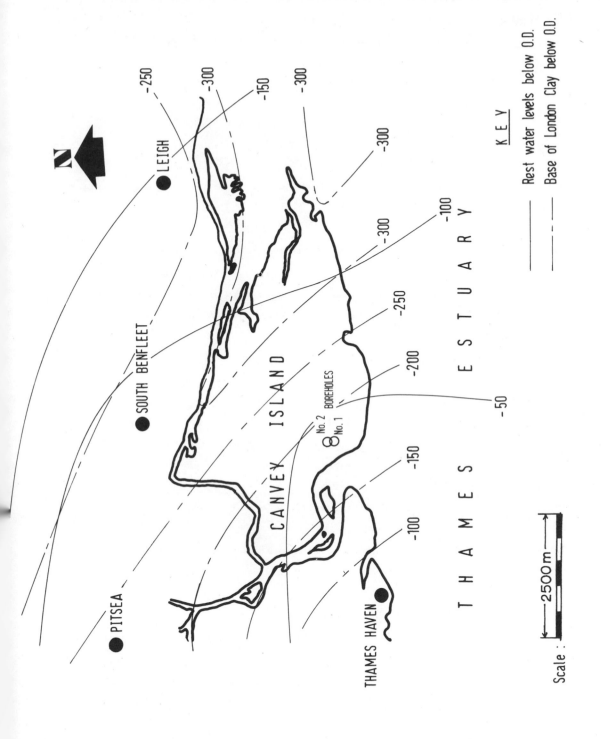

FIG. 82. LOCATION OF HOLE HAVEN BOREHOLES

SALINE UNDERSEEPAGE PROBLEMS AT CHOWILLA DAM
by E.T. Haws,
Engineering and Resources Consultants Ltd., Bracknell, Berks, England

1. POLLUTING MECHANISM

The proposed Chowilla Dam for irrigation storage in South Australia is to be sited within the semi-arid flood plain of the River Murray.

The pressure head created by the new reservoir would force the naturally occurring highly saline groundwater into the Murray River irrigation channel downstream, with accompanying detrimental results. A complete cut-off through the 37 m foundation sands for the 5.5 km length of the dam was found to be uneconomic.

2. DAM SITE AND GROUND CONDITIONS

The dam site lies across a wide flood plain (Fig. 83). The average retained height of water across the valley flats will be 11 m with a maximum of 17 m at the river. In general terms the aquifer through which underseepage will occur, consists of 15 to 30 m of Older Valley Deposits of graded fine, medium and coarse sand underlain by Upper Estuarine Deposits of finer sand; the combined thickness is about 37 m. The aquifer is underlain by Lower Estuarine of alternating very fine sand and clay lenses, which is assumed to be the impermeable base to the aquifer.

Large scale pump tests were used to determine the permeability of the aquifer and unprotected underseepage of $1.8 \text{ m}^3/\text{sec}$ was predicted.

3. CHEMISTRY OF THE GROUNDWATER

Spot sampling from purpose boreholes gave peak groundwater salinities away from the river of 60 000 mg/litre; the average salinity was some 30 000 mg/litre. Only $0.06 \text{ m}^3/\text{sec}$ underseepage could be permitted to enter the irrigation channel downstream.

4. MEASURES TAKEN TO COMBAT POLLUTION

Analogue model tests including Teledeltos, three dimensional resistor, electrolytic tank and Hele-Shaw viscous flow types were used to simulate the seepage system. These analogues were used to devise a system of relief wells intercepting the saline underseepage before flow into the irrigation channel. The wells were designed to allow saline seepage from low levels to be discharged separately from fresh seepage at upper levels. The salt water is to be pumped to evaporation ponds on the abutment. Predictions were made of the period required for flushing out the salt water.

The studies resulted in a proposed design which will deal satisfactorily with the saline underseepage problem permitting full use of the $6200 \times 10^6 \text{ m}^3$ Chowilla irrigation storage.

A full technical paper on the work is being presented by Haws and Legg to the 1973 ICOLD Conference in Madrid.

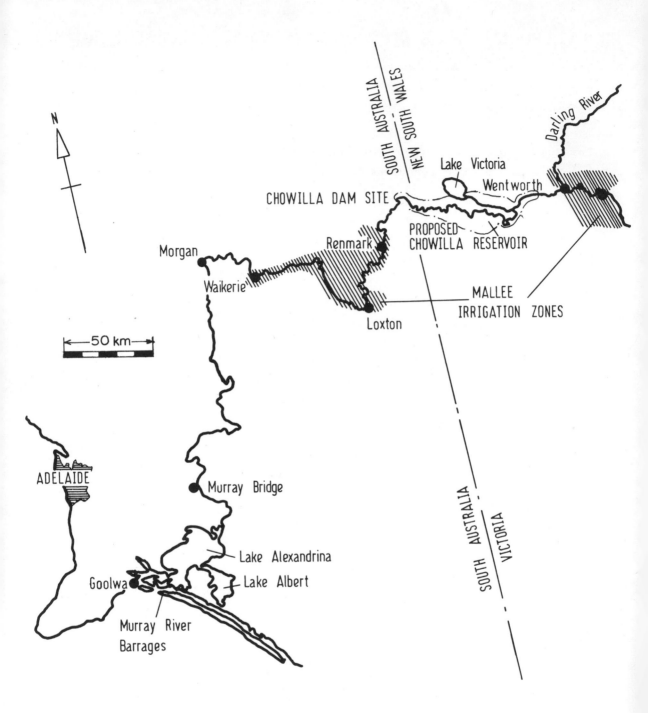

FIG. 83.   LOCATION OF PROPOSED CHOWILLA DAM

268.

NITRATE POLLUTION OF CHALK GROUNDWATER IN EAST YORKSHIRE: A HYDROGEOLOGICAL APPRAISAL*

by S.S.D. Foster and R.I. Crease
Institute of Geological Sciences, London, England and East Yorkshire (Wolds Area)
Water Board, Duffield, Yorks, England

1. THE AQUIFER

The chalk formation underlies some 1800 km^2 (700 sq miles) of East Yorkshire and outcrops as shown in Fig. 84. Groundwater from it provides the bulk of the water supply for a population of over 500 000; the Chalk sources of the public water supply undertakings having an estimated combined drought yield of about 168×10^6 litres/day (37 m.g.d.) (1). The protection of this aquifer from a level of pollution incompatible with its water supply function is thus of regional significance.

2. RECENT INCREASE IN NITRATES FOUND IN GROUNDWATERS

The evidence for nitrate pollution derives from the routine periodic analyses by the statutory undertakings. It is in many respects comparable to that for Chalk sources in North Lincolnshire (2) and in the Eastbourne area (3). During 1960-69, all the East Yorkshire sources had constant and low nitrate analyses in the order of 2.5 to 4.5 mg/litre NO_3-N, but during 1970-72 many sites showed apparently sharp increases and values in the range 8.0 to 11.5 mg/litre NO_3-N are now reported from widely distributed sources. (see Figs. 85 and 86).

An eight-laboratory determination of split samples revealed an absolute variation of more than ± 1.0 mg/litre NO_3-N, with similar variation in repeated determinations by individual laboratories using the rapid selective ion electrode technique. The analytical accuracy factor was not discussed in detail in relation to the published case histories nor is it of special interest to the present authors but it would appear to affect the value of historical data for the establishment of trends. It may also pose problems for the medical authorities when attempting to implement the recommendation that in the long-term, concentrations should not exceed 11.3 mg/litre of NO_3-N. However, the adoption of an intermediate range of tolerable concentrations (4) relieves this situation in the short-term.

3. DISCUSSION OF CAUSES

On balance, the existing evidence from East Yorkshire must be interpreted as indicative of appreciable nitrate pollution. The worst affected part of the aquifer is shown

---

*The paper given in summary form here appears in a full version, under the same title, in: Journal of the Institution of Water Engineers, 1974, 38 (3), 178-194.

in Table 29 to be the escarpment; there has, on the other hand, been some increase in the confined zones. The cause for concern is not so much the present level of NO_3-N concentrations but their widespread and apparently rapid rate of increase in the outcrop area.

TABLE 29

CLASSIFICATION OF SOURCES RELATED TO INCREASE IN NITRATES

Source	No.	Strata	Type	Site	Confined	1960-9 NO_3-N (mg/litre)	1970-2 NO_3-N (mg/litre)	Ratio	Avge. ratio
Filey	1	Corallian	Bore	Dip	Yes	0	0.7	-	
Haisthorpe	3	Chalk	Bore	Dip	Yes	3.4	6.1	1.8	
Burton Agnes	4	Chalk	Bore	Dip	Yes	2.9	7.2	2.5	
North End	7	Chalk	Bore	Dip	Yes	3.4	7.0	2.1	2.1
Hutton	8	Chalk	Bore	Dip	Yes	3.4	6.3	1.9	
Mill Lane	2	Chalk	Bore	Dip	No	3.2	9.3	2.9	
Kilham	5	Chalk	Bore	Dip	No	3.4	8.1	2.4	
Spellowgate	6	Chalk	Bore	Dip	No	3.2	7.9	2.5	2.6
Etton	9	Chalk	Bore	Dip	No	3.2	8.6	2.7	
Newbald	10	Chalk	Bore	Scarp	No	3.4	9.5	2.8	3.4
Springwells	11	Chalk	Bore	Scarp	No	2.9	11.5	3.9	
Millington	12	Chalk	Spring	Scarp	-	2.9	6.8	2.3	2.2
Kirby Underdale	13	Chalk	Spring	Scarp	-	3.4	6.8	2.0	

From the same type of negative reasoning employed by Davey (2), there seems little doubt in East Yorkshire that the pollution has its origin in agricultural activity. The authors have not undertaken an extensive survey of agricultural practices but it would appear that certain processes in specific hydrogeological settings probably lead to the greatest hazard.

In practice, an inter-disciplinary research programme is probably required and within this both the hydrogeochemical and hydrogeophysical aspects need to be represented, since in subsurface environments, oxidation and reduction and the processes of dilution and dispersion in the zones of aeration and saturation, will influence the fate of the infiltrating pollutant. The hydraulics of saturated flow in the East Yorkshire Chalk are well understood (5) and even natural flow rates can be of the order of 50 m/day (164 ft/day), in this fissure-flow formation, characterized by high values of transmissibility and low specific yield. Unsaturated flow in the Chalk is currently a subject of great contention; Smith et al. (6) suggest that intergranular seepage dominates with downward velocities as low as 0.88 m/year (2.9 ft/year) at a site in Berkshire and tritium age determinations of some Chalk groundwaters in East Yorkshire indicate a dominance of pre-1953 precipitation. If this is taken to indicate that the bulk of the nitrate pollution currently being experienced is of a similar age, it can be inferred that further increases could occur because the rate of application of nitrogeneous fertilizers has increased since then.

270.

REFERENCES

1. YORKSHIRE OUSE & HULL RIVER AUTHORITY — Survey of Water Resources. Leeds, Yorkshire Ouse and Hull RA, 1969

2. DAVEY, K. W., — An investigation into the nitrate pollution of Chalk borehole water supplies. Scunthorpe, The Board, 1970

3. GREENE, L. A. and WALKER, P. — Nitrate pollution of Chalk water. Water Treat. Exam. 1970, 19, pp 169-182

4. WORLD HEALTH ORGANIZATION — European standards for drinking waters. Geneva, WHO, 1970, 2nd edition.

5. FOSTER, S. S. D., MILTON, V. A. and CREASE, R. I. — Hydraulic behaviour of the Chalk Aquifer in the Yorkshire Wolds. (in preparation)

6. SMITH, D. B., WEARN, P. L., RICHARDS, J. H. and ROWE, P. C. — Water movement in the unsaturated zone of high and low permeability strata using natural tritium. IAEA Symposium on Use of Isotopes in Hydrology, 1970.

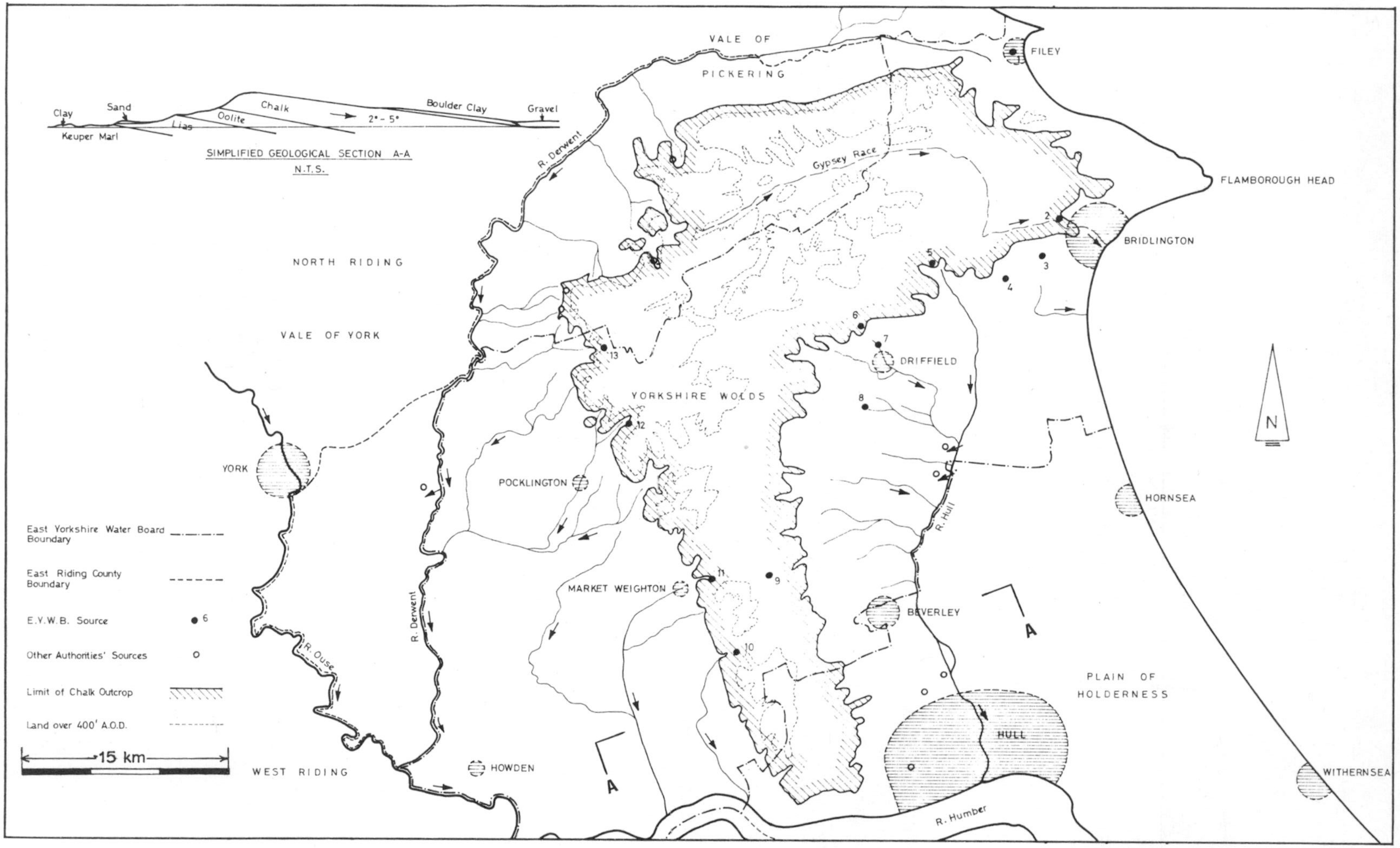

FIG. 84. GROUNDWATER SOURCES - EAST RIDING OF YORKSHIRE

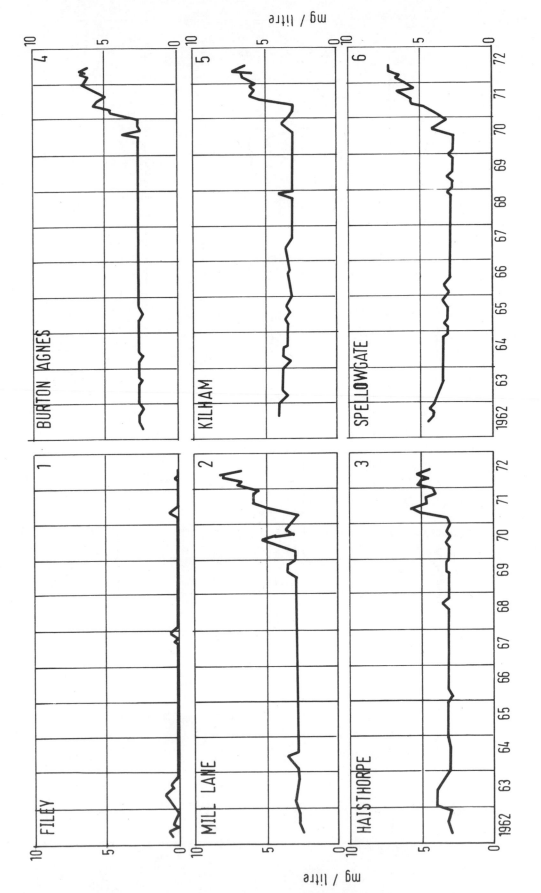

FIG. 85.   GRAPHS OF NITRATES AS N

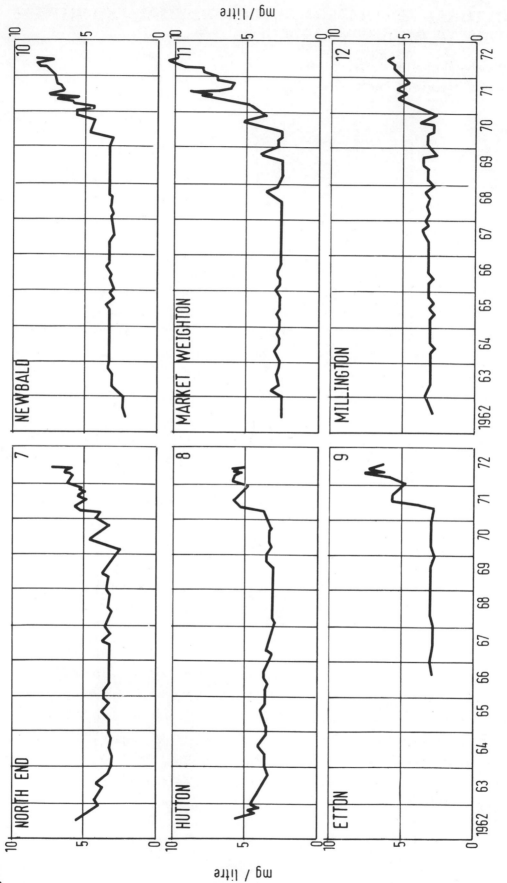

FIG. 86. GRAPHS OF NITRATES AS N

274.

AGRICULTURAL FERTILIZERS, SLURRY DISPOSAL AND NITRATE POLLUTION OF GROUNDWATERS

by J.R. Burford and B. Pain
University of Reading, Berks, England

SUMMARY

Current evidence indicats that the rates of artificial nitrogenous fertilizers applied to crops and pastures on well-managed farms have not had deleterious effects on the nitrate contents of groundwaters. Increases in nitrate contents have occurred, but these are within acceptable limits.

The utilization of slurry as a source of nutrients by spreading onto agricultural land can cause serious pollution of streams due to organic materials contained in run-off waters. The extent to which nitrate derived from slurry contributes to groundwater pollution is not known, but it is possible that much of the nitrate formed could be denitrified under the conditions necessary for leaching.

1. INTRODUCTION

Nitrogen may be added to soil by fixation of atmospheric nitrogen by free-living and symbiotic bacteria, by addition of artificial fertilizers and animal excreta, and by small additions in rain. It may be lost from soil by plant uptake and by reduction of nitrate (or nitrite) to the gases N_2 and N_2O, although it is the loss from the soil by deep leaching of nitrate into groundwaters below the soil profile that has become the primary subject of renewed scientific interest. Deep leaching of nitrate has always been of importance since it results in a loss of a form of nitrogen which is chemically highly available for plant uptake. It is also necessary to know the amounts of nitrate contributed to groundwaters as the concentration of nitrate in drinking water may represent a health hazard.

Although it has been asserted that agricultural fertilizers have been the source of significant nitrate pollution of groundwaters, available data show that there are probably no deleterious effects on well-managed farms, and that significant pollution occurs only when excessive amounts of nitrogenous fertilizer are applied to soil (1)(2)(3). There is, however, a continuing trend towards heavier applications of fertilizers under intensive farming or stocking systems, and therefore an increasing risk of these causing significant pollution. Hence it is important to know the contribution that such increased applications are likely to make to the quality of groundwaters.

2. NITRATE IN THE SOIL SOLUTION

Nitrate present or formed in the surface soil is easily leached due to the high solubility of all common nitrate salts and the lack of significant adsorption by the surfaces of the colloids in most agricultural soils. Nitrate is also the form in which nitrogen is most likely to move down the soil profile.

Nitrate in soils usually represents only a small proportion of the total nitrogen content of the soil, and most of the soil nitrogen is present as organically combined nitrogen. Decomposition of soil organic matter results initially in the production of ammonium ions, which may be assimilated by plants or other microbes, or oxidized to nitrate by autotrophic bacteria.

Nitrate in the soil originates from rain or fertilizer, or by nitrification of ammonium ions. The source of ammonium ions may be ammonium fertilizer, urea or organic matter; the organic matter may be the soil organic matter, freshly added animal excreta and plant debris, such as dead leaves and roots.

Nitrate is removed from the soil solution by uptake by plant roots, assimilation by micro-organisms (i.e. utilization as a nutrient), or dissimilation (i.e. utilization by denitrifying organisms as an alternative hydrogen acceptor to oxygen in situations where there is an oxygen deficiency).

The nett result of nitrate additions to and removals from the soil water determines the amount of nitrate present in the percolating water entering the water table.

3. NITRATE IN GROUNDWATERS

Present evidence indicates that all significant water supplies used for drinking purposes in Britain contain less than 10 mg/litre of NO_3^--N and that they will probably remain at this satisfactory status until the end of this century (4). A level of 10 mg/litre NO_3^--N is acceptable since even the most rigorous limit suggested for the safe consumption of water by infants is 10 mg/litre (and some authorities consider that this limit could be raised to 45 mg/litre). Nitrate pollution of groundwaters in Britain has not yet been generally deleterious to water supplies used for human consumption.

It is difficult to assess accurately the contributions made by agriculture to the levels of nitrate in groundwaters due to the number of transformations involved and variations in weather as well as in soil structural and chemical properties. Analyses of stream waters have indicated that there is not a clear relationship between nitrate concentrations in English rivers and fertilizer usage on adjacent land (5). Cooke (4) has found that "uncontaminated water drawn from reserves under intensively farmed land seems to contain more than 50% more nitrate than similar water did 80-90 years ago" - as indicated by comparisons of recent analyses with those previously reported (about 5 mg/litre) by Warington (6). Similarly, analyses of deep groundwater samples obtained from an intensively fertilized plot and from an adjacent unfertilized plot showed that N fertilizers had caused the NO_3^--N content to increase by 3 mg/litre to about 7 mg/litre.

Localized high concentrations of nitrate in drainage water result from temporary over-loading of the capacity of the soil-plant system to absorb nitrate. These may result from fertilizer application to bare, wet soil (e.g. arable land) when plants are not present or are insufficiently mature for high rates of assimilation of fertilizer N. Grassland is more efficient than arable land at utilizing applied fertilizer N, and there seems to be little direct leaching of fertilizer N from grassland. However, drainage from grassland may contain some transient high nitrate concentrations in spring or autumn due to a 'flush' of mineralization of soil organic matter, dead plant roots etc. Levels of nitrate in drainage waters may rise as high as 50 to 100 mg/litre due to such effects (1).

Increases in the nitrate content of drainage waters in Britain due to fertilizers may therefore result from direct leaching of nitrate or via assimilation by micro-organisms and plants with subsequent release upon mineralization and nitrification. Although concentrations in land drainage waters may rise as high as 100 mg/litre, Tomlinson (5) has found that the nitrate-N concentration in rivers was almost always less than 10 mg/litre, and that the highest level detected was 22 mg/litre (in analyses of eighteen rivers over periods as long as fifteen years). High nitrate concentrations in groundwater are only of temporary localized occurrence due to dilution with other groundwater containing lower levels of nitrate or to losses by denitrification during percolation in saturated strata. Denitrification in sediments on stream bottoms may also be important since lake sediments are known to have significant denitrification capacities (7).

Nevertheless, the occasional reports of high concentrations of nitrate in groundwater, and the reasons for these, are of great importance, and it is also important that the lower levels of nitrate are controlled and their fluctuations understood. Denitrification is known to proceed slowly at temperatures only just above freezing (8), and the long residence time of the nitrate under the water-logged conditions in the groundwater means that even with very slow denitrification rates there should be very little nitrate remaining in the water. The most probable reason for detection of the small amounts of nitrate found in most groundwaters is that the major limitation for denitrification is the lack of available organic substrates. The presence of very high levels of NO_3^--N in a small number of samples often indicates contamination by raw sewage due to poorly designed disposal systems.

4. ANIMAL EXCRETA

The application of artificial fertilizers in amounts as high as 130 to 270 lb N/acre are recommended for optimum production, and it is considered that rates as high as these will not cause significant pollution on well-managed farms (1). Yet, part of good management is even spreading of fertilizers at the correct time such that overloading of the soil's capacity to use fertilizer does not occur. Where animals graze, pastures or crop residues, dung and urine is not uniformly spread but is voided onto the ground surface in discrete patches. Rapid mineralization and nitrification of the N in these materials can give rise to high concentrations of nitrate, and even on well-managed grassland the spatial heterogeneity involved in recycling of nutrients by the animal can increase the possibility of nitrate accumulation in groundwaters. These effects are magnified where stock are congregated e.g. in pathways or gateways. Similarly, accidental spillages from systems where dung and urine are collected (e.g. milking sheds) can cause localized problems.

In recent years, there have been developed 'zero-grazing' systems, in which herbage is removed from grassland and fed to animals in small concentrated areas. This has increased animal production and has removed the deleterious effects of trampling of pastures and poaching of soils when very wet conditions prevail (e.g. in winter), and has also removed the problems associated with localized dung and urine patches in pastures. But it has resulted in the accumulation of dung and urine in the one place i.e. the feeding area, and so has resulted in a concentration of materials with considerable pollutant capacity.

5. THE SLURRY PROBLEM

The increasing size of livestock units and intensification of production is resulting in the accumulation of increasing amounts of animal excreta on farms with a relatively small acreage. This excreta accumulates as slurry, which is potentially a valuable fertilizer and should be returned to the land (4)(9) but excessively high application rates and bad management are known to cause pollution.

Control of the discharge of farm wastes into water-courses or underground strata is provided for in The Rivers (Prevention of Pollution) Acts 1951 and 1961 and The Water Resources Act, 1963. The most important polluting characteristic of wastes is the potential to de-oxygenate water during the biological breakdown of organic matter. More recently, concern has arisen over the pollution of water by chemical nutrients, in particular nitrates, derived from slurries.

6. QUANTITIES OF SLURRY

The quantity of livestock slurries disposed of in such a way that could lead to groundwater pollution is unknown. An idea of the magnitude of the problem can be obtained from the total animal production of undiluted slurry from livestock in the UK (Table 30).

TABLE 30

ANIMAL PRODUCTION OF UNDILUTED SLURRY FROM LIVESTOCK IN THE UNITED KINGDOM (10)

	Numbers of livestock (millions)	Undiluted slurry (million tons)
Cow	3	45
Other cattle	9	50
Pigs	7	10
Poultry	126	6

7. CHEMICAL NUTRIENTS IN SLURRY

Chemical analyses of slurries vary widely with the type of livestock, feed, amount of washing and rainwater added, length of storage etc. Median analyses of samples examined over several years are given in Table 31.

TABLE 31

COMPOSITION OF UNDILUTED SLURRIES (11)

	N $\%$	P_2O_5 $\%$	K_2O $\%$
Cow	0.5	0.2	0.6
Pig	0.4	0.2	0.2
Poultry	1.7	1.4	0.7

8. THE FATE OF NITROGEN IN SLURRY

Nitrogen in slurry is usually contained in organic nitrogen compounds or as ammonium, about 60% of the total being available to plants in the first season (11). Under normal conditions in soil, the organic nitrogen is mineralized to the inorganic ammonium form which may be subsequently converted to nitrate and nitrite nitrogen. Both these forms are subject to leaching and can lead to increases in the concentration of nitrates in the groundwater. Nitrates are readily taken up by growing crops and the risk to water pollution can be reduced by balancing the quantity of nitrogen applied in slurry with crop requirements. Nitrogen consumption varies with the crop, intensive grass production providing for maximum uptake. O'Callaghan et al (12) have estimated the permissible levels of slurry that can be applied to grassland so that nitrate leaching is reduced to a minimum (Table 32). Their calculations were based on an uptake of about 500 lb N/acre/year by a grass sward, and losses of nitrogen by denitrification and fixation in soil were not taken into account.

TABLE 32

PERMISSIBLE LEVELS OF SLURRY APPLICATION TO GRASSLAND (12)

	Available N (lb/thousand gal of slurry)	Permissible level of slurry application (gal/acre/year)
Cattle	42.5 to 82.8	11 800 to 6 100
Pigs	24.0 to 56.0	22 300 to 9 000
Poultry	101 to 159	5 000 to 3 150

O'Callaghan and his co-workers suggest that spreading on areas where no crop is removed should be avoided because of the danger to groundwater pollution by chemical nutrients. Similarly, water pollution is more likely to result from spreading slurry during winter months when rainfall exceeds evapo-transpiration. In practice, many farmers have no alternative to spreading their slurry during this period.

9. PRACTICAL EXPERIENCE

At the time of writing little is known of the contribution of livestock slurries to nitrate concentrations in the groundwater in the UK.

In the USA, the practice of beef production in confined feeding areas is increasing, many units having a capacity of 2000 head, and 10 000 head is not uncommon. Concentration of nitrates in excess of 5 mg/litre were found in groundwater in the proximity of these areas (13). Miller (14) concluded that high nitrate-N levels in groundwater from feedlots were only of localized concern in specific areas. Similar conclusions were reached by Meikle et al (15) in a detailed study of a feedlot in the Platte River Valley. Here, about a foot of organic matter had accumulated on a permeable soil with a fluctuating high water table. Immediately below the feedlot, nitrate-N concentrations reached 39 mg/litre in the top foot of soil but rapidly decreased with increasing depth. It was thought that denitrification occurred beneath the manure pack, but that nitrification was inhibited.

10. FUTURE TRENDS

There is probably very little direct leaching of nitrate-N from inorganic fertilizers applied at rates of up to 130 to 270 lb N/acre in well-managed agricultural enterprises. Yet, mineralization of N in readily decomposable organic matter and plant residues may occasionally result in drainage water containing moderately high levels of nitrate. There seems to be little hope that such mineralization can be completely controlled, and it is not yet of great overall significance since the general concentrations of groundwaters are within acceptable limits.

The fact that the capacity of the soil-plant system to absorb nitrates can be exceeded, indicates that a complacent attitude cannot be adopted, since spillages and 'leaks' can definitely cause excess nitrate accumulation in the soil solution. In addition, there is the danger that economic considerations in some situations may be such that the cost of nutrients is not a great factor in crop production and that rates in excess of requirements are used to maximize crop production. Particular examples are the production of some high value horticultural crops, and the spreading of slurry onto grassland or bare arable land without recognition of its value - or disadvantage - as a source of nutrients.

The difficulty in predicting effects of fertilizers on groundwater nitrate contents in the future is that many of the present conclusions are based on empirical data, and that too little is known about all the transformations involved in the fate of this nitrogen. For example, it is known that, on average, about 30 to 50% of the fertilizer currently applied can be recovered in the above ground portions of the plant (4), but what proportions are assimilated by micro-organisms, leached, or denitrified, and how are the proportions affected by agricultural practices, e.g. increases in rates of fertilizer application?

When excreta from animals housed indoors were allowed to mix with straw to produce farmyard manure, this was a convenient fertilizer due to its straw content. The high C:N ratio of the straw decreased the probability of very rapid release of mineral N from the excreta since there was a very large N assimilation requirement by the microbial population decomposing the straw. The modern development of omitting straw and producing slurry has the disadvantage of the loss of the useful buffering influence of the straw.

The influence of slurry application on agricultural land is not well known. The major immediate concern over the application of slurry in winter at present is the pollution of streams by organic waste transported in surface run-off waters, and recognition of this has been made in the form of legislation restricting waste disposal. While there has been some concern that slurry disposal might cause high nitrate concentrations in drainage waters, some of the assumptions need critical experimental examination. An obvious example is that slurry application in winter need not result in deep leaching of nitrate during this winter period, since soil aeration might be sufficiently restricted (due to high soil moisture contents) such that nitrification is retarded and denitrification stimulated. Initial experiments at present being conducted to examine the fate of slurry-N should provide useful preliminary information about the availability of the N to plants and the losses in drainage and by denitrification (16)(17).

Little is known of the importance of denitrification in British soils and it is relevant to the current problems of slurry utilization as well as being of importance in the N cycle. In particular, if nitrate is leached deep into the soil profile past the surface horizons where maximum removal occurs by assimilation and plant uptake, denitrification is the only significant avenue remaining for the removal of high concentrations of nitrate. Investigations have shown that small but significant losses of nitrate by denitrification occurred in a moderately well-aerated Australian soil (18)(19). Due to the restriction in aeration that must occur in many British soils in winter and spring due to high soil moisture contents, it would seem that only small amounts of readily available substrates would be required in these - and in the water-logged strata below the solum - to stimulate dissimilatory nitrate reduction. To date, however, there is very little work available to show the extent and significance of denitrification in British soils apart from the initial pioneer work of Arnold (20).

Slurry applications to agricultural soil will result in the mineralization of some N to nitrate but the slurry also contains much carbonaceous material, and the leaching of some carbonaceous constituents must also be considered. Leaching downwards of naturally-occurring water-soluble organic materials has been suspected to be the cause of transient very high denitrification capacities in subsoils from some solodized solonetz (21)(22).

Current work has indicated that, for three months after the application of a high rate of slurry to grassland, the oxygen contents of the soil/air were sufficiently low (always < 10%) that there was very little risk of leaching of nitrate (16). However, evaluation is needed over much longer periods since, with time, the return to better aeration in and under the slurry might provide conditions suitable for the production of nitrate but not necessarily for the reduction of nitrate by denitrification. But it is possible that the deep leaching of nitrate formed might not be significant, since the high soil moisture content necessary for leaching might also be suitable for denitrification - as indicated by Elliot and McCalla (23) for the soil under a feedlot in Nebraska.

REFERENCES

1. COOKE, G. W. and WILLIAMS, R. J. B. Water Treatment and Examination, 1970, 19, pp. 253-276.

2. KEENEY, D. R. J. Milk and Food Technol., 1970, 33, pp. 425-432.

3. KOLENBRANDER, J. W. Stikstof, 1972, No. 15, pp. 8-15.

4. COOKE, G. W. Proc. Fertil. Soc., 1971, No. 121.

5. TOMLINSON, T. E. Water Treatment and Examination, 1970, 19, pp. 277-293.

6. WARINGTON, R. J. Chem. Soc. Trans., 1887, 51, pp. 500-552.

7. CHEN, R. L., KEENEY, D. R., GRAETZ, D. A. and HOLDING, A. J. J. Environ. Quality, 1972, 1, pp. 158-162.

8. NOMMIK, H. Acta Agric. scand., 1956, 6, pp. 195-228.

9. RILEY, C. T. Jl Farmers' Club, 1970, pp. 27-37.

10. JONES, K. B. C. and RILEY, C. T. Proc. Farm Waste Symposium, Univ. of Newcastle upon Tyne, 1970.

11. BERRYMAN, C. Proc. Farm Waste Symposium, Univ. of Newcastle upon Tyne, 1970.

12. O'CALLAGHAN, J. R., POLLOCK, K. A. and DODD, V. A. J. Agric. Engng Res., 1971, 16, pp. 280-300.

13. KELLER, W. O. and SMITH, G. E. Geolog. Soc. Am., 164th Meeting, 1967.

14. MILLER, W. D. Water Resources Bull., 1971, 7, p. 5.

15. MEIKLE, L. N., ELLIS, J. R., SWANSON, N. P., LORIMOR, J. C., and McCALLA, T. M. Relationship of Agriculture to Soil and Water Pollution Cornell Univ. Conf. Agric. Waste Management, 1970, pp. 31-40.

16. BURFORD, J. R. Proc. Farm Waste Disposal Conf., Glasgow, 1972.

17. PAIN, B. Farm Waste Disposal Conf., Glasgow, 1972.

18. BURFORD, J. R. and MILLINGTON, R. J. Trans. 9th Int. Congr. Soil Sci., II, 1968, pp. 505-511.

19. BURFORD, J. R. and GREENLAND, D. J. XI Int. Grassld Congr., 1970, pp. 458-461.

20. ARNOLD, P. W. J. Soil Sci., 1954, 5, pp. 116-128.

21. McGARITY, J. W. and   Proc. Soil Sci. Soc. Am., 1968,
    MYERS, R. J. K.   32, pp. 812-817.

22. MYERS, R. J. K. and McGARITY, J. W.   Pl. Soil, 1972, 35, pp. 145-160.

23. ELLIOT, L. F. and McCALLA, T. M.   Proc. Soil Sci. Soc. Am., 1972,
                                        36, pp. 68-70.

284.

DISCUSSION ON NITRATES IN GROUNDWATER

1.   As a background to the whole subject, reference was made to an additional diagram by E. S. Hall (Fig. 87, p. 286) showing that deep infiltration of nitrate into the ground must depend strongly on organic matter present.   J. Burford related this to the chain of oxidations and reductions possible in the soil and subsoil zone.   Then he and B. Pain introduced some new data from their slurry spreading experiments (p. 287).

2.   It was generally acknowledged that there was no time to be lost in pinpointing the causes of major increases in nitrate contents, and remedying these if possible.

3.   Accuracy of nitrate determinations was subject of an inter-laboratory comparison by the East Yorkshire Water Board which is summarized in Table 33, p. 288.

4.   J.K. Marshall remarked that further support for the East Yorkshire nitrate data came from their consistency of trend, most sites starting steep rises in NO_3^- levels at the same date.   In a written contribution, D. M. Milne offered an explanation of how infiltration intensities and spring flow chemistry might be related (see p. 289).

5.   A case history of nitrate pollution in the N. Lindsey chalk was supplied by the local Medical Officer for Health (p. 290) postulating the leaching of fertilizers by spring rainfall as the main source of the high nitrates found.

6.   The nitrogen application rate of 500 lb/acre annually, given as acceptable in Burford and Pain's paper, appeared unduly high by Dutch standards.   (A case history supplied by R. Dorfmeijer of the Provincial Water Board, Gelderland, was cited in this connection: see p. 292 to 294).

7.   Intensive agriculture in north west Germany had been shown to correlate with high nitrates in groundwater (see written contribution by D.A. Gray on p. 295).

8.   There was an inconsistency between average tritium's percolation rates in chalk (based on core samples - see D.B. Smith's paper on p. 377) and the seemingly rapid response of aquifer water quality to increased nitrogenous fertilizer applications in the Yorkshire Wolds area.   (R.J. Crease said that 26 kg/ha of fertilizer, expressed as nitrate, was typical of arable land in the East Riding in 1959.   The annual rate of application had risen almost linearly to 93 kg/ha in 1971).   Some of the inconsistency

vanished if one accepted that some percolation went rapidly down joints and fissures, leaving the rest to follow slowly through the finer pores. To reconcile the abrupt increases in nitrates since 1969 with the nearly linear growth of fertilizer use, since 1954 it would appear that the groundwater nitrates in that area show the same delay as does tritium. Consequently, much worse NO_3^- levels are to be expected in future.

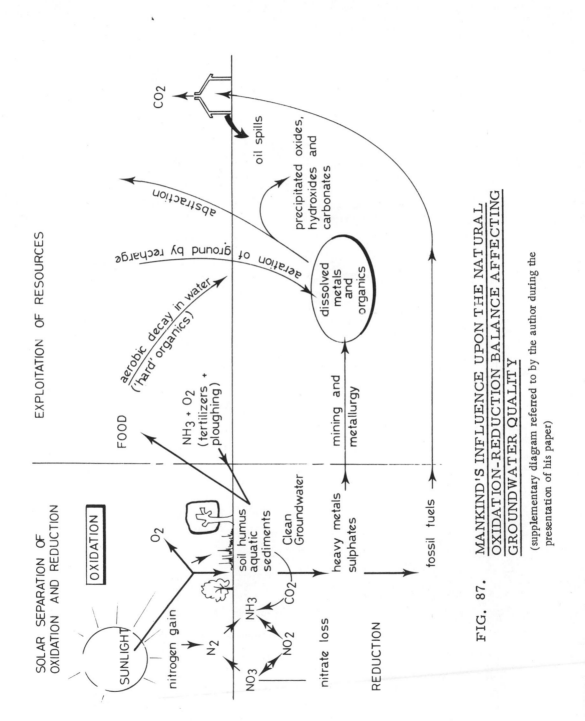

FIG. 87. **MANKIND'S INFLUENCE UPON THE NATURAL OXIDATION-REDUCTION BALANCE AFFECTING GROUNDWATER QUALITY**

(supplementary diagram referred to by the author during the presentation of his paper)

LAND SPREADING OF COW-DUNG SLURRY

by J. R. Burford and B. Pain
University of Reading, Berks, England

In our current experiments on slurry disposal by land-spreading, we are examining the effects of the slurry application on the soil-plant system and the fate of the -N contained in the slurry. Preliminary results show that heavy dressings (220 tons/acre of wet cow-slurry containing 16% oven-dry material applied to grassland in late winter) resulted in restricted aeration of the soil profile for greater than 3 months, as shown by the low oxygen concentration in Fig. 88. During this period, detection of trace amounts of nitrous oxide in the soil atmosphere indicated that denitrification accounted for the small amounts of nitrate formed - no significant accumulation of nitrate-N occurred. It is anticipated that nitrate may accumulate in the surface layers during the summer; formation of such nitrate and its fate (e.g. plant uptake, denitrification, or leaching) will be followed for several seasons to assist in assessments of the long-term effects of slurry disposal on grassland.

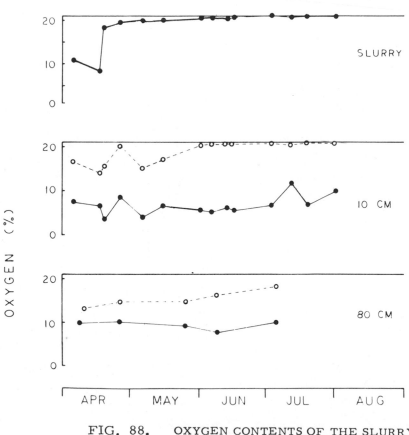

FIG. 88. OXYGEN CONTENTS OF THE SLURRY LAYER (2 CM ABOVE THE SOIL SURFACE) AND OF THE SOIL UNDERNEATH PLOTS NOT TREATED (o ---- o) OR TREATED (o———o) WITH 220 TONS WET COW-SLURRY/ACRE

TABLE 33 (FOSTER AND CREASE)

INTER-LABORATORY DETERMINATION OF ERRORS IN NITRATE ANALYSES ON SPLIT SAMPLES OF CHALK GROUNDWATERS FROM EAST YORKSHIRE

All sampled on 15 February 1972   results in mg/litre NO$_3$-N

Source No.	LAB A METHOD		LAB B METHOD	LAB C METHOD		LAB D METHOD		LAB E* METHOD	MEAN	ABSOLUTE VARIATION	
	DC pda	DC xyl	DC pda	Ion SEl	DC br	UVSp	DC pda	Ion SEl		+	-
9	8.4	8.5	8.5	9.0M	7.8	7.8	8.0	7.7m	8.2	0.8	0.5
8	6.3	7.0M	5.9	6.7	5.7	5.4	5.2m	5.5	6.0	1.0	0.8
5	7.1	7.5	7.4	8.2M	7.0	7.0	7.2	6.9m	7.3	0.9	0.4
11	10.5	10.0	9.1	10.9M	8.6	8.3m	8.8	8.4	9.3	1.6	1.0
10	11.0M	10.5	8.7	9.0	8.9	8.2	8.6	8.1m	9.2	1.8	1.1
3	6.7	7.0M	5.5	6.5	4.9	5.0	4.8m	4.9	5.7	1.3	0.9
Average variation from mean	+0.7	+0.8	-0.1	+0.6	-0.5	-0.7	-0.5	-0.7			

M:maxima   m:minima

METHODS   DC   : Direct colorimetric methods with phenol-2,4 disulphonic acid (pda) or brutine reagent (br) or 2,4-xylenol (xyl)

Ion SEl : Ion-selective electrode (Bunton & Crosby, 1969)

UVSp : Ultra-violet spectroscopy (Hoather & Rackham, 1959)

*mean of five determinations in 5 days with variation of about $^+_-1.0$ mg/litre.

N.B. all samples are organically pure and dilute waters.

TRAVEL TIME FROM SURFACE TO WELL

by D.M. Milne

Sir M. MacDonald and Partners, London, England

Apart from any considerations of time taken for any infiltration carrying a pollutant to reach the saturated 'horizontal' flow zone, the time taken for a given particle to reach a discharging well is dependent on its original distance from that well. If the hydro-geological system at time t_o is assumed to be a previously stored volume together with a vertical recharge rate, then the proportion of a well's discharge formed by vertical recharge originating after t_o should increase gradually as recharge reaches the well after passing along successively longer flow paths, as in the sketch of an admittedly idealised situation:-

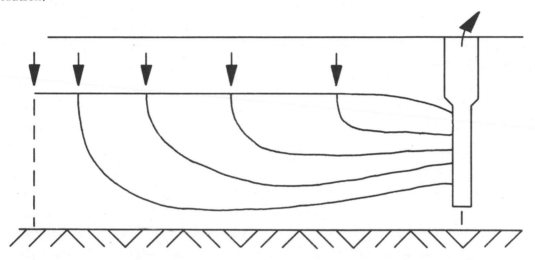

Without raising contentious questions of aquifer parameters and theoretical calculations, surely the simplest single factor affecting the time for a given quantity of dated recharge to reach a well would be the volume of original active storage in the aquifer being affected by flow. One needs an estimate of how these active storage volumes and the total abstractions are related.

If the supposition that a sudden increase in nitrate content of well water was indicative of pollution originating at a given time suddenly reaching a well in large quantities, then the inference must also be that the active volume stored in the aquifer prior to the arrival of the pollutant at its surface, must have been small in relation to the abstractions.

The peakedness of the nitrate concentration/time curves produced in the paper certainly suggests that the period of retention in the saturated aquifer is short indeed.

NITRATE POLLUTION IN NORTH LINDSEY CHALK BOREHOLES

by J.S. Robertson
Lindsey County Council, Scunthorpe, England

The discovery in May 1970 that nitrate levels had risen markedly in chalk sources of the North Lindsey Water Board followed a special analysis which I requested after seeing a cyanosed but otherwise healthy baby. A second blue baby was seen three days later by another doctor. The nitrate concentration revealed by analysis was only 10.5 mg/litre as N, but the mothers of the babies had been using water from kettles in which an unknown increase in salt concentration had been produced by boiling.

Subsequent daily testing showed a steady rise in nitrate level until mid summer, after which the level fell slightly and then stabilized. An investigation by Davey showed that the amount of NO_3 in water vastly exceeded the amount which could have been derived from sewage or animal faeces, that it could not have come from infiltration from the Humber, and that the only possible sources capable of yielding this amount of nitrate in the catchment area were artificial fertilizers or nitrate forming bacteria in soil, and in particular the nitrate fixing varieties which live in association with the pea crop.

Peas are extensively grown in the area for the frozen food industry, and one of the boreholes affected by nitrate is subject to seasonal bacterial pollution due to the bacteria which grow on pea haulm after harvesting. If the major source of the nitrates were the bacteria associated with peas, one would expect the concentration to rise at the same time as the bacterial count, and to be higher at this borehole than at those less subject to pollution from pea growing. In fact, however, the nitrate level is higher at a different borehole, and is unrelated to the pea harvest in time. The most marked rise occurred in the spring and early summer of 1970 when following a very wet period farmers had used unusually high doses of fertilizer to make good leaching losses. Less fertilizer was applied in 1970-71 and in 1971 nitrate concentration remained stable, but in 1972 a slight further rise occurred in the early summer, falling once more at the time when pea harvesting might have led to a rise. It seems highly probable that the amount of fertilizer applied in the spring and the amount of rainfall immediately afterwards may be the major determinants of our nitrate pollution.

One interesting observation is that private bores pumped at lower rates show lower nitrate concentrations than do large public sources nearby.

It seems possible that this is due to stratification, and that the nitrate level is only high in superficial layers of chalk.   Rapid pumping producing a big cone of depression may extract a higher proportion of such superficial nitrate rich water while at low rates of pumping only "old" water from the bottom of the borehole is extracted.

Intensive use of ammonium nitrate has only been practised for a little over a decade.   The annual percolation of water into the chalk is not likely to exceed half the annual rainfall i.e. 11 inches per annum.   The depth of water in the aquifer is almost 250 feet, and if the storage coefficient is 25% it follows that the water in the aquifer represents at least 65 years rainfall.   Much of the water at the bottom of the aquifer must have fallen as rain at least 65, and more probably over 100 years ago, since our observations from pollution tracing show a considerable proportion of the water abstracted to be of very recent origin.   Salt and bacteria reach the boreholes within 3 to 7 days from a site two miles from the borehole.

It follows from these considerations that the quality of the water may be expected to deteriorate slowly as the nitrates borne by rainfall since intensification of agriculture replaces more and more of the "old" water in the aquifer, unless stabilization results from anaerobic bacteria deep in the chalk or there is a change in agricultural practice.

AGRICULTURAL POLLUTION IN THE 'HIERDENSE BEEK' AREA, GELDERLAND, THE NETHERLANDS

by R. Dorfmeijer
Provincial Water Board, Arnhem, The Netherlands

Situation (see also Fig. 89, p. 294)

The "Hierdense Beek" is a brook in the North of Gelderland, which discharges into the lake "Veluwe randmeer", which is very important for recreation purposes. The total catchment area is about 10 000 ha. In the southern part of the area there is a great concentration of calf-fattening farms. On a total of about 250 farms 36 000 calves are being fattened. The soil consists of a mainly sandy upper layer, with some clay lenses at a depth of about 15 metres.

Nature of the problem

Because most of the calf-feeding farms have very little land, they are faced with a liquid manure surplus problem.

Because of the high water content of calf liquid manure (about 98.5%) the costs for transport to other farms are very high. This fact brings about danger for ground and surface water pollution by overmanuring, dumping and direct discharge into surface waters.

The quality of the groundwater is important in the first place, because there are many private wells, which in the past provided water of good quality. Moreover the quality of the water in the Hierdense Beek depends strongly on quality of the groundwater. In this respect especially, those compounds are important which bring about eutrophication of the lake "Veluwe randmeer" such as nitrogen and phosphorus.

Volume of the manure surplus

From the total of 250 calf-farms, 65 have less than 1 ha of land. All farms together have 760 ha of grassland and 180 ha arable land (about 2200 acres in total). According to agricultural standards about 50% of the manure can be used on the calf-farms themselves. As in each calfbox about 3 m^3 of liquid manure are produced a year, this means a surplus of more than 50 000 m^3 a year in this area, corresponding to nitrogen application of about 200 lb/acre.

Consequences for the quality of ground and surface water

Because direct discharge of manure into surface water is legally prohibited and a legal basis for the prevention of dumping and over-manuring is lacking at the moment, the problem concentrates on soil and groundwater pollution.

In 1970, water samples from 56 private wells were taken and analyzed. Also the quality of the surface water at point K 14 was determined four times a year. The results of these investigations can be summarized as follows:

a) Groundwater analyses (56 samples)

	Well depth (m)	NO_3^- content (mg/litre)	NH_4^+ content (mg/litre)	PO_4^{3-} content (mg/litre)
minimum	4	0.1	0.1	0.02
maximum	50	360	70	30
average	12.5	88.5	3.5	0.83

Frequency distribution of NO_3^- contents (mg/litre)

Range	0-30	30-50	50-100	>100
Percentage of samples	23	13	30	34

b) Surface water analyses

Date	NO_3^- content	NH_4 content	PO_4^{3-} content	$B.O.D._5^{20}$
21-9-'70	13.5	1.7	1.1	1
14-12-'70	15.5	2.0	1.0	2
4-3-'71	16.0	2.0	1.2	3
21-6-'71	14.0	6.4	4.6	9

These figures clearly indicate that a severe groundwater pollution is going on in this area and that also the surface water quality is strongly affected by this process. Because of the pollution, most of the private wells have been condemned and replaced by connections to the public water supply, which is not yet affected.

Measures to be taken

1) A legal basis must be provided to prevent overmanuring and dumping.

2) Measures must be taken to achieve a better spreading of the liquid manure over the agricultural area in the neighbourhood.

3) For those cases in which spreading is too expensive, central biological treatment plants must be built with transport of the liquid manure by tankers. The purification process must include tertiary treatment, i.e. nitrogen and phosphorus removal.

Points for discussion

1) Which degree of manure dosing is acceptable from an environmental point of view?

2) Which compounds are important to make a judgement about this problem; organic matter, nitrogen, potassium, phosphorus?

3) Which processes play a role, and how do they work?

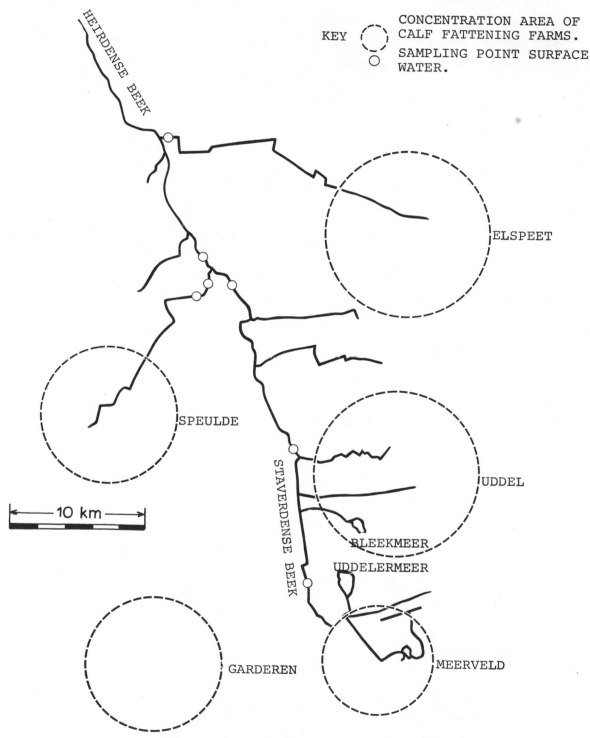

KEY ◯ CONCENTRATION AREA OF CALF FATTENING FARMS.
◯ SAMPLING POINT SURFACE WATER.

FIG. 89. SOUTHERN PART OF HIERDENSE BEEK CATCHMENT AREA, GELDERLAND

NITRATES IN A SAND AQUIFER IN GERMANY

by D.A. Gray

Institute of Geological Sciences, London, England

A recent paper by Groba and Hahn* described their study of groundwater chemistry, including nitrates, over a considerable area in a sand aquifer. The area concerned was characterized by extensive areas of forest, meadows and intensive agriculture. Under the forests and meadows, nitrate contents in groundwater were generally less than 2.5 mg/litre NO_3 to compare with 24-28 mg/litre under small areas of agriculture surrounded by meadows and forests. Groundwater under the intensively cultivated ground, however, is characterized by nitrate values of under 100 mg/litre NO_3. The authors conclude that "If, on the part of agriculture, no measures are taken to stop this development, a strict separation of groundwater and farming land will be necessary in the future".

In the United Kingdom, such division of land use would not be practicable and a serious position would arise if, for example, the extensive Chalk catchment areas were to become characterized by comparable values.

---

*GROBA, F. and HAHN, J.   Variations of groundwater chemistry by anthropogenic factors in North West Germany. Internat. Geol. Cong. 1972, Section 11 - Hydrogeology, 270-281.

MIGRATION OF LIQUID INDUSTRIAL WASTE FROM A GRAVEL PIT

by G.D. Goldthorp and D.V. Hopkin
Great Ouse River Authority, Cambridge, England

Exler's paper (p. 215) is particularly interesting because it describes a very similar investigation to the one recently carried out in Norfolk to study the movement of large quantities of liquid industrial waste deposited in a disused gravel pit. This pit is one of a vast series of old gravel workings dug years ago to provide the sand and gravel for war-time airfields and various buildings. These disused dry pits look, at first sight, attractive to anyone wishing to dispose of contaminated liquids, which for obvious reasons, cannot be discharged direct from factories into the rivers.

Unfortunately these former gravel workings in Norfolk are generally underlain by the most important Chalk aquifer which provides the majority of water undertakings, industrial users, irrigators and domestic users with their water supply.

At a particular site shown in plan in Fig. 90, p. 298, there was known to be over sixty feet of unsaturated Pleistocene glacial deposits separating the liquid waste disposal lagoons from the Chalk water table. The infiltration rate of the glacial deposits was obviously extremely high as there was no difficulty in the waste disposal lagoons absorbing liquids at rates which exceed the average annual natural recharge by over 30 times.

A hydrogeological and chemical investigation was carried out by the Great Ouse River Authority (GORA) because it was considered that the continual depositing of liquid industrial waste constituted a potential threat to the Chalk water supplies. Eight six-inch diameter boreholes were sunk to about one hundred feet deep to sample both soils and liquids at various depths around the tip. In addition, a further twelve three-inch diameter holes were drilled with the River Authority's power auger, down to the first impermeable layer. Permeabilities of the various clays encountered were obtained from laboratory tests of undisturbed core samples, supplemented by in situ tests. Where possible water level recorders were fitted.

From the bores and auger holes, samples of the soil were taken at three-foot intervals and samples of water taken wherever possible. These were analyzed chemically in the GORA laboratories. Amongst other things a large amount of hydrochloric and hydrobromic acids had been tipped into the pit, together with phenolic and odoriferous substances. These materials were therefore chosen as tracers to establish the degree of movement of the waste from the pit.

The soil samples were first examined by smell and initially only those with a strong odour were examined chemically. An aqueous extract of the samples was made and, as stated, examined for chlorides, bromides and phenols. Chlorides being conservative substances and easy to measure, gave us a quick means of establishing the position of the maximum pollution load. Further samples either side of these were then examined to locate the peaks. Bromides, also conservative substances, are not readily found naturally in underground water and these therefore were more specific to the material from the pit. Phenols were also more specific to the material from the pit, but these gave us a measure of the degree of travel of a degradable substance before its effect became insignificant.

By these investigations we were able to show that at the time of the examination the material from the pit had travelled vertically to a depth of 6 to 9 m, and horizontally as a fairly narrow lobe in an easterly direction for a distance of 160 m (see Fig. 90, p. 298).

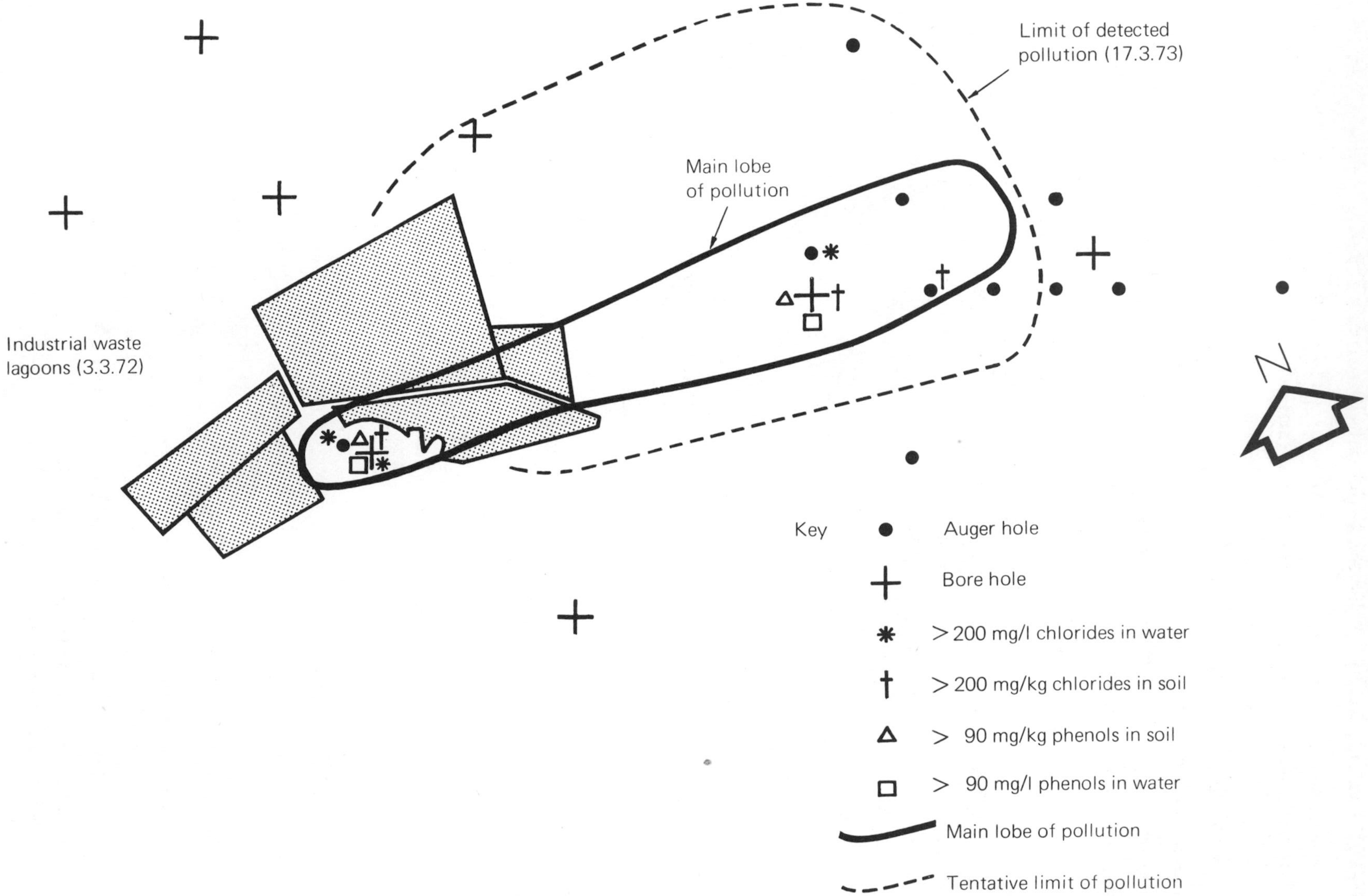

Key

•	Auger hole	
+	Bore hole	
✳	> 200 mg/l chlorides in water	
†	> 200 mg/kg chlorides in soil	
△	> 90 mg/kg phenols in soil	
□	> 90 mg/l phenols in water	
▬	Main lobe of pollution	
- - -	Tentative limit of pollution	

Limit of detected pollution (17.3.73)

Main lobe of pollution

Industrial waste lagoons (3.3.72)

FIG. 90.   INDUSTRIAL WASTE DISPOSAL IN A GRAVEL PIT

PHENOLIC POLLUTION IN CHALK AT BECKTON, ESSEX
by R. Aspinwall *
Essex River Authority, Chelmsford, Essex, England

1. <u>BECKTON CHALK GROUNDWATER POLLUTION</u>

   1.1. <u>Background information</u>

   In 1954, phenolic pollution was discovered in industrial chalk boreholes in Lower Roding Valley. The contaminant was identified as carburretted water gas tar, or in one analysis, 85% gas oil, 15% coal tar oil. Specific gravity slightly > 1. No means of estimating total volume of pollutant. Pollutions persisted to present day and spread to two other industrial borehole abstractions.

   Figs. 91 and 92 show basic geology, groundwater contours and flow lines (1967) in relation to abstraction points, affected boreholes and suspected points of entry of contaminant.

   1.2. <u>Argument for source of contaminant</u>

   Contaminant product of gas manufacture, therefore point of entry to ground likely to be in vicinity of two gas works G_1 and G_2.

   Reasons G_1 not prime suspect:

   (i)   No evidence that water gas was product of works.

   (ii) Underlying geology known to be Drift deposits underlain by 56 ft of relatively impermeable London Clay and 71 ft of inter-bedded semi-permeable and permeable Lower London Tertiaries. Access to Chalk of large quantities of contaminant therefore most likely through boreholes.

   (iii) Bores on this site were not affected until 1963 whilst other bores up the hydraulic gradient were affected before this.

   Reasons G_2 prime suspect:

   (i)   Major manufacturers of water gas.

   (ii) No record available of means of disposal; likely to be on waste marshland at site. Subsequent excavations have encountered extensive deposits water gas tar in alluvium and made-ground near site.

   * now Consulting Hydrogeologist, Oswestry, Shropshire.

299.

(iii)   Some evidence of bomb damage to areas of works.

(iv)   Base of low permeability London Clay outcrops to north of probable disposal area which is underlain by variable alluvium over Lower London Tertiaries, making entry to Chalk more probable than at G_1.

(v)   Piezometric head in Chalk reduces northwards up Roding Valley, primarily determined by heavy pumping from public water supply sources, therefore flow path of any contaminant entering Chalk at G_2 well defined.   Chlorion concentrations of water from bores at B and C indicating saline infiltration give additional evidence for northerly flow.

(vi)   Bore A at site of G_2 affected in 1915, bores at B affected 1954, bore at C slightly affected 1959, bore at D affected 1963, therefore path of contaminant consistent with groundwater flow through fissured aquifer.

Contaminant in bores at B and C found low in aquifer, consistent with specific gravity, but tar content raises question of how flow took place through granular deposits without clogging occurring and suggests that entry to Chalk might be via boreholes or excavations at G_2.

Note that flow velocities are small making flow times from disposal to pollution large.

Pollution at A to bores at B = 29 years = average velocity of 170 ft/year
Bores at B to bores at D   = 9 years = average velocity of 190 ft/year

1.3.   Measures to combat pollution

(i)   Site A

Some disused boreholes (12 on this site) located and sealed.
Producing well cleaned.

(ii)   Site B

Borehole linings checked to water level, found to be sound.
Continuous pumping to clear contaminant, successful on short-term basis only.

(iii)   Site D

Existing pump raised by 100 ft, subsequently lowered 50 ft because sucking air.
Scavenger bore sunk, only 4000 gallons contaminant removed.

(iv)   General

Consideration given to creating a pumping trough at right angles to flow. Very expensive and decided that because of fissured aquifer might be unsuccessful.
Scavenger pumps at depth would remove oil but disposal problems acute.

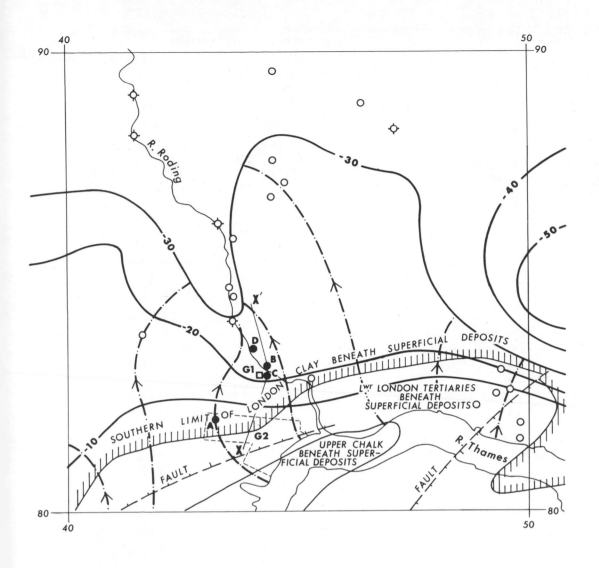

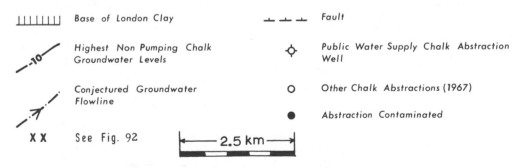

⊥⊥⊥⊥⊥⊥	Base of London Clay	
—·—·—	Highest Non Pumping Chalk Groundwater Levels	
—·—·— →	Conjectured Groundwater Flowline	
X X	See Fig. 92	
—·—·—	Fault	
⟡	Public Water Supply Chalk Abstraction Well	
○	Other Chalk Abstractions (1967)	
●	Abstraction Contaminated	

⊢—— 2.5 km ——⊣

FIG. 91.   BECKTON CHALK GROUNDWATER POLLUTION

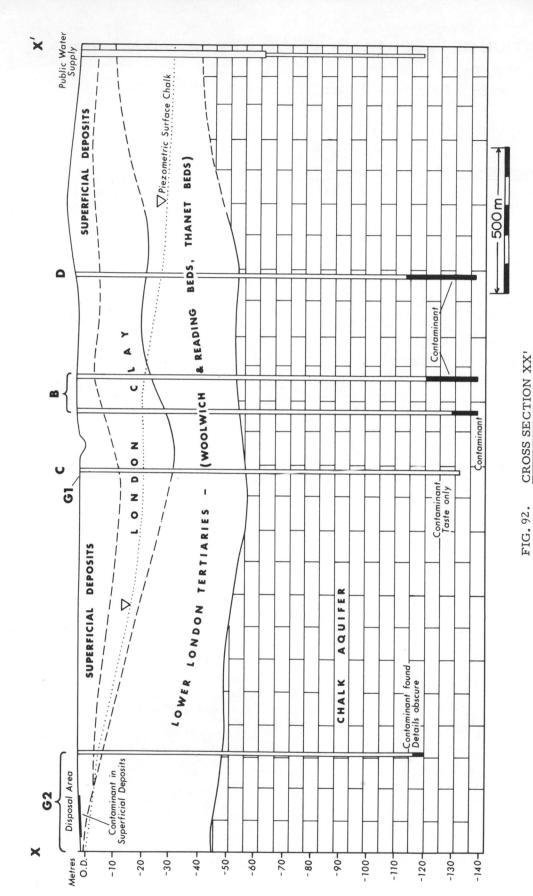

FIG. 92. CROSS SECTION XX'

POLLUTION OF FLOOD PLAIN GRAVELS BY GAS WORKS WASTE

by Mrs. H.P. Toft
Lee Conservancy Catchment Board, Waltham Cross, Herts, England

1. INTRODUCTION

Gas works wastes had been deposited in ponds formed in an old pit excavated in the Flood Plain Gravels in the lower Lee Valley adjacent to the Pymmes Brook, from 1905 until 1967 (Fig.93). Wastes included tar acids, sludge from sulphate settlement pits and oil from gas holders. These liquids have moved through the gravels causing groundwater pollution over a wide area.

The gravels overlie London Clay which varies considerably in thickness being 2.3 at well number 70c, and 21 m at well number 613 (Fig.94). The London Clay cover is absent in an area 3.5 km to the south.

TABLE 34

ANALYSES OF POLLUTED GROUNDWATER FROM VARIOUS EXCAVATIONS

Date	Site	Analysis	
22.8.1963	A	Hydrocarbon oil sp. gr. 0.887 phenol present	
23.8.1963	Effluent pond	Hydrocarbon oil sp. gr. 0.867 phenol present pH = 2	
10.8.1965	C	Heavy tarry material phenol - 16 mg/litre	
22.6.1966	Well 666 - used to dewater gravel workings		mg/litre
		Suspended solids	382
		Chloride as Cl	182
		Ammoniacal nitrogen as N	29
		Nitrate nitrogen as N	0.13
		Albuminoid nitrogen as N	6.5
		Permanganate Value	43.0
		Biochemical Oxygen Demand in 5 days at 20°C	16
		pH	7.6
		Odour	gas liquor
		Phenol	0.6

2. POLLUTION HISTORY

Pollution of the groundwater in this area has been known since 1935 when oil emerged at the surface of playing fields between the Pymmes Brook and the River Lee Navigation.

This oil caused a fire on the fields in 1943. An analysis indicated gas works liquor as the source.

Polluted water from excavations made in 1958 when the Pymmes Brook was channelized contained high sulphates.

Oily water was encountered during culverting of watercourses immediately west of the ponds in 1961, and at site E in foundation works.

Pymmes Brook was polluted in 1965 by oily liquids which emerged from the ground in the vicinity of the effluent ponds when the water table in the gravels rose after heavy rains. In the same year, discharges of polluted groundwater from the gravel workings caused pollution of the brook and the Lee Navigation. (See Table 34 analysis).

Oil from the effluent ponds moved through the gravels and emerged at weep holes in the channel wall and from expansion joints in the Pymmes Brook in 1966.

3. SITE INVESTIGATION

Boreholes sunk through made ground and gravels, to depths of 6 m to the London Clay surface, on the site of an old backfilled effluent pond showed the ground to be polluted down to the clay.

4. REMEDIAL MEASURES

Remedial measures were confined to prevention of pollution of watercourses. Liquid wastes were removed, ponds infilled, weep holes plugged at bed level and upper weep holes connected to a manifold collecting pipe draining to a sump.

No attempt has been made to remove the oil by pumping.

Construction of a bentonite clay mixture cut off wall contiguous with the Pymmes Brook was considered.

The gravel aquifer is not exploited in this area, most abstractions being from the Chalk aquifer.   There is no evidence of pollution of the Chalk.

The Flood Plain Gravels consist of sands and gravels with interbedded silts and clays. Over parts of the area they have been excavated and refuse used to backfill.   The random nature of these deposits make it impossible to predict the groundwater flow.

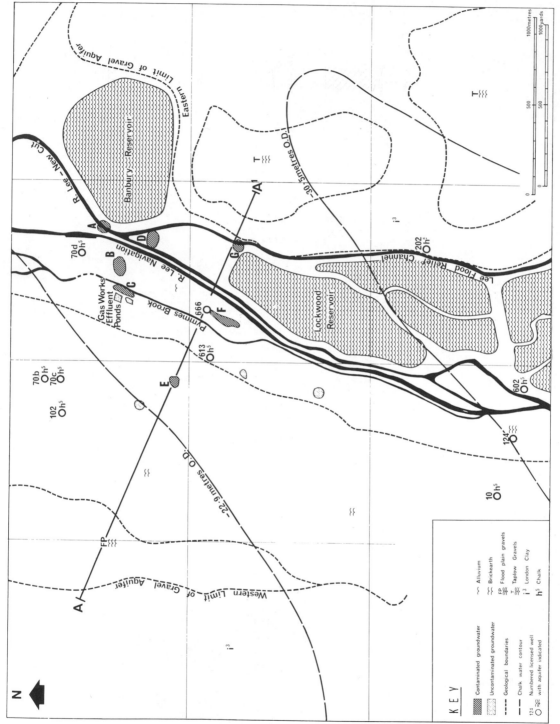

FIG. 93. MAP SHOWING AREAS OF CONTAMINATION BY GAS WORKS WASTE – TOTTENHAM AREA

K E Y

Contaminated groundwater	
Uncontaminated groundwater	
Geological boundaries	
Chalk water contour	
Numbered licensed well with aquifer indicated	

Alluvium
Brickearth
fp Flood plain gravels
T Taplow Gravels
i^3 London Clay
h^5 Chalk

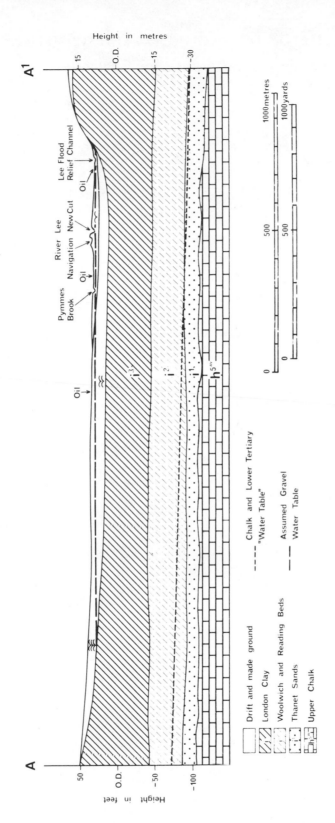

FIG. 94. CROSS SECTION INDICATING HYDROGEOLOGY – TOTTENHAM AREA

ENDURING POLLUTION OF GROUNDWATER BY NITROPHENOLS
The Essex Water Company, Romford, Essex, England

1.   NATURE OF POLLUTANT

  Picric acid, nitrophenols, in wastes from explosives manufacture 1914 to 1918. Wastes called liquids by contemporary consultant.   Thresh, Beale and Suckling "The Examination of Waters and Water Supplies, 7th Edition, p. 75".

2.   MAP OF AREA

  Water table: 5 ft OD.   Seasonal fluctuations.   Numerous local old chalk pits worked to about this level.   See Fig. 95, p. 309, for:

  i.   Grays Well (floor level 17 ft OD depth 36 ft) for various reasons not used for public supply since 1932.

  ii.   Stifford Well (floor level 17 ft OD depth 140 ft) present abstraction 6 m.g./week for public supply.   Treatment: iron removal and chlorination.

  iii.   Little Thurrock - location of explosives factory.

3.   POLLUTION

  Grays well: yellow tint to water first noted early 1920's.   Samples 1939 to 1955 all tinted, Hazen values up to about 35.   1955 sample matched against picric acid indicated 0.7 mg/litre.   Pumping machinery removed 1955, sampling discontinued.

  Private deep borehole at West Thurrock: About 1 mile west of Grays.   1942 sample had marked yellow tint.

  Stifford Well: no abnormal colour ever detected.

  Laboratory tests: heavy chlorination, coagulation, ineffectual for removing colour. Activated carbon treatment only partially successful.

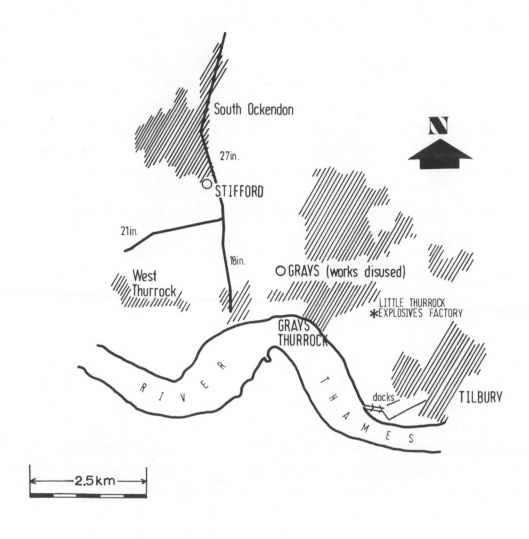

FIG. 95.   LOCATION OF GRAYS WELL, STIFFORD WELL
AND LITTLE THURROCK EXPLOSIVES FACTORY

PROTECTION OF AQUIFERS AGAINST PETROLEUM POLLUTION IN SWITZERLAND

by Th. Dracos
Federal Technical University, Zurich, Switzerland

SUMMARY

The protective measures against contamination of aquifers are described by means of two examples. In the shunting yard near Dietikon-Kilwangen, lying within the area of an extended groundwater stream, a compacted delaying layer under the ballast has been set up, which will allow the dredging of the contaminated soil materials before the oil can reach the groundwater. At the Refinery in Cressier a system of wells is continuously in function. These form a depression of the groundwater table towards the interior of the Refinery area and prevent any contaminated groundwater from flowing out of this area.

A further protective measure is described that has often been successfully applied to already existing objects which do not come up to the protective regulations.

1. INTRODUCTION

In Switzerland the most usable aquifers can be mostly found in the alluvia at the bottoms of the valleys. They are fed by infiltration of rain water and snowmelt by underground inflows coming from the valley slopes and by infiltration from surface waters. They are generally aquifers with a free water surface. They are neither very extended nor very deep and the free water surface of these aquifers often lies only a few meters below the ground surface. In spite of their small extent the groundwater deposits are for the moment the most important sources of drinking water supply. The topographical conditions which have led to the formation of these aquifers also cause the concentration of agglomerations of industries and of means of communication in the areas, where groundwater deposits exist. Particularly in Switzerland these deposits are endangered by infiltration of harmful substances, especially of petroleum products. This danger is constantly growing because the increasing motorization and the popularity of oil heating have led to a larger use of petroleum products.

Seeing this, the Federal Office for the Protection of Environment, acting for the Department of the Interior, issued in 1967 severe technical regulations in order to protect the waters against pollution caused by fuel and by other stored liquids, which could endanger the water. These regulations prescribed special protective measures for constructions, which could be a danger to groundwater and which should be built in zones, where there is usable groundwater or from which usable aquifers are fed. The same state organization also asked the Swiss Federal Institute in Zurich to analyze the propagation of petroleum products in aquifers.

The results of these analysis have been published (1)(2)(3). They correspond to the results of the analysis made by Schwille (4)(5) and confirm to a great extent the theoretical paper of van Dam (6). They show that the migration of the oil in bulk occurs mainly in the capillary zone and that the migration stops after a residual saturation has been reached. We can be happy with this result, because it shows that in case of an oilspill the pollution remains within certain limits. It must be kept in mind, however, that depending on the product this pollution lingers on for several years, if not for decades and that it can affect the quality of the water through its soluble elements. It is therefore easy to understand why special protective measures must be taken in the zones, where precious aquifers are in danger. Some examples of these measures are discussed below.

2. SHUNTING YARD LIMMATTAL

The Swiss Federal Railways (SFR) are now building a large shunting yard near Dietikon–Kilwangen. This station covers an area of about 850 000 m^2. It is just on this area that a large groundwater stream can be found. It flows on the left side of the Limmat Valley (Fig. 96). This stream is fed by affluents coming from the left side of the valley and also by infiltrating meteoric water and is an important source of water supply for the adjacent municipalities.

These groundwater deposits are now in great danger of being polluted, because in this station tanker wagons containing fuel and other chemical products are shunted. For this reason the competent authorities have imposed severe conditions of the SFR regarding the protection and preservation of groundwater in the shunting yard area. These conditions can be summed up as follows

a) Under no circumstances should the groundwater level be reached by fuel.

b) The precipitation falling on the station area should be able to infiltrate just the same, thus charging the groundwater stream.

The groundwater level lies at least 5 m below the surface. Besides the special problems arising from the construction of an impermeable layer over such a great area, the second condition imposed is one of the reasons, why such a layer cannot be built. Further studies led to a combination of active and passive protective measures with which one hopes to fulfil the conditions imposed.

The active protective measures depend on having suitable technical equipment ready to be used and on the existence of an organization which can put this equipment quickly and effectively into action. In this way the area of the accident should be cleared quickly and, if possible the spilled oil products should be recovered before they infiltrate into the soil; the tracks should be removed from the polluted spot in the shortest time possible.

According to the railway authorities these operations should take 2 to 3 hours.

In order to prevent the infiltration of the spilled liquid down to the groundwater level, the polluted soil should be removed layer by layer as quickly as possible. Since the above mentioned preparations need a few hours, it is desirable to slow down the infiltration. For this reason it is necessary to build a delaying layer (i.e. a scarcely permeable layer) under the railway ballast in order to delay the infiltration of the liquid. To fulfil the condition of the precipitation infiltration, the permeability of this layer must not be less than 2.10^{-5} cm/sec. With regard to the practical execution of this layer, we assumed an average permeability of 4.10^{-5} cm/sec with a vertical variation between 2.10^{-5} cm/sec and 10^{-4} cm/sec measured over the whole thickness of the layer.

The following parameters were decisive in ascertaining the thickness of this layer.

a)   The maximum quantity of the pollutant infiltrating per m^2.

b)   The relation between depth of infiltration, saturation and time.

c)   The delay time between spillage and moment, when the dredging begins.

This third point has already been answered. In order to answer the remaining questions, an experiment of large-scale has been made. The disposition of this experiment is represented in Fig. 97. On a fill made up of the material which would be used to prepare the delaying layer, three infiltration experiments have been made with oil and one with petrol. Those with oil were conducted in order to relate the saturation distribution of the oil products in the ground as a function of time. The results of the measurements made are represented schematically in Fig. 98. Based on the results obtained by Youngs (7) and by Childs (8) a simplification was made assuming that saturation remained constant throughout the contaminated depth. With the aid of these measurements it was possible to relate the saturation redistribution as a function of time. The corresponding function is represented in a non-dimensional form in Fig. 99

and allows for an estimate of the influence of a change in conductivity k or of the viscosity μ of the liquid in this relation.

The experiments with petrol were made in order to estimate the infiltrated quantity per square meter. In this case petrol is decisive because of its low viscosity which leads to a maximum of infiltration. The most important thing was to find out up to which point the ballast prevents the spreading, causing a concentrated infiltration. The result of this experiment was, that the horizontal spreading of petrol is not essentially hindered by the ballast. This determination, together with the experience made form various accidents, brought us to the assumption that an infiltration of 50 litre/m^2 petrol corresponds to the highest quantity possible. This all the more after seeing that the infiltration experiments made with oil converted to the infiltration of petrol gave the following results: 50 litres/m^2 petrol take about 1 hour to infiltrate in a layer which has a conductivity of 4.10^{-5} cm/sec. Based on this assumption and on the results of the infiltration experiments made with oil, it was possible to calculate the dredging capacity necessary to remove the contaminated material before the oil reached a certain depth. The result of this calculation is represented in Fig. 100. The curves reproduced show the various depths of the delaying layer as a function of the dredging capacity in m^3 per hour and m^2 with the maximum infiltration depth as parameters. In order to determine the real dredging capacity this value must be multiplied by the extent of the polluted zone in m^2.

Aided by this experimental data, the engineering firm in charge of the project fixed the thickness of the compacted delaying layer at 1.50 m and in collaboration with the Federal Railways determined the technical equipment to be kept ready in order to satisfy the condition mentioned above.

In the meantime this layer was put in. The resulting reached conductivity is slightly less than the required average and its dispersion lies within the imposed limits. One of the most important tasks of the railway administration will be to train its personnel in the correct usage of the available equipment is case of an accident, if the above described protective measures are to be successful.

3. REFINERY IN CRESSIER

Special protective measures for groundwater had to be taken for the refinery in Cressier which lies between the lake of Neuchatel and the lake of Bienne. Its area is bounded in the south east by the Zihl-canal that connects the two lakes and crossed by the old course of the Zihl, a section of which is now earth-filled (Fig. 101). Between the Zihl-canal and the old Zihl the soil consists of fine partially permeable lake sediments.

North of the old Zihl, the lake sediments slowly change to the delta sediments of a small stream. The permeability of these sediments increases with the distance from the old Zihl. On the north western and northern margin of the refinery zone, extensions of gravel-sand deposits can be found, which are usable aquifers. One of these aquifers is used by a food industry which requires a good water quality. Compared with the neighbouring ground the fill in the old Zihl is very permeable. With regard to the adjacent surface and sub-surface water, which are important reserves for the water supply of the whole region, the federal and cantonal authorities demanded that besides normal protective measures, additional steps be taken to prevent the outflow from the refinery area of oil or of water contaminated with oil.

The used water and the polluted surface water are collected and led through a purification plant. Where there is danger of infiltration, leak-proof collecting basins are built to protect the groundwater i.e. a concrete basis under the production plant, leak-proof collecting basins around the standing tank deposits which lie on permeable areas, concrete basins below the valve station, etc.

In spite of these measures, a groundwater pollution in the refinery area cannot be ruled out entirely. In order to prevent the subterranean outflow of the possibly contaminated groundwater from the refinery area van Dam (6) worked out the following proposal which has been accepted by the appropriate authorities of the Confederation and of the Cantons.

Van Dam's proposal is based on the results of a thorough geophysical investigation of the refinery area and of its surroundings and completed by observations with piezometers made on a net of test-borings. These examinations have clearly shown that most of the refinery area was drained by the old Zihl before the refinery was built. To enable the construction of the refinery, the canal was earth-filled on the refinery area. Because of the relatively great permeability of this fill, it is supposed that the draining effect of the old Zihl has not changed after having been filled. This effect had to be strengthened by taking adequate measures. A row of seven wells was built in the fill of the old Zihl, four of which are always in operation. In this manner it is possible to keep the groundwater table in the filled Zihl-bed within prescribed limits. These limits have been chosen in such a way that at all stages of groundwater level most of the groundwater contained below the refinery area will certainly drain into the old Zihl. In this way the contaminated water will not leave the refinery area in an uncontrolled fashion. In case of a larger oil infiltration, the artificially increased gradient of the groundwater table in the direction of the Zihl fill hinders an oil migration towards the outside of the refinery. To keep the pumping rate down to capacity on a tolerable level, great quantities of water from the old Zihl must be prevented from infiltrating down to the lowered groundwater of the filled segment of the Zihl. For this reason two impermeable screens were built on each side of this fill. They reach the less permeable layers of the surrounding ground both laterally and in depth.

The satisfying effect of the lowered groundwater level obtained by pumping and the effectiveness of the diaphragms can be clearly seen on the groundwater map in Fig. 102 and 103. These maps represent the evaluation of control measurements carried out on a dense net of observation tubes every 15 days. It can be seen that some marginal zones of the refinery area remain outside the influence of the lowered groundwater level caused by the four wells. In case of an accident, the groundwater table should also be lowered in the inside of the refinery in these zones. For this reason more wells were built and equipped with pumps which operate in case of an emergency. The effectiveness of this additional row of wells can be seen on the representation of the groundwater contour line in Fig. 104. A last protective measure had to be taken along the north-eastern boundary of the refinery area. The watershed between the drainage of the refinery and the depression of the well belonging to the food industry lies along this boundary. In order to prevent the outflow of oil towards this well an impermeable screen was built which goes down to about 2 m below the groundwater table. These measures combined with a severe constant groundwater control have worked very well up to now and hopefully will do the same in case of an emergency.

4. FURTHER PROTECTIVE MEASURES

The examples described are unfortunately the rare cases in which it was possible extensively to examine and elaborate protective measures in advance i.e., during the planning phase. It is much more difficult to advise protective measures for installations already existing that do not come up to the prescribed regulations. A protective measure often successfully applied in such areas is sketched in Fig. 105. An impermeable screen generally made out of concrete with bentonite added is constructed in the ground all around the area which endangers the surroundings. If a low-permeability layer exists at a small depth, this screen will be fixed in it. (Fig. 105a). The result is a nearly permeable underground basin which hinders the outflow of meteoric water and oil that infiltrate in the area. In this case an artificial drainage with subsequent control of the water level must be foreseen. If the permeable layer is deep and if the water table does not lie very far below the surface of the ground, the screen can be driven down to 3 m below the lowest measured ground water level. The screen forms a barrier against the spreading of oil but lets the groundwater pass. A special drainage of the protected object is not necessary in the latter instance. In both bases, wells must be constructed inside the area surrounded by the screens which can be put in operation in case of an emergency. In this way most of the oil that infiltrated into the ground can be fetched out again. These wells also serve to observe the groundwater.

In spite of all these measures it is not possible to prevent completely the infiltration of petroleum products caused by accidents. In most cases it is very difficult to restore the area after such accidents occurred. Generally one tries to remove the contaminated soil as quickly as possible. Unfortunately this measure cannot always be applied, especially in cases where the infiltration lies near buildings. In such cases one tries to recover at least part of the spilled oil products by means of recovering wells and to keep its spreading within certain limits. Such measures are expensive and last too long. In Switzerland the situation is particularly difficult, because generally the soil is not homogeneous, which often makes it very hard to form a reasonable estimation of how the oil spreads after an accidental infiltration.

REFERENCES

1. DRACOS, Th. — The behaviour and movement of immiscible fluids in the subsoil. S.V.G.W. Monthly Bulletin 48, no. 10, 1968.

2. DRACOS, Th. — Survey and results of model trials of protective and restorative measures for oil spillages. Plan, no. 3, 1968.

3. DRACOS, Th. — Experimental investigations on the migration of oil products in unconfined aquifers. De Ingenieur, 1969, 81, No. 51/52.

4. SCHWILLE, F. — Petroleum contamination of the subsoil. A hydrological problem. DGM 10, 1966, (6).

5. SCHWILLE, F. — Petroleum contamination of the subsoil. A hydrological problem. The joint problems of the oil and water industries. Elsevier, 1967.

6. VAN DAM, J. — The migration of hydrocarbons in a water bearing stratum. The joint problems of the oil and water industries, Elsevier, 1967.

7. NYDEGGER, H. — Protective measures for groundwater in the construction of the railway shunting yard at Zurich- Limmattal. Wasser-Boden-Luft, Special publication, no. 16, May 1969.

8. YOUNGS, F.G. — Redistribution of moisture in porous materials after infiltration. Soil Sci., 1958, 86.

9. CHILDS, E.C. — Soil Water Phenomena. Wiley, 1969, pp 255-261.

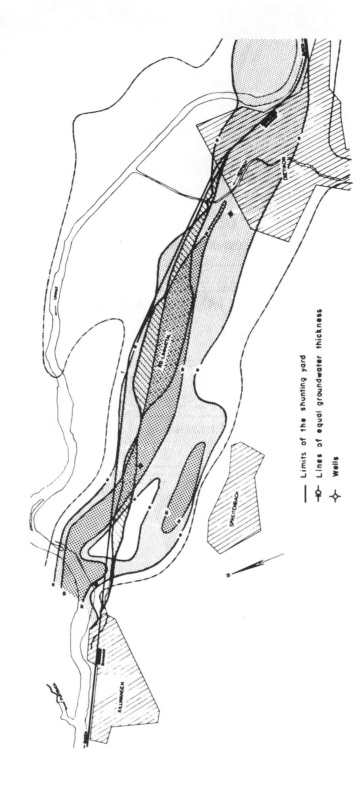

Limits of the shunting yard
Lines of equal groundwater thickness
Wells

FIG. 96   PLAN OF THE SHUNTING YARD AT LIMMATTAL

318.

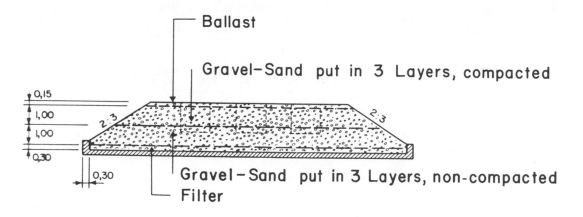

Ballast

Gravel–Sand put in 3 Layers, compacted

0,15
1,00
1,00
0,30

2:3   2:3

0,30

Gravel–Sand put in 3 Layers, non-compacted
Filter

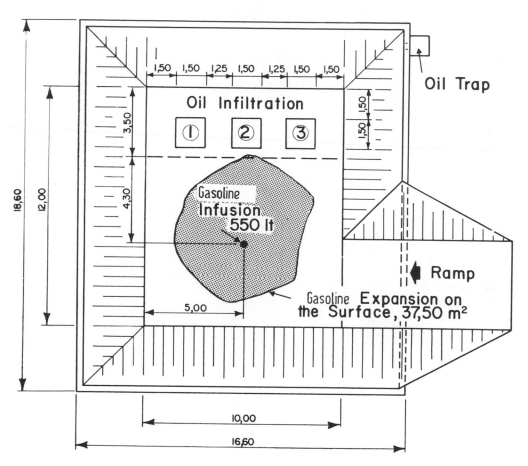

1,50 1,50 1,25 1,50 1,25 1,50 1,50

Oil Infiltration

① ② ③

Oil Trap

3,50

1,50
1,50

4,30

18,60

12,00

Gasoline
Infusion
550 lt

Gasoline Expansion on
the Surface, 37,50 m²

Ramp

5,00

10,00

16,60

FIG. 97.   DISPOSITION OF THE LARGE SCALE EXPERIMENT

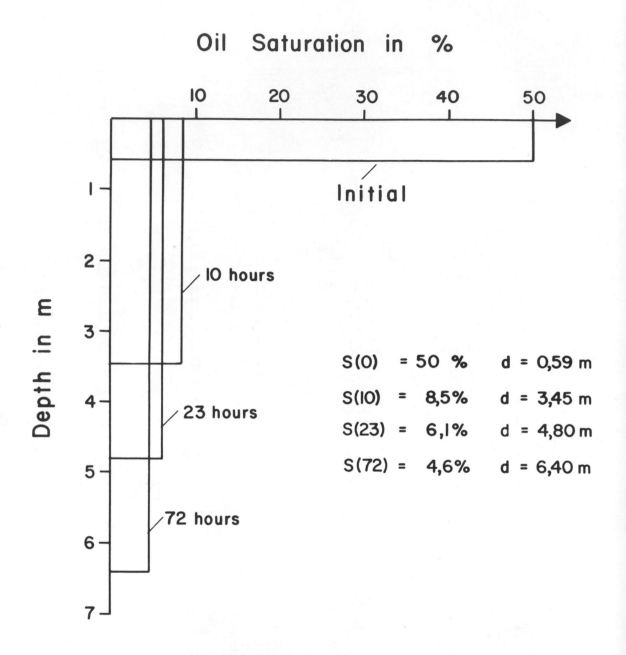

FIG. 98.   OIL SATURATION DISTRIBUTION AS A FUNCTION OF TIME

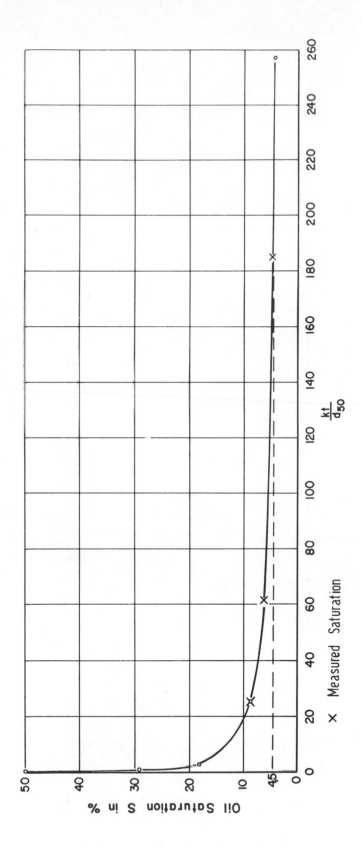

FIG. 99. RESULTS OF THE LARGE SCALE EXPERIMENT

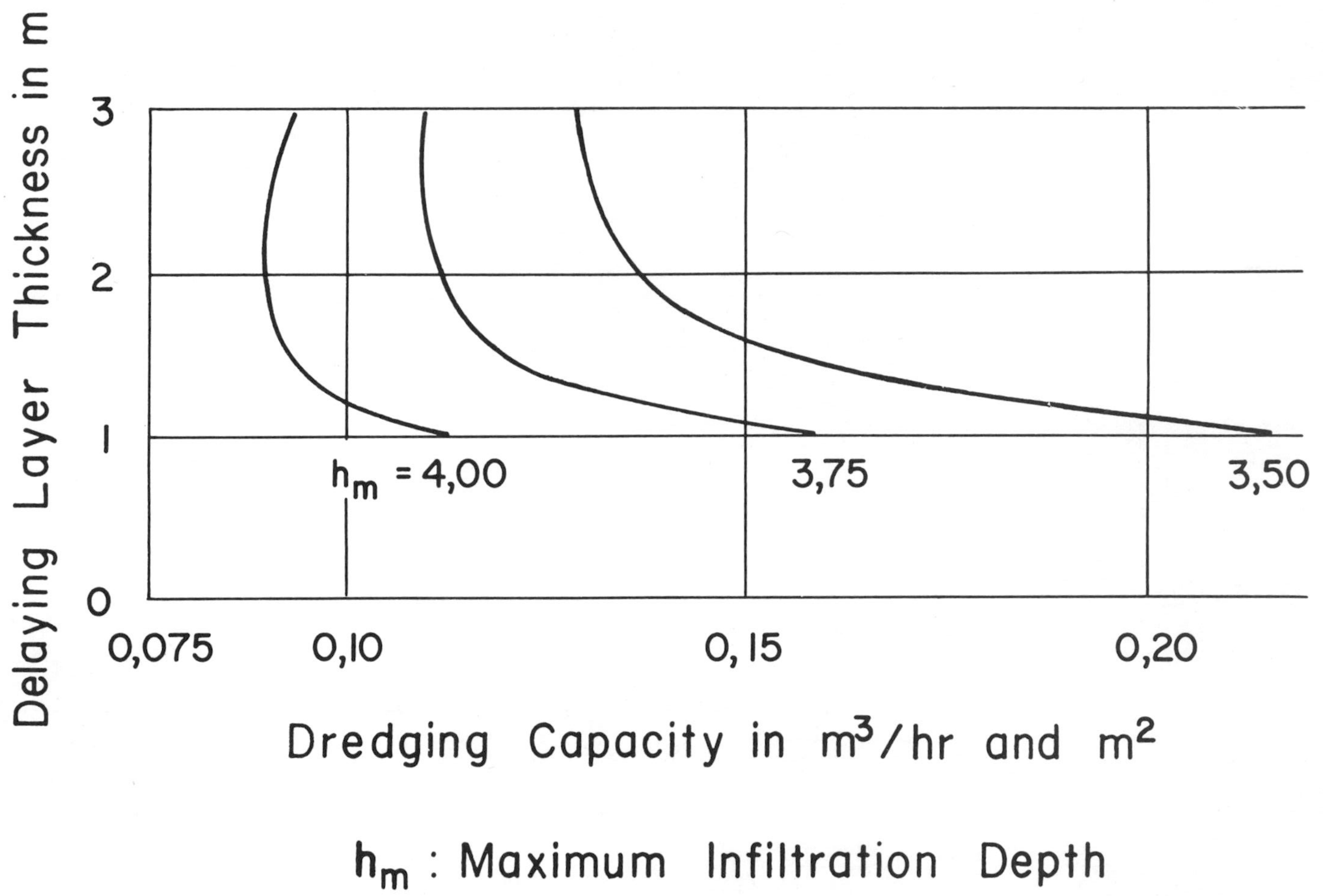
Delaying Layer Thickness in m
3
2
1
0
0,075
0,10
0,15
0,20
h m = 4,00
3,75
3,50
Dredging Capacity in m3/hr and m2
hm : Maximum Infiltration Depth

FIG. 100. BASE FOR THE SIZING OF THE DELAYING LAYER

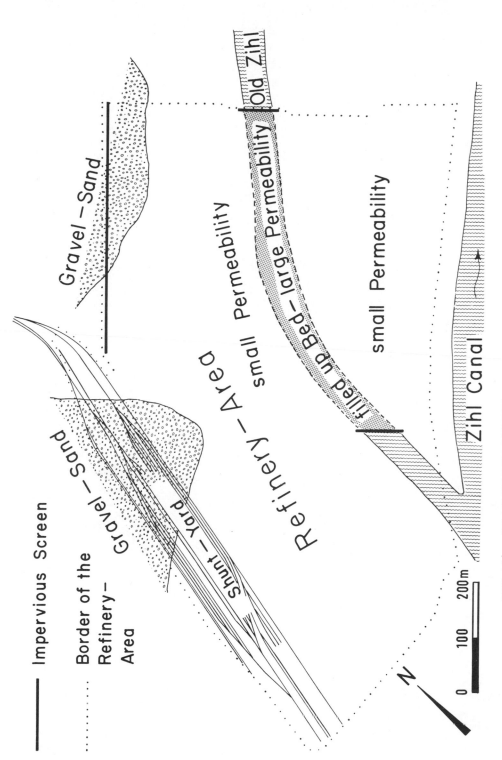

Impervious Screen

Border of the Refinery-Area

Gravel—Sand

Gravel—Sand

Refinery—Area small Permeability

filled up Bed—large Permeability

small Permeability

Old Zihl

Zihl Canal

Shunt—yard

N

0   100   200m

FIG. 101.   PLAN OF THE REFINERY AT CRESSIER

323.

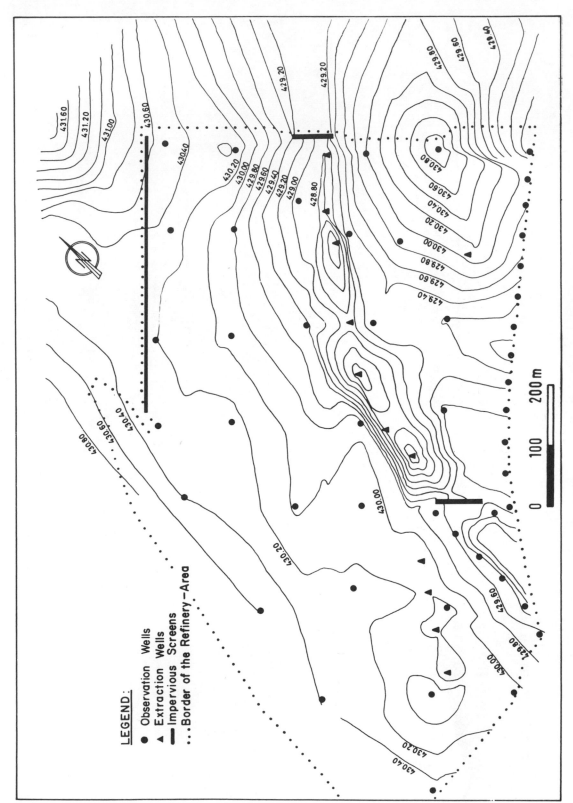

LEGEND:

• Observation Wells
▲ Extraction Wells
▬ Impervious Screens
··· Border of the Refinery-Area

200 m

100

0

FIG. 102. GROUNDWATER MAP, LOW GROUNDWATER LEVEL

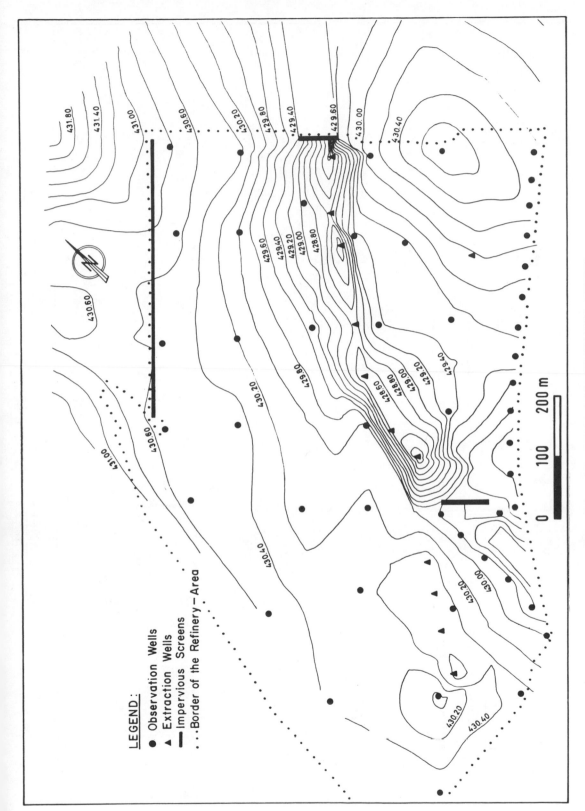

LEGEND:
● Observation Wells
▲ Extraction Wells
— Impervious Screens
···· Border of the Refinery-Area

200 m

100

0

FIG. 103. GROUNDWATER MAP, HIGH GROUNDWATER LEVEL

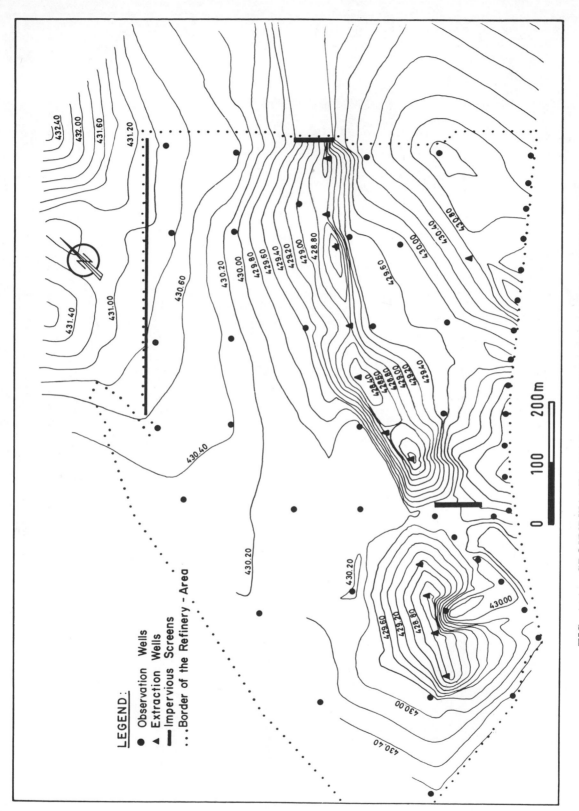

LEGEND:
● Observation Wells
▲ Extraction Wells
▬ Impervious Screens
⋯ Border of the Refinery - Area

0   100   200m

FIG. 104. GROUNDWATER MAP, HIGH GROUNDWATER LEVEL WITH AUXILIARY WELLS OPERATING

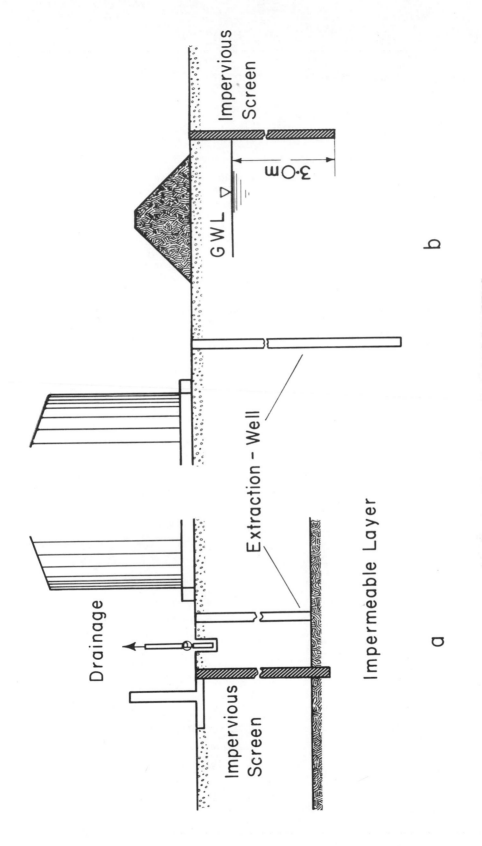

FIG. 105. PROTECTIVE MEASURES USING IMPERVIOUS SCREENS

a) SCREEN FIXED IN AN IMPERMEABLE LAYER

b) SCREEN IN DEEP PERMEABLE LAYER

SAFEGUARDING UPPSALA FROM OIL POLLUTION

BY J. Sidenvall

Department of Public Works and University Institute of Geology, Uppsala, Sweden

1.   OIL POLLUTION, (CAUSES)

After World War II, wood and coal were used less and less for heating purposes in Sweden. Instead oil and electricity came. Soon oil tanks were common and in Sweden they were generally stored in basements or in the ground. Nobody thought of the risks of oil leakage.

During the sixties the different authorities in Sweden realized the seriousness of the situation. There were several hundreds of thousands of tanks in the ground. The authorities began to issue new especially designed laws. The latest one came during May 1972: all of them have one main reason: to stop oil pollution.

As the urban areas of Uppsala lie on or near the esker, there are many risks of oil pollution of groundwater. Therefore the screens of the groundwater wells have been put deep under the groundwater table. During the sixties, oil accidents happened often. The most common reason for spilt oil is the overflow of oil tanks. Burst tanks are also common. These accidents usually lead to the spilt oil flowing into the River Fyris or to the sewage works. In Uppsala there is a dual sewage system. Such oil accidents are fairly easy to handle and the oil can be quickly controlled.

The accidents caused by corrosion either of the underground tanks or of their pipes are much worse. Several such accidents occur every year. These tanks lie both on the esker and on the clay plains alongside. We have therefore had long experience of these accidents and now know how to combat them.

2.   OIL POLLUTION OF ESKERS

When oil percolates through homogeneous layers of sand or gravel the oil body will sink in a 'pear' like form until it reaches the capillary zone. Here the body will spread out on top of the water filled capillaries. Oil will now follow the groundwater flow. Unfortunately such an ideal stratification is rare. Generally in the esker there are layers of coarse materials surrounded by layers of fine materials and so on. There are often thin layers of silt or clay too. These different layers make it almost impossible to forecast how oil will act after an accident. The thin layers of fine materials lead oil in unpredictable directions.

When such accidents have to be handled, the first step is to try to stop the leakage. The next step is to ascertain where all different water pipes, cables, etc. lie in the ground. Oil frequently flows towards pipes and cables where it collects easily, but there is the risk that oil will follow the pipes a long way, which is generally the case when oil is free in clay areas.

When the amount of oil spilt is known an estimate can be made of how far it has penetrated the ground. It is fairly well known how much oil will be permanently attached to sand, silt, etc. which is important when the method to be used to get rid of the oil has to be chosen.

Generally, most of the oil-polluted soil, i.e. that soil that is saturated with oil, is dug away. Then the polluted soil around houses, pipes, cables etc. is dug away. Any soil that has less than two or three percent oil is left. That percentage is generally a threshold percentage and means that oil is permanently fixed in the soil matrix.

If large quantities of oil have been spilt and have already reached the groundwater table the best method is, in combination with the first one, to dig a ditch or push a tube well down to the water table. From there so much water is pumped that the groundwater flow towards the ditch or the well is altered. A small pump is put on the water table and this will suck up oil and water, which is conveyed to a tank where the oil settles out(Fig. 106). This method means the directions of all groundwater flows in the area must be known.

The largest danger with these accidents is that there are constituents in oil which will be dissolved by water, but when the oil has been identified the risks for nearby wells can be estimated. It is important to know this when deciding how much oil and, if necessary, how much oil polluted water, must be pumped away. There is also the question of how to handle all that polluted water but the risks must not be overestimated since the screens of the well galleries lie very deep under the groundwater table.

3. OIL POLLUTION ON CLAYS

This is one of the more difficult types of contamination to handle. Luckily the worst of them all - oil in rock crevices - is seldom encountered.

In clays leakage of oil often happens. There are several reasons for that. First many tanks were laid right down in the clay. There are different sorts of clays. This can cause galvanic action, which promotes corrosion of tanks and its pipes as in the case of one tank that was isolated according to the laws, but after one year "there were just holes left". Nowadays there must be a 0.5 m thick layer of sand around the tank.

Another reason is that tanks are laid down in excavated pits in the clay and the pit is filled with sand. After a few years the tank has several holes around at a certain level. The reason for this is that the pit becomes filled with water too, because wet clay is impervious. The holes show exactly where the local groundwater table has been and indicate the presence of galvanic action just around groundwater tables.

To combat oil accidents in clay areas is a difficult task. It is known that clay layers that always have been wet are practically confined. Dry clays on the other hand have many cracks. These will spread oil faster than in sand.

When oil has to be cleared from clays the first step is to stop the leakage and then all waterpipes, cables etc. must be located. After that an estimate is made of how much really must be done. If it has happened near the esker it is necessary to dig. Otherwise it is generally enough to remove the top soil layers, i.e. vegetation layers, but if there are cables and such things close by it will be necessary to dig around them.

When excavating clays for this purpose care must be taken. The capillary forces may be broken if too much is dug at once with the result that more soil may be polluted by oil quite unnecessarily. Therefore we dig slowly down to the top of the capillary zone, if needed. Often oil will now flow out in this pit and can be collected. If we are forced to enlarge the pit it may at first appear that there is no more oil. The cracks can be completely dry but are suddenly full of oil again, enough to destroy a water well.

4. OIL-POLLUTED SOIL

All these oil accidents mean that there will be a great deal of oil-polluted soil to dispose of. This has been a nuisance but the problem is solved in Uppsala where all oil-polluted soil is transported to a new refuse disposal plant.

This plant consists of a pulverization unit, different dumping areas and a specially constructed purifying plant for polluted water. This sewage works has both biological and chemical treatment of polluted water from the dumping areas.

At the dumping area, the oil-polluted soil will be spread out on top of the refuse. The soil depth must be less than 0.2 m.

Due to bacteria in the refuse, high temperature in the rows of refuse (around + 70 to + 80°C) and aeration, the oil molecules will be broken down in less than six months. After that time it is impossible to notice any traces of oil.

If there is loose oil in the oil-polluted soil it will be absorbed by the underlying refuse.

If barrels of waste oil were to be dumped at the dumping area (which is forbidden), oil would reach the ditches for polluted water and the compensation reservoir. If that ever occurs a special oil separator will take care of it there. If it is necessary to pump up liquid oil it is transported to a special receiving station for refuse oil. From this station, oil is transported at intervals to special factories.

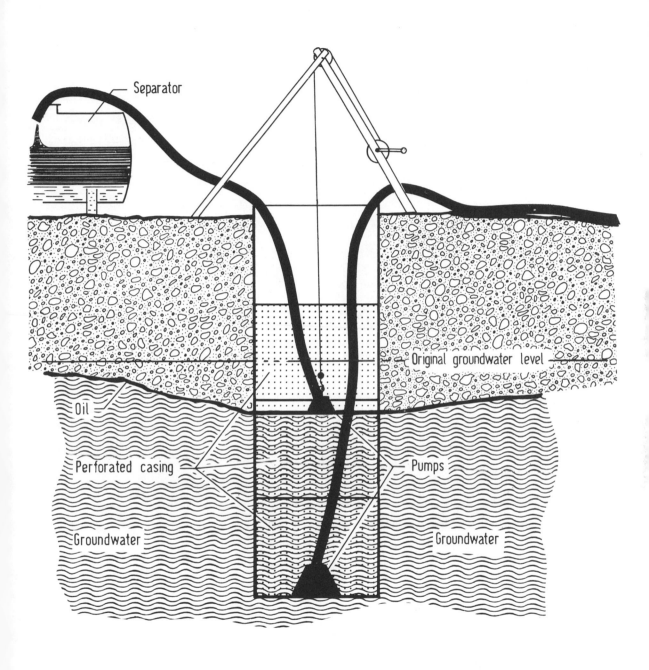

FIG. 106.   REMOVAL OF OIL POLLUTION BY USE OF A BOREHOLE

331.

STUDIES FOR THE PREVENTION OF OIL POLLUTION NEAR BRATISLAVA, CZECHOSLOVAKIA

by A. Hunter Blair
The Water Research Association, Medmenham, Bucks, England

1. BACKGROUND

The majority of the water supply for Bratislava is obtained by induced infiltration of the River Danube. However, in 1960, due to increasing pollution of the Danube by petroleum wastes and sugar refinery wastes, it became necessary to reduce the number of infiltration wells and seek a second major water supply source.

This could be conveniently situated on the island formed south of Bratislava between the Danube and Little Danube rivers. However the situation was complicated by the increasing size of an oil refinery positioned on the divergence of the Danube and Little Danube (Fig. 107).

A description of measures investigated to prevent groundwater pollution by oil leaks and spills from the refinery is given. Consideration is also given to prevention of groundwater pollution caused by induced recharge of the oil polluted River Danube.

2. MEASURES TAKEN TO COMBAT POLLUTION

Several methods for the prevention of oil pollution were studied separately and conjunctively on analogue and digital models:

(a) Two rows of abstraction boreholes were situated in the middle of the oil refinery. Cones of depression were formed under the site by the abstraction. The oil therefore flowed under gravity along the surface of the groundwater cone of depression into the top of the abstraction boreholes. Intermittent pumping abstracted a mixture of oil and water from the top of the boreholes. This water was subsequently passed through separators and recycled within the plant (Fig. 108 a and b).

(b) To prevent pollution occurring by natural groundwater flow to the south-east, a row of twenty boreholes was to be situated to the south-east of the plant. The cone of depression which they produced caused the oil to flow into the boreholes where it was abstracted from the surface of each borehole as in (a) above. The water was abstracted at 580 litre/sec and following separation was used for cooling purposes; after further treatment it was returned to the Danube (Fig. 109).

(c)   The above scheme was then tested at 750 litre/sec abstraction rate. Due to inconsistencies in the bedrock, a uniform cone of depression was not obtained. The boreholes were therefore pumped at different rates   about 15 litre/sec at one end and 40 litre/sec at the other.   A uniform drawdown of about 3.5 m was thus obtained.

(d)   The stratum beneath the oil refinery has on the east side an impermeable bedrock some 13 to 22 m below ground surface.   On the west side is sand and silt which is in hydrological connection with the water table.   A scheme was therefore investigated in which an impermeable wall or membrane was constructed to the bedrock on the east side, and in which a borehole system abstracting 670 litre/sec was situated in the centre of the west side (Fig. 110).

(e)   A variation of (a) above was tested in which the borehole was moved to the south-west corner of the refinery and pumped at 670 litre/sec. (Fig. 110).

(f)   To prevent contamination by induced recharge from the River Danube, a line of 15 abstraction wells was placed near the Danube and between the Danube and the main water supply source, then pumped at 200-225 litre/sec.

The final solution to prevention of pollution from the oil refinery and from oil pollution in the Danube will be obtained from a study of all the above data in various combinations.   The final scheme may be one in which the abstraction boreholes within the refinery, the line of abstraction boreholes to the south of the refinery and the line of abstraction boreholes between the main water supply source and the Danube are all used.

3.   ACKNOWLEDGEMENTS

Grateful acknowledgement is made to Dr. A. Sikora, Director of Vyskumny Ustav Vodneho Hospodharstva, Bratislava, for his permission to present these data. Thanks are also due to Ing. M. Bartolcic and Ing. I. Brachtl for providing the above information.

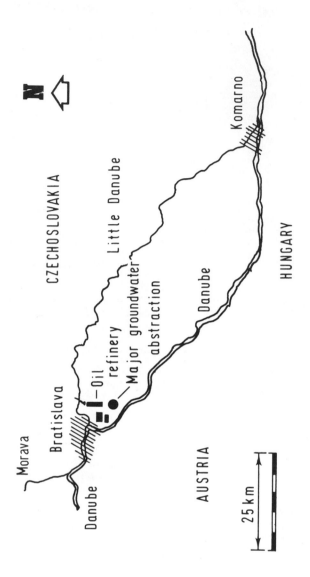

FIG. 107.   SITE PLAN

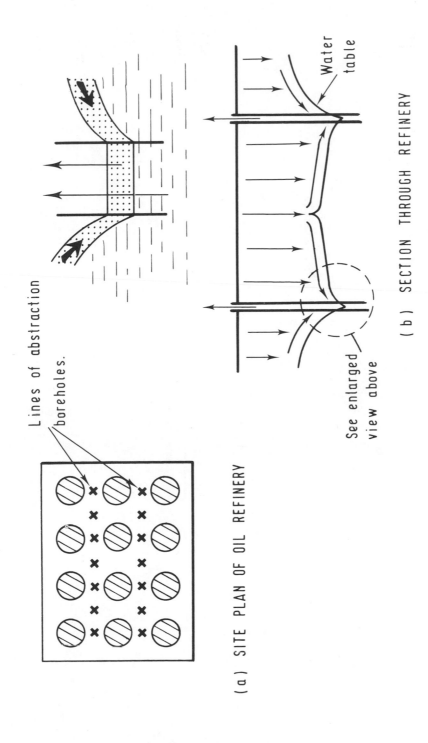

Lines of abstraction boreholes.

(a) SITE PLAN OF OIL REFINERY

Water table

See enlarged view above

(b) SECTION THROUGH REFINERY

FIG. 108.   SOLUTION (A) WITHIN OIL REFINERY

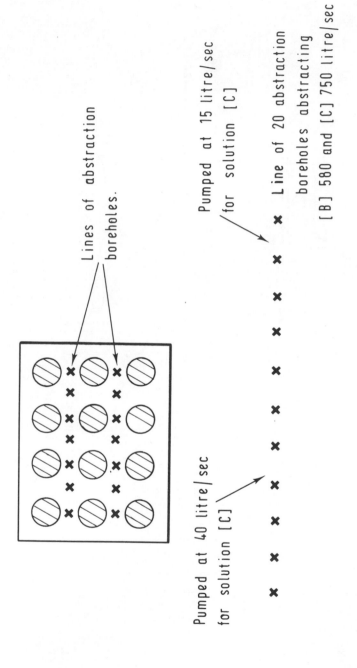

Lines of abstraction boreholes.

Pumped at 15 litre/sec for solution [C]

× Line of 20 abstraction boreholes abstracting [B] 580 and [C] 750 litre/sec

Pumped at 40 litre/sec for solution [C]

FIG. 109.   SOLUTIONS (B) AND (C) WITHIN AND OUTSIDE REFINERY

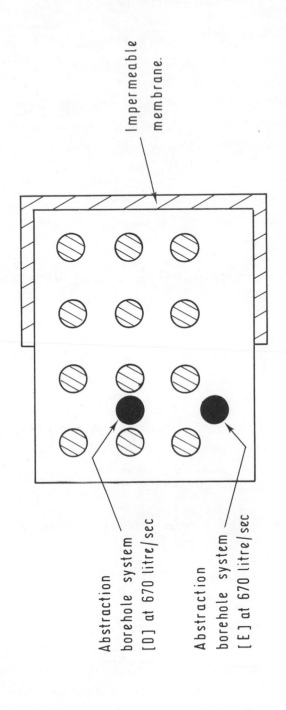

Impermeable membrane.

Abstraction borehole system [D] at 670 litre/sec

Abstraction borehole system [E] at 670 litre/sec

FIG. 110.   SOLUTIONS (D) AND (E) WITH IMPERMEABLE
MEMBRANE AND LARGE ABSTRACTION SCHEME

337.

NAPHTHA LEAKAGE AT PURFLEET AND SPILLAGE OF LEADED FUEL AT SOUTH STIFFORD, ESSEX

by R. Aspinwall [*]

Essex River Authority, Chelmsford, Essex, England

1. PURFLEET CHALK GROUNDWATER POLLUTION

1.1. Background information

In June 1969, approximately 700 tons of naptha were reported to have been lost from petroleum storage area buried in dry Chalk. Loss was eventually traced to one particular leaky installation but only after extensive tests were carried out.

Chalk of area outcrops in east-west ridge from Purfleet to Grays and is important unconfined fissured aquifer in this area. Heavy industrial groundwater abstractions take place creating persistent cone of depression of water table below river level and saline infiltration has ensued.

Fig. 111 shows groundwater contours, abstraction bores, leakage point and points at which contaminant was identified.

1.2. Sequence of events

(i)   June 1969. Leakage reported, naptha appearing as large globules on top of water surfaces, groundwater survey initiated by Essex River Authority (ERA) groundwater flow directions predicted.

(ii)   July 1969. Predictions proved correct by appearance of contaminant at edge of a shallow lake in disused chalk quarry. (Q in Fig. 111). Appearance of contaminant in south of quarry suggested some product might have flowed along fissure systems to south of quarry. ERA recommended the following:

(a)   Boreholes b_1 to b_6 to be sunk to determine extent of pollution and to act as early warning stations to the south-east and east of the quarry.

(b)   Monitoring of 30 wells, boreholes and excavations in area to be implemented on a daily basis.

(c)   A 2.4 m deep trench 200 m in length (T) to be immediately constructed at southern edge of quarry Q in quarry floor and continuously pumped to lower groundwater levels locally in effort to induce flow of contaminant. Pumped water discharged over embankment into area of open water surface on which contaminant could evaporate safely.

(iii)   August 1969. 100% explosive mixture reported in ducts carrying electrical installations and in abstraction well W_5 at Cement Factory (Fig. 111).

* now Consulting Hydrogeologist, Oswestry, Shropshire.

Ventilation system installed to de-gas ducts. Boreholes sunk to south-east of quarry encountered contaminant. Borehole b_3 encountered contaminant on water table within made-ground overlying Chalk. This borehole subsequently enlarged into sump 3m x 3m x 2m deep and continuous pumping into quarry trench T employed to induce flow of contaminant - method successfully removed contaminant from this vicinity. Evidence of contaminant at well W_5 and under Cement Factory gradually lessened.

(iv) March 1971. Contaminant finally ceased to flow into quarry trench. Pumping discontinued. Table 35 indicates the wide range of velocities computed for this leakage.

TABLE 35

FLOW RATES OF CONTAMINANT

Shortest time (days) leakage area to		Longest time (days) leakage area to Quarry	Distance (m) leakage area to			Flow rate of contaminant m/day
Quarry	Cement Works		Quarry		Cement Works	
			Max	Min		
69			640			9
				330		5
	89				1900	21
		669	640			1
				330		0.5

1.3. Effectiveness of measures

It was fortunate in this case that the quarry lay on flow path of contaminant. Groundwater exposed in base of quarry, therefore pumping trench could be contemplated and evaporation practised. Quarry is remote from residential areas; security reasonably good. Measures taken considered to have been highly effective but drawback in not being able to quantify amounts of contaminant recovered because of evaporation method used for disposal.

Pollution could have had very serious consequences if quarry had not been fortuitously sited.

Note should be made that only legislation which could be used was vested in local authority as the petroleum storage licensing authority.

Therefore, fortunate in one sense that contaminant was not, say, gas oil which is not subject to licence.

Good liaison was achieved between local authority, petroleum company and River Authority and this is considered of prime importance.

2. SOUTH STIFFORD GRAVEL/CHALK GROUNDWATER POLLUTION

2.1. Background information

July 1966. 350 tons leaded petroleum spirits spilled in rail sidings (A in Fig. 112).

October 1966. 40 tons leaded petroleum spirit leaked as result of corroded pipeline (B in Fig. 112).

August 1968. 36 tons leaded petroleum spirit spilled (C in Fig. 112).

September 1969. Explosion in basement of house at D (Fig. 112) initiated a site investigation into the hydrogeology of the area and whether petroleum was present in the strata beneath the area.

Fig. 112 shows site investigation borehole positions, groundwater contours in flood plain gravels and in Chalk, general geology and position at which explosion occurred. Cross section YY' shows general geological relationships.

No groundwater abstractions take place in immediate vicinity. Problem one of public safety in short-term, water supply in longer term.

2.2. Results of site investigation

(i)   At time of investigation, Chalk groundwater levels were confined by overlying gravels. Upward leakage from chalk to gravels taking place. Locally clay lens within gravel supports perched water table. Groundwater table very flat.

(ii)   Contaminant found in boreholes indicated in Fig. 112 within gravels.

(iii) Petroleum odour zone of 2.4m to 3m encountered during boring above water level.

(iv)   Possible that petroleum gases above water table had migrated into basement of house probably as result of rising water levels after abnormal rainfall on 14 and 15 September 1968 when 200 mm fell in 48 hours, causing rises in level in local chalk quarries ranging from 0.7m to 2.4m.

2.3. Measures taken to combat pollution

Cofferdam, 12m x 1.6m x 6.5m deep constructed in gravels at Z. Drainage slots in piles from 2.1m down to allow drainage of contaminant and water into cofferdam. Liquids then pumped away. Largely unsuccessful due to low permeability of saturated gravels and decreasing saturated thickness of gravels with time as effects of abnormal meteorological conditions on groundwater levels died away.

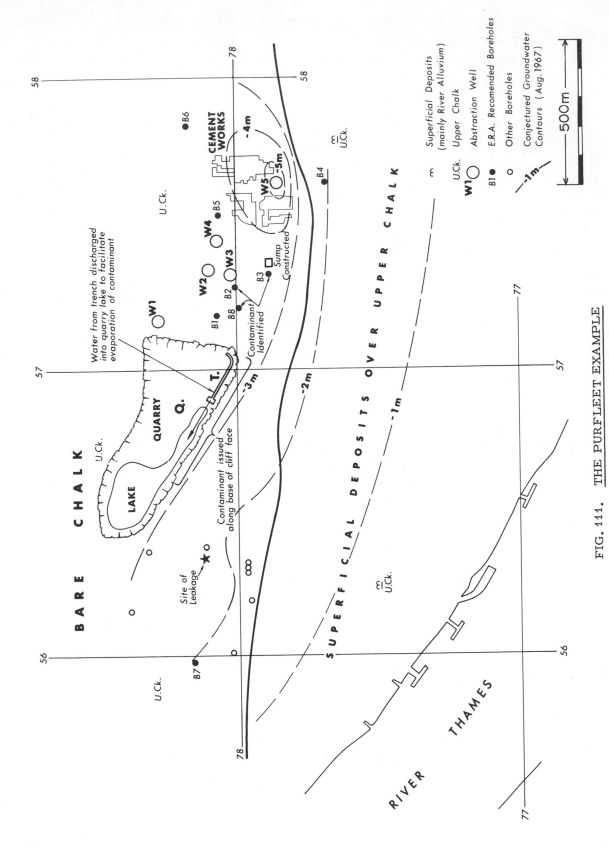

BARE CHALK

Water from trench discharged into quarry lake to facilitate evaporation of contaminant

Contaminant issued along base of cliff face

Site of Leakage

Contaminant Identified

Sump Constructed

CEMENT WORKS

SUPERFICIAL DEPOSITS OVER UPPER CHALK

RIVER THAMES

W1 ○ Abstraction Well
B1 ● E.R.A. Recomended Boreholes
○ Other Boreholes
—1m— Conjectured Groundwater Contours (Aug. 1967)

Superficial Deposits (mainly River Alluvium)
U.Ck. Upper Chalk

500m

FIG. 111. THE PURFLEET EXAMPLE

342.

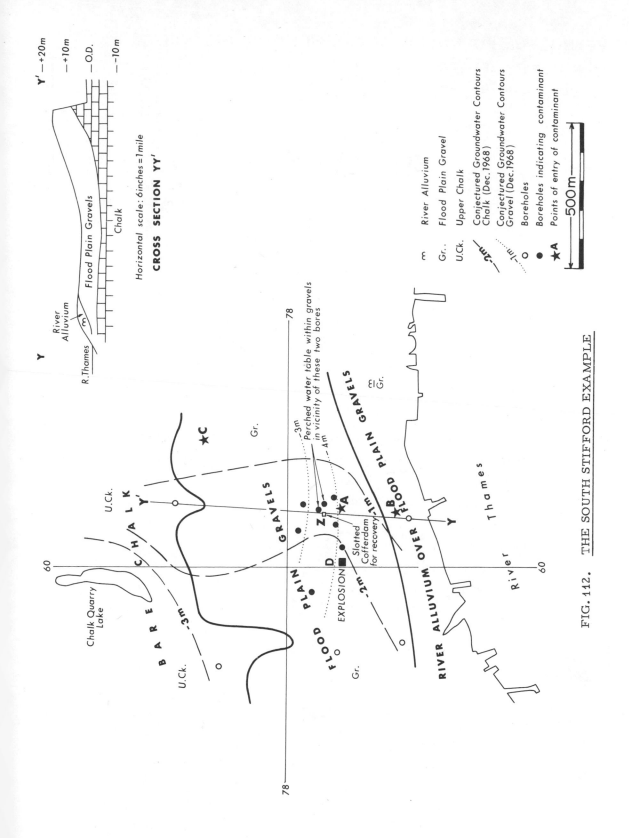

CROSS SECTION YY'

Horizontal scale: 6 inches = 1 mile

Y Y'

River Alluvium

R. Thames

Flood Plain Gravels

Chalk

+20m
+10m
O.D.
−10m

m — River Alluvium
Gr. — Flood Plain Gravel
U.Ck. — Upper Chalk
⌇ — Conjectured Groundwater Contours Chalk (Dec. 1968)
⌇ — Conjectured Groundwater Contours Gravel (Dec. 1968)
○ — Boreholes
● — Boreholes indicating contaminant
★A — Points of entry of contaminant

500m

Chalk Quarry Lake

U.Ck.

BARE CHALK

U.Ck.

★C

Gr.

Y'

Perched water table within gravels in vicinity of these two bores

−3m

−4m

GRAVELS

Z

★A

Slotted Cofferdam for recovery

−1m

Gr.

FLOOD PLAIN GRAVELS

★B FLOOD

FLOOD PLAIN

D ■ EXPLOSION

−2m

Gr.

Y

RIVER ALLUVIUM OVER

River Thames

60

60

78

78

FIG. 112. THE SOUTH STIFFORD EXAMPLE

343.

PETROLEUM POLLUTION AT GREAT HOSPITAL, NORWICH

by B.M. Funnell and G.S. Boulton
School of Environmental Sciences, University of East Anglia, Norwich, England

NATURE OF POLLUTANT

Light petroleum distillate, refractive index 1.4861 to 1.4782, specific gravity 0.872 to 0.859, boiling point range $54^{o}C$ to $184^{o}C$, flash point $114^{o}F$, aliphatic predominating over aromatic hydrocarbons.

MAP OF AREA

See Fig. 113.

VERTICAL SECTION

See Fig. 114.

HISTORY OF DISCOVERY

January 1951 explosion in Great Hospital boiler house located in 2 m deep concrete-lined pit, later in 1951 gas oil discovered in pit dug for new drainage system nearby. November 1954, and again after remedial action December 1954, renewed percolation of oil into boiler house. September 1970 further oil discovered in pit dug 3 m deep not far from boiler house. Samples analyzed November 1970. History of percolation of oily liquid into cellar of nearby public house in 1943.

MEASURES TAKEN

Boiler house rebuilt in 1954 and refloored in 1954; open pit pumped out in 1951. It appears that the pollutant responsible for the incidents in 1943, 1951 and 1954 is unlikely to have been of the same type as that analyzed in 1970; the use of light petroleum distillates (for gas production) only dates back to about 1960. No remedial action yet taken, responsibility not yet determined at law, but proposals include cutting drain across course of pollution stream thereby intercepting flow so that oil may be pumped off.

Immediate danger not to water supply, but of fire risk, which is not considered too serious in general case because top of water table generally 1 to 2 or more metres below surface; flashpoint corresponds roughly with paraffin.

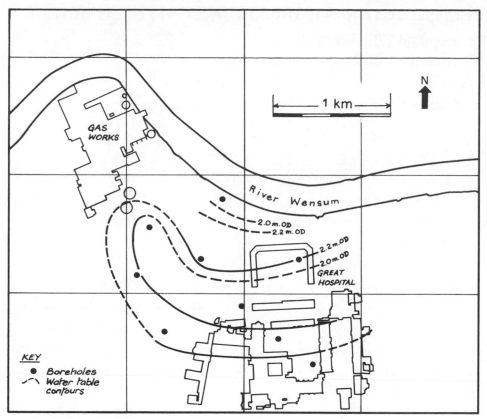

TG 2309 SE (part)

FIG. 113. MAP OF AREA AFFECTED: GREAT HOSPITAL

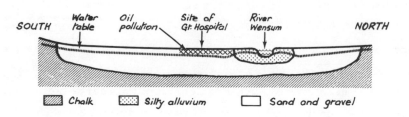

FIG. 114. VERTICAL SECTION: GREAT HOSPITAL

A RECENT CASE HISTORY OF GROUNDWATER POLLUTION BY ORGANIC SOLVENTS

by J.B.W. Day
Institute of Geological Sciences, London, England

1. DISCOVERY AND SAMPLING OF THE POLLUTED WATER

Towards the close of 1969 a burst water main on a factory site led to investigations which revealed the presence of severe pollution to groundwater in close proximity to public supply sources. (Institute of Geological Sciences' Annual Report for 1970, p. 89).

The factory concerned was built some years previously upon made ground which overlies several feet of Superficial Deposits, consisting mainly of flint gravels in a clay matrix, in turn overlying the solid Chalk. Just outside the factory fence and within 200 m of the burst main, lay the nearer of two public-supply wells sunk into Chalk, but lined out through the superficial deposits. Both wells normally contributed a total of 1.5 m.g.d. to the local town's supply. Wells and factory are situated in a groundwater discharge area, the site of springs, now in conduit.

The sides of the excavation made to repair the main were seen to be exuding an aromatic liquid, analysis of which revealed the presence of organic solvents, mainly ketones and alcohols, at a total concentration of 1 part in 200. Sodium and potassium contents were also abnormally high. Although the solvents concerned were not thought to be toxic in low concentrations, the nearby public supply wells were immediately closed down, but not until samples had been taken. Initially, both boreholes were thought to be free of contamination, but the conventional analytical methods used could not detect solvent at low concentrations. As a sample from the nearer borehole was thought to have a detectable odour, further samples were analyzed by gas chromatography, still with negative results. This method, however, was subsequently found to be capable only of detecting solvent concentrations in excess of 10 mg/litre. Meanwhile, it was remembered that the meter located on the delivery main from the nearer borehole had been removed recently and found to contain an unusual slime of a kind thought to be associated with bacilliary and algal growths capable of feeding on organic liquids at low concentrations.

Twenty-two further pumped borehole samples were taken and of these, nineteen had detectable (by colorimetric methods) traces of ketones in concentrations up to 8 mg/litre and averaging 2.2 mg/litre. Subsequently, refinement of gas liquid chromatography techniques enabled solvents at very much lower threshold values to be determined.

2. DIAGNOSIS

Examination of the factory premises revealed that the source of pollution was probably leaking underground effluent drains. The Institute of Geological Sciences was asked to assist at this point, and the resulting investigation established that severe pollution was restricted to the vicinity of one particular factory block, near which was a number of organic liquid storage tanks in bunds. Shallow pits and auger holes into the drift provided sampling points. Groundwater levels, after pumping from the nearby supply wells had ceased, recovered throughout the factory area to about 3 ft 6 in. below surface, to a level within the made ground and above that of the drains which were about 10 ft below surface. Whilst the nearby pumps were operating, however, water levels had been below drain level, so that the leaking drains had been effluent with respect to surrounding ground. Subsequent examination of a drain revealed that the main source of trouble was probably fracturing due to compaction and subsidence of made ground in which the drains were laid.

3. SCAVENGE PUMPING

After the defective lengths of drain had been bypassed, suction scavenge pumping was carried out from eight pits in the affected area in an effort to clear the body of polluted groundwater which appeared to lie mainly at relatively shallow levels. There was of course, a tendency for the natural groundwater discharge to have a clearing effect, and it was hoped that bacterial degradation would play a part. One pit was sunk to the underlying Chalk (the remainder were not, in an effort to prevent ready access of pollutants to a fissured formation) and a small sample of Chalk was obtained and centrifuged. The resulting liquids were analyzed for potassium, which was found to be present in concentrations of 35. 8 mg/litre in liquid from portions of the outer skin of the sample, and 22. 0 mg/litre in liquid from the interior.

Scavenge pumping from pits within the most severely contaminated area continued for a period exceeding one year during which further defective drains and tanks came to light. The solvent content of the local groundwater fluctuated widely as a result, but the total solvent concentrations, at first measurable in thousands of milligrams per litre, gradually reduced until values consistently less than 100 mg/litre were obtained. At the end of this period, during which the supply source remained out of commission, pumped samples from the two wells revealed no contamination.

At the factory, suspect bunds were repaired, in some cases with butyl rubber membranes, and all underground effluent drains put out of use in favour of a pumped overhead effluent system.

347.

DISCUSSION ON OIL AND PETROCHEMICAL POLLUTION

1.   To supplement his paper, Professor Dracos showed a composite slide of the successive stages of movement of an oil spill down through an unsaturated layer, then laterally spreading on reaching the capillary zone above a water table.   It was stated that although the soluble constituents of oils were not enough to change the chemical composition of groundwaters significantly, they were obnoxious from taste and odour aspects even at minute concentrations.

2.   As site excavation was not always feasible, when oil pollution has occurred near a building or highway, the costlier and less effective alternative was to sink scavenger wells and maintain pumpage to waste, with interception of undissolved oil.

3.   J.B.W. Day mentioned that his paper (p. 346-7) had originated from pollution caused by ketones and alcohols escaping from pipe drains into a superficial alluvial deposit. The water in that deposit was nearly stagnant, so the polluting chemicals could only be cleared by scavenge pumping.   An original concentration of some thousands of mg/litre was reduced to less than 100 mg/litre in a matter of months.   The public water supply well by the site had to be abandoned, to counter the unknown risks of low-level ketone toxicity, and was still out of action.   About £200 000 had had to be spent on new effluent disposal arrangements at the site, and over £50 000 had already been devoted to the scavenge pumping operations.   The public had not been alerted to the situation.

4.   The question of oil pipelines bursts under pressure was raised.   Given that any immediate action for protecting the public was the police and fire authority's responsibility, it then fell to the pipeline company to make the necessary urgent repairs, to reinstate ground damaged by an oil spill and to sink scavenger wells where appropriate.   Some companies were offering training in emergency procedures to police, fire, river and water authority personnel, and good liaison was essential.   There did not appear to be a universal chain of command for pipeline emergencies, however, so priorities for action would vary regionally.

5.   River and water authority delegates spoke well of the effort devoted by oil pipeline operators in the UK to consultation with water interests, following procedures laid down to Department of Trade and Industry specification under the provisions of the Pipelines Act, 1962.   A similar aim was in view in the USA where an oil and water industry working group was drafting a code of practice.

6.   Special anti-pollution measures were being designed into the multiple pipeline route, connecting Rotterdam and Antwerp.   The proposals included concentric pipes, cathodically protected externally and with sand in the annular space inside; this sand can be monitored for electrical conductivity changes caused by leakage and can be drained also.   Another proposal is for the pipelines to be laid over a bed of sand holding down a PVC sheet, draped in a shallow trench to form a safety trough.   Eventually, about 50 pipelines will be traversing an aquifer along $1\frac{1}{2}$ km of the route, so the cost of protecting the latter will be considerable.

7.   Central storage of heating oil was a potential risk, which had given troubles in the Lee valley.   Such storage is not permitted in Switzerland.

THE IMPACT OF ORGANIC PESTICIDES AND HERBICIDES UPON GROUNDWATER POLLUTION

by B.T. Croll
The Water Research Association, Medmenham, Bucks, England

1. INTRODUCTION

The groundwater pollution problems which could arise from the use of pesticides and herbicides may be judged from the concentrations likely to reach groundwaters under adverse conditions. Thus, if an area of land were sprayed at the rate of 2 kg/ hectare active ingredient and 25 mm of rain fell in 24 hrs, then the water percolating through the soil could contain 2 kg/250 m^3 or 8 mg/litre of the chemical. Such a concentration of some materials could present a very great hazard to receiving rivers or domestic consumers; the organo-chlorine insecticides Endrin and Endosulphan, for instance, being toxic to fish at levels down to 0.0005 mg/litre. In practice, levels reaching groundwaters are likely to be considerably lower than those calculated above, due to adsorption on soil and plants, and degradation by soil bacteria. Levels reaching rivers or domestic consumers are likely to be even lower than those entering the groundwater, due to dilution.

Organic pesticides and herbicides have been detected in many parts of the environment, and organo-chlorine insecticides in particular occur almost universally in environmental samples (1). Other types of pesticides and herbicides are normally only found within the localized environments in which they have been released, although organic mercury fungicides appear to have contributed to more widespread pollution especially in Sweden (2). With particular reference to the aqueous environment, organo-chlorine insecticides have been found in rain at levels of up to 230 µg/litre (3) and in surface waters at similar levels. Other types of pesticides and herbicides have been found sporadically in various types of water.

The above mentioned levels of organo-chlorine insecticides may be regarded as background levels arising mainly from agricultural uses, although other factors such as malaria control are undoubtedly important. Contamination of water by other pesticides and herbicides has not usually been caused by their correct use in agriculture, but by misuse, accident or as industrial effluent. Accidental contaminations have probably been the largest problem numerically, and the reports of the River Authorities in England and Wales, and the River Purification Boards in Scotland, give numerous examples covering all types of pesticides. Pollution due to industrial effluent is less frequently encountered, but is generally more serious when it occurs. In surveys of levels of organo-chlorine insecticides in England and Wales (4) (5), the highest levels encountered were due to industrial effluents, particularly the use of Dieldrin in the mothproofing of textiles.

Data on the pollution of underground waters by pesticides and herbicides are very much less abundant than those for surface waters and the present paper is an attempt to outline the problems of this type of pollution and to collect together the published information.

2.   THE EFFECTS OF PESTICIDES AND HERBICIDES ON WATER

2.1.   Toxicity

Toxicity is of prime importance when considering effects on humans, and both chronic and acute toxicities (6) must be taken into account. Compounds having low acute toxicities, but which accumulate in the body, may have much lower long-term acceptable intake levels than ones of much higher acute toxicity which do not accumulate in the body. These materials may be taken into the body from air, food or water by ingestion or skin adsorption; in many cases residues in water may not be the prime source of the compound. Acceptable intakes of toxic compounds are assessed from long-term feeding and dermal application experiments on those animals found to be most sensitive to the toxicant. The transference of these data to acceptable daily intakes to man is naturally a job for expert toxicologists and involves the setting of a suitable safety factor. This safety factor is necessary because man may be more sensitive to the toxicant than the most sensitive test animal, and because individuals may vary in their reactions. The size of the safety factor, which is determined by the nature of the compound, may be of the order of 1000. In setting acceptable levels of materials in potable water not only man must be considered but also any animals, plants etc. which might use the supply. For instance, farm animals, domestic pets, tropical fish and garden plants (irrigation), all might come into contact with the water supplied by an authority to a consumer. In this respect, re-use of the water for irrigation is of great importance when considering acceptable levels of herbicides in domestic water supplies. Residues likely to be found in air and food must also be taken into consideration.

Where the groundwater is entering a river, overall ecological effects are of paramount importance and such factors as the build up of lipid soluble materials, for example organo-chlorine insecticides, along food chains can reduce acceptable levels to extremely low values.

2.2. Taste and odour

Many organic chemicals can give rise to unpleasant tastes and odours in water when present at levels considerably less than 1 mg/litre. Waters contaminated in this way will be unacceptable to the consumer even though they are perfectly safe from biological and toxicological aspects. Even pleasant tastes and odours will be rejected, as the water is obviously 'contaminated'. It is particularly important to consider complete pesticide or herbicide formulations rather than active ingredients with respect to taste and odour, as formulation solvents may have very low threshold odours. For instance, diesel oil which is used as a solvent, particularly for 2,4 5-T ester formulations, has a threshold odour (a minimum detectable level in water) of 0.0005 mg/litre (7). The pesticide or herbicide or its formulation materials may react chemically with other materials present in, or added to, the water and produce or enhance odours. A particular example of this is the reaction of phenol with chlorine to produce much more odorous chlorinated phenols.

2.3. Other effects

Some highly coloured compounds can give rise to a colour in the water at low concentration, for instance the nitrophenols DNOC and DNBP (Dinoseb) give rise to a yellow colouration at levels down to 0.1 mg/litre. The possible effect of herbicides on the treatment processes used at a waterworks must also be considered. Although it does not seem likely, the normal processes could be disrupted by the presence of organic residues.

3. LEVELS OF PESTICIDES IN GROUNDWATERS

3.1. Organo-chlorine insecticides

This class of compounds is extremely important because of the persistence and chronic toxicity of many of its members. DDT can persist for longer than 20 years in the soil (8), and any problems associated with these materials could last for many years despite the measures that have been taken to control their usage in many parts of the world. Their ability to accumulate in body fat means that very low intake levels can have toxicological significance. The maximum acceptable levels of the various compounds in potable waters have been quoted as lying between 0.001 mg/litre for Endrin and 0.6 mg/litre for Lindane (9), where mammalian toxicity was the prime consideration. However, the same source quotes the toxicity of Endrin to fish (TLm Bluegills) as 0.2 µg/litre and that the recommended maximum level in surface waters be one hundredth that level for periods longer than 24 hours.

The very persistence of the organo-chlorine insecticides in the soil and their low water solubilities (9) indicates that only very low concentrations are likely to be found in water percolating through soil treated with them. However, concentrations as low as those indicated as permissible maxima in the previous paragraph could easily be reached. In a study of the leaching of eleven insecticides through eight soil types, Bowman and his co-workers (11) found that only Lindane was eluted from a soil high in organic matter. TDE and pp'DDT were eluted only from a very sandy soil. The other insecticides were eluted from the varying soils according to the clay and organic matter content of the soils. Thus, the higher the clay and organic matter the smaller the leaching of these insecticides. The only leaching that Beran and Guth observed in similar experiments was of Lindane in sandy soils (12). Tsapko (13) was not able to detect leaching of Lindane or Heptachlor at the 0.1 mg/litre level when applied at 40 to 60 times the field doses. His methods of analysis were somewhat insensitive, however, and leaching at lower levels may have occurred despite the lack of taste and odour, or toxicity to <u>Daphnia magna</u> in the soil column effluents. Eye (14) studied the leaching of Dieldrin through various soil types and concluded that"...from the evidence obtained in this investigation residues of Dieldrin cannot be transported through soils into subsurface waters in significant amounts by infiltrating water. A period of several hundred years would be required for Dieldrin to be transported in solution at a residual concentration of 20 μg/litre to a depth of 1 ft (0.305 m) in natural soils. Thus it appears reasonable to conclude that residues of Dieldrin in or on the upper layers of the soil do not pose a threat to the quality of groundwaters at the assumed permissible concentration of 20 μg/litre." However, it seems possible that lower levels toxic to fish and microfauna could leach through shallow soils and cause problems, as indicated by other workers.

Once in groundwater, the organo-chlorine insecticides might be expected to persist for long periods of time. There have been no studies of persistences in groundwater, but these might be expected to be as long as, or longer than those for soil or surface waters. As mentioned previously, DDT may persist in soil for longer than 20 years showing 95% disappearance in 4 to 30 years with an average of 10 years (8). The persistences of other materials, apart possibly from Endrin, are shorter, with Lindane being probably the shortest. Eichelberger and Lichtenberg (15) studied the persistence of pesticides in river water. Of the twelve organo-chlorine insecticides studied over two month periods, seven were completely unaffected while four were changed to stable chlorinated compounds. Only one, Telodrin, was decomposed to materials not detectable with the use of electron-capture gas chromatography.

Once in an aquifer there is very little information on the movement of pesticides through the strata. Scalf (16) showed that there was no penetration of DDT to nearby boreholes when it was injected into mixed river deposit aquifer through a recharge bore.

Much of the DDT was recovered as suspended material when the recharge bore was used for abstraction.

In practice there have been no reports outside the USSR of detectable background levels of organo-chlorine insecticides in underground waters. Croll (4) examined 74 samples of underground water from wells and springs in Kent, England and was unable to detect organo-chlorine insecticides in any of the samples. Kent, which has extensive fruit growing, hop growing and market gardening areas, is considered to have the heaviest pesticide applications of any British county. The waters sampled are shown in Table 36. The majority of the samples were taken monthly over a period of one year at sites 1, 2, 3 and 4. The remainder were taken at the other sites at the beginning of the year and then at sporadic intervals until the end of the period.

TABLE 36

SAMPLING SITES FOR UNDERGROUND WATERS IN KENT

Site No	Well	Aquifer	Depth	Grid reference
1	Farleigh	Greensand	Spring	TQ 707525
2	Boarley	Chalk	Spring	TQ 766594
3	Goudhurst	Ashdown Sand	4 bores 150-280 ft	TQ 715372
4	Barham	Chalk	2 bores 200 & 220 ft	TR 196510
5	Ospringe	Chalk	200 ft	TR 001602
6	Godmersham	Chalk	2 bores 200 ft	TR 070512
7	Throwley	Chalk	3 bores 340 ft	TQ 994558
8	Harty Ferry	Chalk	250 ft	TR 013647
9	Sir Thomas A'Beckett Springs	Chalk	Spring	TQ 933632
10	Matts Hill	Chalk	300 & 400 ft	TQ 818628
11	Highsted	Chalk	200 ft	TQ 910614

Sites 1, 2, 3 and 4 were chosen as those most likely to show contamination, and Site 1 is a shallow spring whose catchment is planted entirely to orchard and hops. This catchment had been treated regularly and intensively with organo-chlorine insecticides for many years. Under these conditions it might have been expected that some insecticide would have penetrated into the groundwater. It seems likely that the rate of disappearance of the insecticide from the soil, probably by volatilization, has been greater than its rate of leaching downwards, thus avoiding groundwater contamination. Alternatively, the pesticides may not have leached to sufficient depth in the soil, or they may have broken through at levels too low to detect with the methods of analysis employed in the exercise. Subsequent to the work in Kent, some twenty samples of chalk well waters from various parts of south eastern England have been analyzed at the WRA. None had detectable levels of organo-chlorine insecticides. All the above analyses were performed using electron-capture gas chromatography and the limits of detection of the method ranged from about 0.5 ng/litre for α BHC to about 10 ng/litre for pp'DDT. The limits of detection of the other thirteen insecticides investigated lay between the above figures. In surface waters in Kent, three organo-chlorine insecticides, namely α BHC, γ BHC (Lindane) and Dieldrin, were detected in all of some 200 water samples at average levels of about 20, 60 and 20 ng/litre respectively. In April and May of the year, pp'DDT was also found in the surface waters at levels of up to 164 ng/litre. Grasso and his co-workers (17) obtained similar results and were unable to detect any organo-chlorine insecticides in 35 samples of Italian well waters.

In the USSR, Vrochenskii (18) detected maximum levels of Lindane and DDT of 0.025 and 0.01 mg/litre respectively in well waters. The average levels were 400 and 500 ng/litre respectively. In a further article (18) dealing with well waters from state farms and collectives he states that DDT was detected in 13.8% of the samples and Lindane in 10.3%. The worst sample contained 0.01 mg/litre DDT and 0.02 mg/litre BHC. The maximum levels are much higher than those normally found even in polluted surface waters, except in the USSR, and would certainly be toxic to fish (9) although acceptable on mammalian toxicity considerations (9).

Two specific instances of groundwater pollution by organo-chlorine insecticides have been found, the first reported by Jettmar (20) where a shallow well in Switzerland was contaminated by soil applied materials. The second is in England, where the Chew Valley reservoir was contaminated by organo-chlorine insecticides used in sheep dips (21). This second incident occurred in a region of fissured limestone and is particularly interesting. Used insecticide solutions from a sheep dip operated by several local farmers had been spread on an adjacent field, and percolated downwards into an apparently stagnant natural underground reservoir in the limestone. This effluent accummulated over a period of years until some natural phenomenon, possibly unusually high water levels or a cavern collapse, caused it to be discharged to an underground stream. The insecticides were then transported some distance underground before surfacing in a stream and then progressing to the reservoir. In the reservoir they caused a massive mortality of Daphnia which in turn allowed an algal species on which the Daphnia grazed, to multiply to such proportions that severe treatment problems were encountered for about two months.

3.2. Organo-phosphorus insecticides

Unlike the organo-chlorine insecticides, which have similar properties, the organo-phosphorus insecticides exhibit a wide range of toxicities, persistence and water solubilities. Accordingly each compound must be judged separately as a potential pollutant.

In general, the organo-phosphorus insecticides do not build up in the body and therefore although some of them are more acutely toxic than the organo-chlorine insecticides the allowable levels in water can be somewhat higher. A maximum of 0.1 mg/litre total organic phosphates plus carbamates (cholinesterase inhibiting compounds) has been recommended for potable water supplies (9). However, when considering environmental contamination, although allowable levels are often higher than organo-chlorine insecticides, the limits on several organic phosphorus compounds are extremely low. Carbophenothion, for instance is recorded as being toxic to Daphnia magna at 9 ng/litre and allowable levels should be one tenth this figure (9). Sub-acute effects have been noted in some species at very low levels of organo-phosphorus compounds in water.

The organo-phosphorus compounds are generally much less persistent in soil than the organo-chlorine, being measured in days or weeks rather than years (9). However, some, for instance those used as sheep dips, must have a persistence on the animal of several months and so similar figures could apply to soil. Again, the phosphorus compounds are generally more water soluble than the chlorinated insecticides, some being very soluble. They would therefore be expected to leach through soil more readily than the organo-chlorine insecticides.

McCarty and King (22) studied the movement of six organo-phosphorus insecticides, namely Parathion, Methyl Parathion, Thimet, Disyston, Ethion and Trithion in four soils. They found that all leached through the soils, the rates of leaching being slower with soils of high clay content. Four of the insecticides eluted rapidly in fairly narrow bands, but Ethion and Trithion eluted more slowly as broad bands. These effects were related to the water solubility and the adsorption properties of the various insecticides. The recoveries of insecticides varied from less than 1 to about 50% depending upon the degree of breakdown in the soil, and the interval between application and leaching. It was thought that biological activity was the most important factor in determining the breakdown of many organic phosphates. Parathion was studied closely and this hypothesis was found to hold true. In 6 inch (15 cm) long soil columns, over 100 ft (30 m) of water and more than one year were required to complete the elution of the recoverable pesticides.

No data are available concerning the movement or persistence of organo-phosphorus insecticide in underground waters. Eichelberger and Lichtenberg (15) have studied the persistence of nine organo-phosphorus insecticides in surface waters. Parathion, Methyl Parathion, Malathion, Trithion, Fenthion and Merphos were significantly degraded within one week and were not detectable after eight weeks. Ethion and Dimethoate were reduced to 50% of the original concentration after eight weeks and Azodrin was recovered unchanged.

The only recorded instance of groundwater pollution by organo-phosphorus insecticides that has been discovered is that recorded by Nicholson and Thoman (23). Well waters 125 to 185 ft (40 to 60 m) deep were contaminated to a level of 1 μg/litre by Parathion used in agriculture. Grasso (17) was not able to detect any organo-phosphorus insecticides in well waters in Italy.

3.3. Carbamate compounds

The carbamate compounds are used both as herbicides and insecticides and are generally of low mammalian toxicity (6). They do not build up in the body. Their toxic action is identical to that of the organo-phosphorus compounds and they are included with the phosphorus materials in the level of 0.1 mg/litre total cholinesterase inhibiting compounds allowable in potable waters.

These materials are relatively new and very little is known about their environmental effects. The most widely used of these compounds, Carbaryl, is toxic to a species of stream invertebrae (9) at a level of 1.3 μg/litre. This would limit its environmental level to one tenth that figure. The other properties of these materials are also not well documented collectively and no information is available concerning levels leaching through soils, although it is known that the depth to which certain members of the group leach in soil depends upon their solubility in water, and the clay and organic content of the soil (24). In this respect they resemble the other types of insecticides in their properties.

No residues of carbamates have been recorded in groundwaters even due to pollution incidents. Eichelberger and Lichtenberg (15) investigated the persistence of seven carbamates, namely Carbaryl (Sevin), Zectran, Matacil, Mesurol, Baygon, Monuron and Fenuron, in surface waters. All were signficiantly degraded within one week and only Baygon showed any residue (5% of original level) after eight weeks. Aly and El Dib (25) obtained similar results with Sevin and Baygon and recorded no hydrolysis of Pyrolan and Dimetilan after eight weeks at pH values varying from 4.0 to 10.0.

3.4. Phenoxyalkyl acid herbicides

The problems associated with the phenoxyalkyl acid herbicides are of taste and odour and not toxicity when potable supplies are being considered. The active compounds give rise to tastes and odours in water at levels of 0.01 to 0.1 mg/litre and they may degrade to chlorinated phenols (27) whose threshold odours in water can be as low as 0.002 mg/litre. In the United Kingdom a maximum level of 2,4-D of 0.05 mg/litre has been recommended for irrigation water (26). The levels toxic to aquatic fauna are generally somewhat higher than this last figure (9).

Some of the phenoxyalkyl acids are applied as esters. These rapidly break down into the corresponding acids (27) which are extremely water soluble. Thus the acids might be expected to leach from soil into water, and several instances of this happening in surface waters close to the point of application have been recorded (28) (29), at levels up to 2 mg/litre of 2,4-D. The persistence of these compounds in soils varies from a few days, up to several months for 2,4,5-T. The exact persistence depends upon the previous history of the soil to which the herbicide is applied. Soils with previous histories of treatment break down these herbicides more quickly than soils which have had no previous contamination. Similar considerations apply to the persistence of these materials in water and no degradation may be seen after several months in some waters (30).

No background levels of phenoxyalkyl acids in either underground or surface waters have been detected (31). Two notable incidences of extensive groundwater pollution due to industrial effluent have been recorded.

The first of these pollutions occurred at Montebello, California (32). A small plant manufacturing 2,4-D discharged a poorly reacted batch of chemicals to the sewers, and this eventually found its way to the Rio Hondo River. Further downstream the river charges the aquifer from which Montebello derives its potable water. The contamination of the aquifer with 2,4-D and the 2,4-dichlorophenol used in its manufacture led to all the city's 11 wells having to be taken out of operation. These were put back after chlorine dioxide treatment was installed to remove the chlorophenolic tastes. However, these taste problems persisted for 4 to 5 years. The second incident is less clear cut and involved wastes from the Rocky Mountain Arsenal in Colorado USA (33). Although no 2,4-D was being manufactured at this plant it appeared that waste chemicals had reacted in a lagoon to form 2,4-D. This waste had then percolated into groundwater and subsequently caused crop damage after the water was used for irrigation. Although it was certain that wastes from the lagoon had caused the crop damage, 2,4-D was not isolated from the irrigation water.

3. 5. Other pesticides and herbicides

None of the remaining pesticides and herbicides have been recorded as contaminating underground waters. Of these materials the only important group whose toxicity could cause concern are the nitrophenol herbicides. These compounds have high acute toxicities and sometimes exhibit cumulative effects. As they are extremely water soluble they could easily contaminate underground waters.

Other types, such as the triazine herbicides, are so strongly adsorbed onto soil that they are not likely to be leached into underground waters. As with other types of pesticides and herbicides discussed previously, the soil adsorption is often related to organic and clay contents of the soil and leaching problems are most likely on sandy soils.

Water solubility is very important and, generally, the more water soluble the compound the more likely it is to be leached through the soil. There are exceptions to this generalization, for instance the bipyridyl herbicides, which, despite their reasonable water solubilities, are very strongly adsorbed into clay and are very unlikely to leach from any but extremely sandy soils.

4.   REMOVAL OF PESTICIDES AND HERBICIDES FROM WATER

Most organo-chlorine insecticides are not affected by normal water treatment methods (4). Laboratory experiments have shown that DDT may largely be removed by coagulation (34). Treatment with activated carbon is the only effective method of removing the majority of these materials from water. The same conclusion may be drawn for the phenoxyacetic acid herbicides (34)(35). Despite these herbicides being acidic they are not removed effectively at low mg/litre concentrations by ion exchange resins (36).

Because of their diverse properties, methods for the removal of organo-phosphorus insecticides must be tailored to the individual compounds. Although none have been recorded as being effectively removed by coagulation, oxidative treatments with $KMnO_4$, chlorine, chlorine dioxide or ozone are effective in removing some compounds. However, oxidation can sometimes result in the more toxic oxons being produced from thiophosphorus compounds. Activated carbon treatment is sometimes effective (34), but the high doses likely to be needed to remove the more water soluble materials would prove prohibitive on cost considerations.

No information has been published concerning the removal of other types of pesticides and herbicides from potable waters, except for the removal of rotenone when used as a fish poison. Activated carbon was again the most effective method of removal (37).

5. CONCLUSIONS

There is no evidence to suggest the widespread contamination of underground waters with any type of pesticide or herbicide. This contrasts with the presence or organo-chlorine insecticides in practically all surface waters. Similarly there seems little danger from the correct agricultural use of many pesticides and herbicides, however, problems could arise with the more water soluble materials, especially under unusually heavy rainfall conditions. In all cases the information is scanty if it exists at all and much more work is necessary if reliable predictions are to be made concerning the likelihood of agricultural chemicals reaching underground waters. The greatest dangers would appear to arise from the incorrect use of the chemicals, the disposal of unused chemicals, waste chemicals and containers, or from industrial effluents. In these cases, the concentrations of pesticides or herbicides applied to the soil may be very great in terms of quantity per unit area, thus increasing the likelihood of leaching. Alternatively, the chemicals may be introduced directly into the aquifer. Again much work is needed on the movement of this type of material within the aquifer.

ACKNOWLEDGEMENTS

The author wishes to thank the Director of the WRA for permission to publish this paper.

REFERENCES

1. EDWARDS, C.A.   Persistent pesticides in the environment. London, Butterworths,1970, vi, 78p

2. WATER POLLUTION RESEARCH LABORATORY   Mercury and water pollution Notes on Water Pollution,1971,(55) pp 1-4

3. TARRENT, K.R. and TATTON, J.O'G   Organo-chlorine pesticides in the British Isles Nature,1968, 219, pp 725 - 727

4. CROLL, B.T.   Organo-chlorine insecticides in water part I Wat.Treat.Exam. , 1969, 18, pp 255-274

5. LOWDEN, G.F. and others   Organo-chlorine insecticides in water part II Wat.Treat.Exam.,1969, 18, pp 275-287

6. JONES, K.H. and others   Acute toxicity data for pesticides (1958) World Rev.Pest.Control, 1968, 7, (3), pp 135 - 143

7. INESON, J. and PACKHAM, R.F.   Contamination of Water by Petroleum products Proc.Symp.Joint Problems of the Oil and Water Industries, Brighton, 1967, London,Inst. Wat.Engineers,1967, 12 p.

8. EDWARDS, C.A.   Insecticides residues in soils Residue Reviews, 1966, 13, pp 83 - 132

9. UNITED STATES FEDERAL WATER POLLUTION CONTROL ADMINISTRATION   Water Quality Criteria Washington, U.S.G.P.O. ,1968.

10. GUNTER,F.A. and others   Reported solubilities of 738 pesticide chemicals in water. Residue Reviews, 1968, 20, pp 1-148

11. BOWMAN, M.C. and others   Behaviour of chlorinated insecticides in a broad spectrum of soil types. J.Agric.Food Chem. ,1965,113,pp 360-365

12. BERAN, F. and GUTH, J.A.   Organic insecticides in various soils with particular reference to possible groundwater pollution. Pflanzenschutz.Ber. ,1965,33, (5/8), pp 65-117

13. TSAPKO, V.V. and others   Possible penetration of Hexachloran and Heptachlor into ground waters (in Russian) Gig. Naselennykh Mest. ,1967, pp 93-5 Chemical Abstracts, 1968, 69, 109644

14. EYE, J.P.   Aqueous transport of Dieldrin residues in soil J. Wat.Pollut.Control Fed. ,1968,40, (8), pp R316-R332

15. EICHELBERGER, J.W. and LICHTENBERG, J.J.
Persistence of pesticides in river water
Environ. Sci. Technology, 1971, 5, (6), pp 641-544

16. SCALF, M.R. and others
Movement of DDT and nitrates during groundwater recharge.
Wat. Resour. Res., 1969, 5, (5), pp 1041-52

17. GRASSO, C. and others
Detection of pesticides in underground waters.
Ann. Sanita. Publ. 1968, 29, (4), pp 1029-1032

18. VROCHINSKII, K.K. and others
Contamination of water reservoirs with pesticides.
Gig. Sanit., 1968, 33, (11), pp 69 - 72

19. VROCHINSKII, K.K.
Pesticide levels in water of reservoirs and sources of water supply.
Gig. Naselennykh. Mest. 1969, 8, pp 112-114

20. JETTMAR, H.M.
Pollution of groundwater by substances difficult to decompose.
Oesterr Wasserwortsch., 1957, 7, p. 56

21. BAYS, L.R.
Pesticide pollution and the effects on the biota of Chew Valley Lake.
Wat. Treat. Exam., 1969, 18, pp 295-326

22. McCARTY, P.L. and KING, P.H.
The movement of pesticides in soils.
Proc. 21st Ind. Wastes Conf.
Purdue Univ. 1966, pp 156-171

23. NICHOLSON, H.P. and THOMAN, J.R.
Pesticide persistence in public water their detection and removal. In Research in pesticides.
New York, Academic Press, 1965, p 181

24. KOREN, E. and others
Absorption, volatility and migration of thiocarbamate herbicide in soil.
Weed Sci., 1969, 17, (2), pp 148-153

25. ALY, O.M. and EL DIB, M.A.
Studies on the persistence of some carbamate insecticides in the aquatic environment - hydrolysis of Sevin, Baygon, Pyrolan and Dimetilan in waters.
Wat. Res. 1971, 5, pp 1191-1205

26. AGRICULTURE FISHERIES AND FOOD, Ministry of
Chemical compounds used in agriculture and food storage. Recommendations for safe use in the UK.
London, HMSO.

27. KEARNEY, P.C. and KAUFMAN, D.D. (Eds)
Degradation of Herbicide.
New York, Marcel Dekker, 1969.

28. ALDOUS, J.R.

2,4-D residues in water following aerial spraying in a Scottish forest.
Weed Res.,1967, 7, (3), pp 239-241

29. FRANK, P.A. and others

Herbicides in irrigation water following canal-bank treatment for weed control.
Weed Sci., 1970, 18, (6), pp 687-692

30. ROBSON, T.O.

Some studies of the persistence of 2,4-D in natural surface waters.
Proceedings of 9th British Weed Control Conference, 1968, pp 404-408

31. LICHTENBERG, J.J. and others

Pesticides in surface waters of United States - 5 year summary 1964-1968.
Pestic.Monit. J.1970, 4,(2),pp 71-86

32. SWENSON, H.A.

The Montebello Incident.
Proc.Soc. Wat. Treat. Exam.,1962, 11,(1), pp 84 - 88

33. WALTON, G.

Public Health Aspects of the contamination of ground water in the vicinity of Derby, Colorado
Cincinnati, RA Taft Center,Technical Report W61-5,1961, pp 121-125

34. ROBECK, G.G. and others

Effectiveness of water treatment processes in pesticide removal.
J.Am. Wat. Wks. Ass.,1965,57, p 181

35. FAUST, S.D. and
ALY, O.M.

Water pollution by organic pesticides
J.Am. Wat. Wks.Ass.,1964, 56, pp 267

36. ALY,O.M. and
FAUST, S.D.

Removal of 2,4-D acid derivations from natural waters.
J.Am. Wat. Wks.Ass.,1965, 57, (2),pp 221

37. COHEN, J.M. and others

Effect of fish poisons on water supplies part 3 - field study at Dickinson.
J.Am. Wat. Wks Ass.,1961, 53, pp 233-246

DISCUSSION ON PESTICIDES AND HERBICIDES IN GROUNDWATER

1.   B.T. Croll was asked about the sensitivity and specificity of the analytical methods
available for chlorinated hydrocarbons.   For gamma BHC, DDT his results had shown
standard deviations of the order of 10 µg/litre when present at 150 µg/litre and 0.5 µg/litre
when 1.5 µg/litre.   To avoid the possible confusion with paradichlorobenzene, the extract
from the carbon columns was normally run on three GLC columns of different polarity,
to obtain maximum selection against paradichlorobenzene;   occasional confirmatory tests
are made with loose layer and thin layer chromatography.

2.   Phenoxyacetic acid, a common 'hormone' weed killer, was mentioned as being one
of the less degradable chemicals in common use, whose presence in groundwaters should
be watched for.

3.   Polychlorinated biphenyls (PCB) had emerged as environmental contaminants of
late: Dr. Croll had examined his chromatograms for PCB traces in the Kentish and
Essex waters sampled in 1966 and found no detectable quantity, i.e. any amounts present
were at concentrations less than 1 microgram/litre.

4.   It was generally agreed that organo-chlorine compounds had little chance of
getting through the adsorption barrier of an agricultural soil, but risks were recognized
in respect of direct entry to fissured strata.   A gross example of the latter was cited,
where surface run-off had conveyed Dynoseb (a dinitro-phenol, used as a herbicide)
into a swallet, and reappeared some miles away.

INFILTRATION FROM FARM SILAGE

by H.J. Richards
Water Resources Board, Reading, Berks, England

1. INTRODUCTION

A number of river authorities have discussed with the Water Resources Board problems which may arise as a result of contamination of groundwater by agricultural effluents. Several authorities have received applications for permission to discharge such effluents into aquifers and these have almost invariably been turned down. Some effluents are sprayed onto grasslands and apparently improve sufficiently before infiltration into an aquifer. Silage produces a highly polluting liquor and the possibility of this contaminating groundwater supplies has been of concern to some river authorities and statutory water undertakings. The Board undertook to study a site with the Avon and Dorset River Authority in order to determine if there was the possibility of contamination which could restrict the quantity of good quality groundwater available for development. Any findings might help in providing advice to farmers regarding the disposal of silage liquor. There was also the possibility that rates of infiltration of the effluent, or any change in chemistry, such as a breakdown of the phenols during their passage through the unsaturated Chalk, could be determined.

2. THE SITE

The River Authority and the Board chose to study an isolated site so that the extent of local contamination might be determined. It was difficult to obtain agreement to the disposal of measured quantities on a new site and so an existing silage clamp site has been utilized. A site (NGR SU 012 113) was chosen on the outcrop of the Upper Chalk near Gussage All Saints in Dorset, where the unconfined water table fluctuates annually between 20m and 26m below the surface and slopes southwards at 1 in 200. The Chalk is a soft, white limestone and is fairly uniform in composition at this site. Its porosity is about 2 per cent and there is some evidence of fine, near-vertical cracks within individual beds.

The nearest licensed groundwater sources are bores 1.5 km away to the north east, and the nearest public supply well lies more than 3 km to the south. Cut grass is laid on the bare Chalk. At the study site the soil has been removed and the Chalk excavated so that a flat area of approximately $185m^2$ forms the base of the silage clamp. The back wall stands 2m high and the side walls slope forwards to the front edge. Two

200 mm diameter bores were drilled early in 1970, No. 1 at a position 7. 5m and No. 2 at 15m in front of one corner of the front edge of the clamp. Both bores were drilled to a depth of 38m below surface and lined with solid steel casing to 5m. An autographic water level recorder has been installed on the bore nearer the clamp (No. 1) since April, 1970.

Within the Chalk groundwaters of the region the total hardness ranges from 215 to 310 mg/litre, sulphate concentration averages 8 mg/litre and total dissolved solids 280 mg/litre; pH averages 7. 8.

3.   THE EFFLUENT

Silage is made seasonally by placing cut grass into the clamp over a 6 to 8 week period in June and July. This feed is used during the winter and the clamp is normally clear during April and May. Following a wet spring 1. 02 Tonnes (1 ton) of unwilted silage could produce 273 litres (60 gal) of strong, polluting effluent. Silage clamps holding 305 Tonnes (300 tons) and more are not uncommon. The silage liquor effluent is highly noxious, having a BOD in the range 12,000 to 60,000 mg/litre, due to the presence of sugars and other carbohydrates. There is also a high phenol content derived from the hydrolysis of plant phenolids, and their detection would be a fairly clear indication of the movement of some pollutant from the surface to the main body of groundwater. The mono-hydrate phenols are said to break down in a matter of hours but some of the higher phenols are persistent.

This particular silage clamp has been in use since the early 1960's. Silage was not made in 1970 but grass was placed during the first two weeks of June 1971 and during June 1972.

If phenols or high BOD were to enter groundwater from an effluent then it could probably not be used for supply for some time. In particular the phenols react with any chlorine which might be used in treatment and give rise to an unpleasant taste. When present at levels even below 0. 001 mg/litre the phenol is obnoxious and of concern to the water supply chemist.

4.   SAMPLING AND ANALYSIS

Samples of groundwater have been obtained from the two observation wells once a month since April, 1970 and weekly during June and July, 1970 and June to September, 1971. Two litres of water were obtained using a depth sampler, and three samples obtained from each bore, at 2m and 8m below the water table and from near the base of the bores. The uppermost sample was taken first in each bore.

366.

Following the advice of the Water Pollution Research Laboratory these groundwater samples have been analyzed for:

Phenols	Taste
Sulphate	Odour
Total hardness	Total dissolved solids
pH	Alkalinity
BOD	Permanganate value
COD	

COD is usually determined for grossly polluted waters and therefore we were advised not to put reliance on low values in the groundwater but to obtain permanganate values, which are more conventional in this context and which give an arbitrary, approximate measure of the organic content of a sample.

5. TRENDS OF ANALYSES

For every sample collected the taste has been recorded as satisfactory, and no odour has been reported. Some of the persistent constituents such as phenols and sulphate have been detected in the groundwater samples. Their concentration seems to have some relation to seasonal movement of groundwater level and to differences between periods of infiltration and of groundwater recession.

Phenols

This constituent was recorded at its highest concentration of 0.3 mg/litre in December, 1970; from January to September, 1971 at values between 0.02 and 0.2 mg/litre; and then only in April, 1972, of samples subsequently collected to July, 1972. The silage making process in 1971 did not appear to have increased content of phenol in subsequent samples as compared with that reported from samples earlier in the year.

In December, 1970 phenols were recorded in samples from the middle levels of both bores, about 8m below water surface and also from the top of No.1, 2m below water surface. Three weeks earlier phenols were absent from all six samples. Presumably the phenols are persistent enough to infiltrate to the water table but so far there is not enough information to be able to see a pattern of movement.

Sulphate

This constituent was less than 9 mg/litre in the samples collected in November, 1970, but exceeded 30 mg/litre in samples from the top and bottom of borehole No.1 a month later following the start of the winter recharge period. The biggest gross concentrations of sulphate were found in the samples collected in April and June, 1971, 28 mg/litre being recorded early in June, 1971 when a high alkalinity of 575 mg/litre was also measured. Subsequent analyses resulted in values between 6 and 25 mg/litre but from September 1971 to March and April, 1972 the values were generally lower in the range 1 to 12 mg/litre. The content increased to 25 mg/litre in samples taken in June, 1972, before grass was placed in the clamp, and when water levels had been falling since April.

Total hardness

This constituent varied between 250 and 450 mg/litre, with five analyses of 500 to 600 mg/litre in the top sample from No. 1 bore during the winter of 1970-71, July 1971 and February to May, 1972. From September, 1971 until March, 1972 the content appears to be lower in general as compared with most of 1971. This variation is followed closely by the analyses for Total Dissolved Solids.

pH

In general terms this seems to have been between 6.8 and 7.5 during the autumn and winter of 1971/72 as compared with the range from 6.4 to 7.2 during most of 1971. The total range is between 6.25 and 7.6 with the values above 7 tending to occur between October and January and lower values in the summer months. The pH of the groundwater samples from bore No. 2 is almost always slightly greater by 0.2 or 0.3 than samples from the bore nearer the clamp (No. 1). It is not yet clear whether these changes are within normal seasonal changes or not.

COD

Generally low values from 1.5 to 5 mg/litre were recorded for samples during the whole of 1971 but a range of 9.5 to 28 mg/litre was recorded for March 1972, falling to 6 to 7 mg/litre in July.

BOD

These determinations followed the same general pattern as COD, but with more variation in values, with higher levels from 3 to 11.5 mg/litre from March to July, 1972 than the range of 0.5 to 8 mg/litre throughout 1971.

Permanganate value

This ranged from 0.2 to 4.2 mg/litre. There was a general fall from 1.3 to 1.8 mg/litre during the winter of 1970/71 to 0.2 to 0.7 mg/litre before increasing to 1.8 to 4.2 in February to April, 1972.

Alkalinity

This constituent ranged from 150 to 850 mg/litre. A gradual fall from a range of 400 to 550 mg/litre in the winter of 1970/71 to 200 to 400 mg/litre in June 1971, with an isolated set of values in the range 500 to 850 mg/litre in mid July, 1971. Levels fell to 200 to 400 mg/litre during the winter of 1971/72 but one series of samples in May, 1972 contained 500 to 700 mg/litre when water levels were still falling.

Consideration of several of these constituents, therefore, indicates that following periods of infiltration, or of natural recharge to groundwater, there is an increase in the level of phenols, sulphate, total hardness and a fall in pH in the groundwater samples analyzed. COD and BOD values appear to increase as water levels rise and seem to persist for several months after infiltration ceases. A selection of values of some constituents is given in Table 37.

BOD figures of 2 mg/litre or more presents some problems to the water supply chemist as does a COD figure in excess of 20 mg/litre. Both of these levels have been exceeded in samples obtained from these two bores. Phenols too have been determined at concentrations in excess of that level (0.001 mg/litre) above which taste problems will occur in chlorinated water.

6. COMMENTS

During the two seasons over which samples have been obtained and analyzed there is no clearly shown relationship between the quantity of the contaminants reaching the water table and the rates of natural infiltration through the unsaturated zone of the Chalk. Increases occur in the concentrations of several constituents following natural infiltration and the resultant rise in groundwater level. There are many instances, however, where concentrations of some constituents rise even when groundwater recession occurs. There is no pattern showing a consistent rise in constituents in any one sample in a bore, or any but general agreement between samples from the same levels in the two observation bores.

The vertical movement of the effluent through 20m and more of the Chalk is probably very slow, perhaps only 0.8m/year through the bulk of the rock. There is more rapid movement through the plexus of fine fissures, and this path is the more important for immediate problems of local contamination. There is doubt about the proportions of the effluent which move through these fine fissures and through the intergranular space within the Chalk. The constituents of the effluent are moving slowly downwards in the unsaturated zone of the strata and when the water table rises, following natural infiltration from precipitation, the effluent within the unsaturated Chalk is taken into the groundwater and begins to move laterally under the prevailing regional hydraulic gradients. If the bulk of the downward movement were through the fine fissures there would be a rapid movement of pollutant to the water table, but there is no evidence of such quick movement even within a period of several weeks.

It is likely that the unsaturated zone beneath the silage clamp contains persistent effluent constituents on their way to the water table, some of them moving selectively through fine fissures to levels below the water table. These give rise to the higher concentrations sometimes measured in samples from the lower levels within the bores. A rising water table takes in a bigger volume of available effluent constituents and leads to higher concentrations during, or soon after, the peaks in the groundwater hydrograph.

In the zone of aeration improvements are brought about in polluting effluents by natural processes of filtration, adsorption and degradation. Natural filtration of insoluble organic matter retains the latter on the surface to be broken down. Adsorption by clay minerals removes certain constituents, including possibly dissolved organic matter in a relatively hard groundwater. Biological degradation breaks down organic constituents in an aerobic situation, although other processes in anaerobic situations break down nitrates and sulphates. Removal of the soil and the disposal, within a period of a few weeks, of a substantial volume of effluent on the area of a silage clamp probably leads rapidly to an anaerobic condition immediately below the clamp. Clay minerals in the Chalk possibly do play a part in reducing the dissolved organic matter, but sulphates persist through the particular conditions at the study site.

In order to relate movement of an effluent such as this information is needed on rainfall and evaporation so that the influence of natural infiltration can be determined. The mineralogy of the strata and the local hydrogeology need to be related to the possible effects on the chemistry of the pollutant and the likelihood of bringing about changes either before reaching the water table or before the regional groundwater flow has transported the pollutant any distance.

7.   SUMMARY

There is evidence of pollution of groundwater at the study site from disposal of silage effluent. It is not yet known how far pollution may have been carried, beyond the bore 15m from the clamp by the regional groundwater flow. The persistence of the contaminants such as phenols, sulphate, BOD and COD is not known.

Clearly, it would be wise for any sites which are likely to lead to pollution of existing groundwater sources to be monitored, and that new silage clamps should not be made on the outcrop of an aquifer until the persistence of some of these pollutants has been more clearly measured. However, it must be borne in mind that silage-making leads to a heavy local concentration of polluting liquid of comparatively small volumes up to 91,000 litres (20,000 gal), and that such limited quantities of pollutant can be diluted in very large volumes of groundwater. The small volumes of aquifer rendered unusable may have to be accepted provided that there is no hazard to other users, but if the pollutants are persistent the groundwater resource may be unavailable for decades.

Are there layers of material which could be laid beneath a silage clamp to improve the effluent to acceptable standards before it is allowed to infiltrate to groundwater? Can this be achieved with material readily available to the farmer and at an acceptable cost? Some of the problems of rates of flow vertically through unsaturated Chalk are under investigation with the Atomic Energy Authority at a nearby site, using radioactive tracing methods. The question of the persistence of some of the pollutants appears to need a substantial programme of field sampling and laboratory analysis, or laboratory investigations under controlled conditions.

8. ACKNOWLEDGEMENTS

All arrangements for the site were carried out by the Avon and Dorset River Authority and they have carried out many of the analyses of water samples. Other samples have been analyzed by the Reading Public Analyst.

This note is provided by permission of the Director, Water Resources Board, but the opinions expressed do not necessarily reflect those of the Board.

REFERENCES

1. AVON AND DORSET RIVER AUTHORITY   Survey of Water Resources and Future
    Demands.
    Avon & Dorset R.A., 1970

2. HEM, J.D.   Study and Interpretation of the Chemical
    Characteristics of Natural Water. U.S.
    Geol. Survey. Water-Supply Paper 1473.
    1959.

3. HOLDEN, W.S.   Water Treatment and Examination.
    (Editor)   Churchill. London, 1970.

4. SMITH, D.B.,   Water movement in the unsaturated zone
    WEARN, P.L.,   of high and low permeability strata using
    RICHARDS, H.J. and   natural tritium.
    ROWE, P.C.   IAEA Symposium on Use of Isotopes in
    Hydrology.
    Vienna, 1970.

TABLE 37.

DATES OF MAXIMUM AND MINIMUM VALUES OF SELECTED CONSTITUENTS IN GROUNDWATER SAMPLES

Borehole no.	Sample depth (m. below water surface)	Phenols		Sulphate		Permanganate value		pH	
1	2	0.28	(21. 1. 71)	31.4	(16. 12. 70)	1.9	(2. 2. 72)	7.5	(15. 12. 70)
		0		0.8	(7. 12. 71)	0.2	(6. 7. 71)	6.2	(11. 5. 71)
	8	0.30	(15. 12. 70)	25.2	(28. 4. 71)	1.4	(8. 3. 72)	7.4	(21. 7. 71)
		0		1.0	(2. 2. 72)	0.25	(6. 7. 71)	6.4	(30. 6. 71)
	38	0.20	(8. 9. 71)	32.2	(16. 12. 70)	1.75	(16. 3. 71)	7.3	(10. 11. 71)
		0		0	(10. 11. 71)	0.08	(10. 5. 72)	6.4	(16. 3. 71)
2	2	0.20	(8. 6. 71)	24.1	(17. 2. 71)	2.05	(8. 3. 72)	7.5	(10. 11. 71)
		0		3.0	(17. 8. 71)	0.1	(10. 5. 72)	6.4	(11. 5. 71)
	8	0.12	(6. 7. 71)	20.1	(23. 6. 71)	2.8	(12. 4. 72)	7.6	(21. 1. 71)
		0		0.4	(10. 11. 71)	0.15	(7. 12. 71)	6.4	(11. 5. 71)
	38	0.06	(6. 7. 71)	28.0	(14. 4. 71)	4.2	(2. 2. 72)	7.4	(21. 1. 71)
		0		0	(10. 11. 71)	0.2	(28. 7. 71)	6.4	(28. 7. 71)

(Units are mg/litre except for pH)

373.

POLLUTION OF A SPRING SOURCE BY SILAGE LIQUOR
Bristol Waterworks Company

1. INTRODUCTION

The Company's Stoke Bottom Treatment Works is in a valley in the eastern part of the Mendip hills, about $3\frac{1}{2}$ miles north-east of Shepton Mallet at grid reference ST/657482. Water from several springs in the vicinity are piped to the works, and Fig. 115 shows the Ashwick Grove spring sources. These sources, in Carboniferous limestone at 500 ft AOD, have been shown by lycopodium spore tracing to be fed by underground drainage from swallet depressions approximately along the 700 ft contour and also in the Carboniferous limestone.

2. POLLUTION: AN INCIDENT TAKEN FROM THE COMPANY'S RECORDS

In June 1969, the Middle spring rising was polluted by silage liquor discharged through a land drain into the swallet depression at Blakes Farm. Table 38 shows the average chemical and bacteriological analyses for this source for 1969. During the pollution incident the raw water smelled of silage and showed a maximum phenol content of 0.04 mg/litre as phenol (after distillation and diazotisation). The first hint of pollution was a chlorphenol taste in the chlorinated water. Field inspection of known swallets and farms led to discovery of the cause of pollution and the discharge of silage liquor was stopped. Farmers are now requested to contain silage liquors and spread over a wide area of land away from known swallet depressions.

A comment on methods of silage liquor disposal is reported on p. 164-5.

TABLE 38.

ASHWICK GROVE SPRINGS

RESULTS OF EXAMINATION: AVERAGE 1969

Chemical

Total hardness, as $CaCO_3$	247 mg/litre
Alkalinity, as $CaCO_3$	232 mg/litre
pH	7.2
Oxygen absorbed (N/80	
$KMnO_4$, 3 hours at $37^{o}C$)	1.30 mg/litre
Phenols, as phenol	< 0.001 mg/litre

Bacteriological

Plate count, per ml.	
$37^{o}C$, 2 days	524
Presumptive coliform organisms	2800
per 100 ml.	
E. coli , per 100 ml	1700

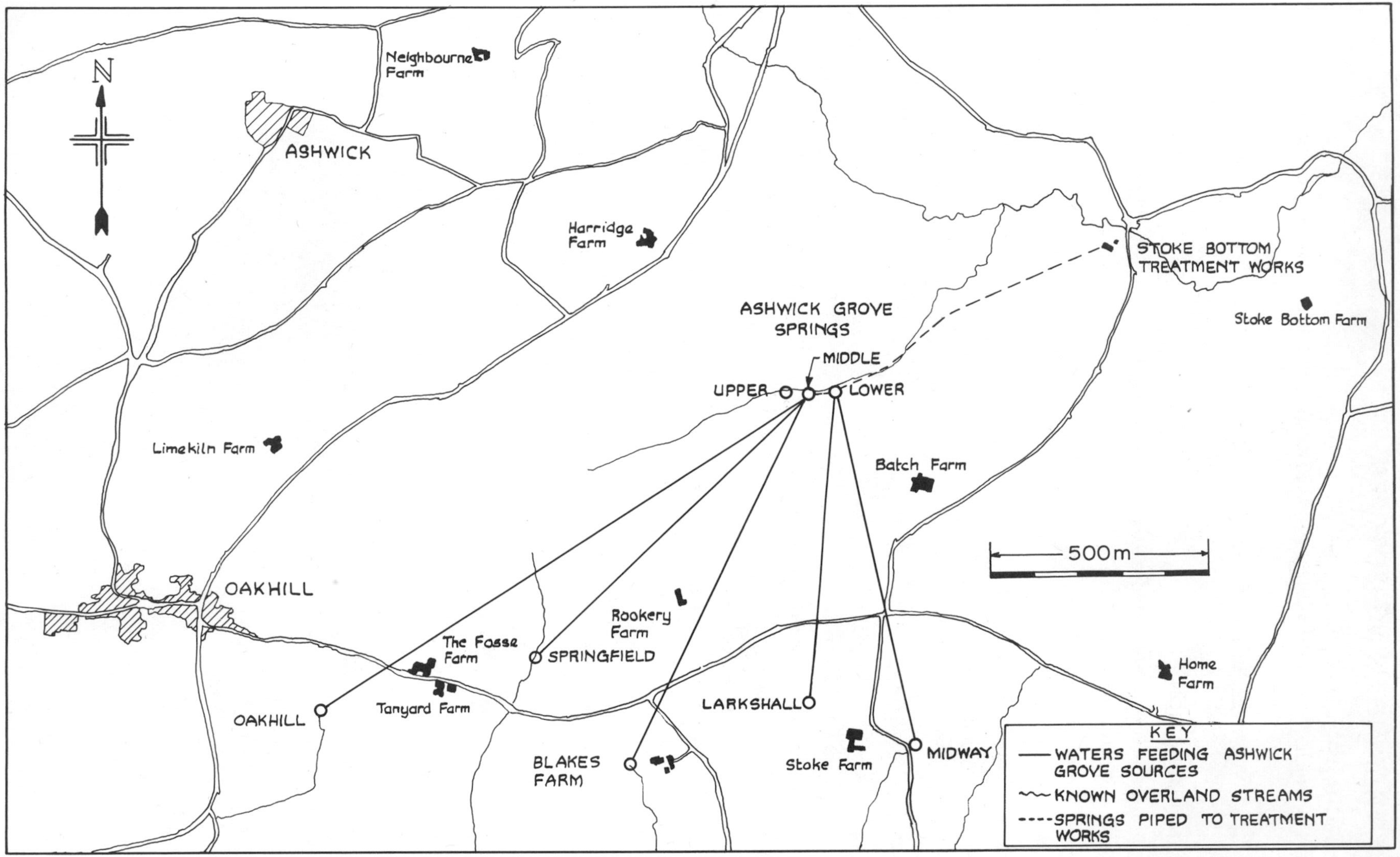

FIG. 115. ASHWICK GROVE SPRING SOURCES, STOKE BOTTOM

FLOW TRACING USING ISOTOPES
by D.B. Smith
United Kingdom Atomic Energy Authority, Harwell, Berks, England

1. INTRODUCTION

Pollution of groundwater usually originates from mans' activities on the ground surface. The pollutant is first transported by water into the unsaturated zone of the strata where it may be retained by adsorption or ion exchange. If it reaches the water table, transport is then controlled by the hydraulic gradient and permeability. The concentration of pollutant remaining in the groundwater when it is abstracted or discharged by springs is of prime importance. Radioactive isotopes provide valuable and unique techniques for investigating many of the different facets of underground water and pollutant movement.

The movement of the water controls the transport of all the soluble pollutants and if its movement is known, then an upper limit can be set to any pollution transport with the additional safeguard that the pollutant may be removed from the groundwater by interaction with the strata or by chemical breakdown. Tritium is unique in providing an almost perfect tracer for groundwater movement.

The movement of a specific pollutant may be traced using a radioactive tracer. The chemical form of the pollutant would have to be established and, as is often possible, an identical radioactively-labelled compound prepared. The movement of this material would then include the effect of retention of the pollutant by the ground. Such an investigation would need to give careful consideration to the choice of the radioactive label to ensure that the tracer continued to follow the hazardous portion of the material in the event of chemical breakdown.

Flow and movement studies using radioisotopes can be divided into two distinct groups of techniques. The first uses added tracer to follow movement or to measure movement using instrumental techniques. The second uses the 'natural' radioactivity in the water and interprets the results of very low activity measurements of tritium or Carbon-14 which occurs in rain, surface and groundwater as a result of cosmic radiation in the atmosphere and as a result of thermonuclear testing.

2. ADDED RADIOACTIVE TRACERS

2.1. Selection of tracer

Tritium, being an isotope of hydrogen, is incorporated within the water molecule. It is not subject to adsorption or other chemical loss and provides an excellent tracer for water. Unfortunately, it only emits low-energy β particles (max. energy 18 keV, half-life 12.26 years) and hence it cannot be detected in the field using portable equipment. Samples must be taken and analyzed in the laboratory. This is inconvenient, although the analysis of trace quantities is easily carried out in the laboratory using an automatic liquid scintillation detector.

Since in-situ measurement of tracer in groundwater is very desirable, a great deal of research has been carried out in an endeavour to find a good radioactive groundwater tracer which emits γ radiation.

The most generally applicable is the Bromide ion, using Bromine-82 which emits a series of high and medium energy γ-rays with a half-life of 36 hours. This is far too short for many investigations and Iodine-131 (half-life 8 days) is an alternative but more toxic isotope. It is also less likely than bromide to form a satisfactory tracer in clayey conditions. Both tracers are more effective when non-active 'carrier' bromide or iodide is added to the tracer to increase the quantity of the anion which is present.

A search for alternatives has indicated that Chromium-EDTA (Chromium-51, half-life 28 days, weak γ emitter) and Potassium Cobalticyanide (Cobalt-60, half-life 5.3 years or Cobalt-58, half-life 71 days) are fairly good chemical forms of Chromium and Cobalt for groundwater tracing although both have only been used under fairly limited conditions.

None of these tracers can be relied upon to follow groundwater through clayey strata or through areas of high ion exchange capacity. When a 'negative' answer, showing the traced water has not come to a particular sampling point is of great importance, then tritium is virtually the only reliable tracer to use.

2.2. Tracer and instrumental techniques

Information on groundwater movement can be obtained by using several different radioactive tracer techniques or by using these methods in conjunction with simple field instrumentation.

2.2.1. Water tracing and interconnection of water supplies.

The high sensitivity of radioisotope detection enables large-scale problems to be studied. Gamma emitting tracers can be measured in-situ. Tritium provides the only practical tracing method for many groundwater studies where it is essential to the interpretation that tracer loss must not occur in the strata.

2.2.2. Borehole investigations.

Labelling the complete column of water in a borehole followed by logging of the tracer can be used to identify horizons at which flow occurs. Similar techniques using a pulse of tracer in a borehole can be used to measure the vertical flow velocity and identify points of entry and exit of the water. Such a method can work at very low velocities with the tracer density accurately matched to the water (which is not possible with most chemical tracers) and with no lower velocity limit such as that set by current meter measurements.

2.2.3. Single well technique.

This can be used to provide quantitative values for groundwater velocity. The tracer is injected into a borehole, allowed to move with the natural velocity of the groundwater for a predetermined time and then recovered by pumping the injection borehole. From the distribution of the tracer concentration observed during recovery, the groundwater velocity can be calculated.

2.2.4. Filtration velocity and direction.

To measure the horizontal groundwater velocity at a specific horizon, a short length of perforated (or unlined) borehole can be isolated with packers. A radioactive tracer solution, accurately matched to the water density, is stirred into the isolated borehole section and the rate of disappearance of the tracer is measured with a detector incorporated in the instrument. In a mixed system, the radioactivity disappears such that:

$$V_a = \frac{V}{Ft} \quad Ln \frac{C}{C_o}$$

where V_a = apparent groundwater velocity

$\quad\quad V$ = volume of the isolated section of the borehole

$\quad\quad F$ = cross section of the isolated section perpendicular to the water flow

$\quad\quad t$ = time interval between concentration measurements C_o and C

This method is useful but several problems occur in practice, errors arising from the vertical velocity component of the water and the distortion of the natural flow due to the presence of the borehole or the gravel packing surrounding it.

The direction of flow can be investigated in two ways. An isotope is required which adsorbs onto the ground or onto a mesh forming part of the instrument. In one method, the isotope is carried into the strata and a collimated detector is used to observe the direction of transit. In another, a mesh of iron gauze is placed at the periphery of the borehole and chromium chloride (Chromium-51) solution is released in the centre some time after inserting the instrument. The mesh is oriented by rigid drill rods and is retrieved to locate the area on which the isotope has deposited, thus establishing the flow direction.

2.3. Examples of tracer techniques

The following examples are intended to illustrate practical applications of radioactive techniques, carried out principally in the UK.

2.3.1. Simple near-surface water movement tracing.

In one investigation in a sand-clay alluvium, an auger hole was drilled to 4 m depth, 2 m below the water table. Sodium iodide 'carrier' solution was added to the hole followed by 100 mCi of Iodine-131 solution. Observation holes were drilled at a radius of 60 cm and 120 cm round the central bore and were periodically logged and sampled for Iodine-131. The water moved between north and north east at a rate of less than 30 cm/week.

Similar techniques have been used to establish the rate of water movement in peat deposits. Following injection of 40 mCi of tritium tracer, sampling was undertaken using an auger in a direction down the hydraulic gradient from the injection site. The sampling pattern was set out on a grid and showed the three dimensional movement of the tracer in addition to its movement into the boulder clay beneath the peat. In one investigation, the mean movement was 5.5 m in 18 months on a hydraulic slope of 10%. This indicated a permeability expressed as the velocity in cm/sec under a hydraulic gradient of 1 cm/cm of 1.4×10^{-4} cm/sec. Another series of investigations on blanket peak produced a value of an order of magnitude lower than this and showed the peat to have a similar permeability to clay.

2.3.2. Near-surface water movement in the unsaturated zone.

The only successful radioactive tracer which has been used to study water infiltration through the humic layer and into the weathered layer of the ground has been tritium. The method involves watering tritiated solution evenly over an area of say $50 \, m^2$ at a rate of about 0.1 mCi tritium/m^2 using a predetermined amount of water to simulate the type of rainfall which is being investigated. Sampling is carried out by distilling water from short lengths of core and assaying for the tritium. Results of limited work show how

irregular (and, initially rapid) is the penetration and movement of the water in the
inhomogeneous weathered layer. In fine grained material, the water displaces that
water which is held at field capacity in the strata but movement in weathered rock
is also by fissure or crack transport.

2.3.3. River supply to groundwater.

Tritium was injected at a constant rate for a month into a small river during
a dry period. Samples were taken from a series of observation wells in the surrounding
alluvial gravels. The movement of tritium into the gravels, which were in hydraulic
continuity with the chalk beneath, was observed. In the same investigation, the river
flow was obtained by measuring the tritium concentration in the river and by comparing
these results with current meter measurements, it was possible to estimate that there
was a 15% loss of the river flow to the gravels.

2.3.4. Lake bed leakage.

A small artificial lake situated on Oolitic limestone was leaking through the
silt bed. A solution of sodium chromate (7 mCi Chromium-51) was spread over the lake.
The lake bed was monitored after 8 days when most of the tracer had seeped away using
a scintillation counter which had previously been used to establish the natural background
and areas of high count rate were observed where the water had leaked through the bed
and the radioactive tracer had stuck to the silt. Grouting of these areas cured the
major leaks in the lake.

2.3.5. Movement of water in mine workings.

An investigation was carried out in collaboration with the Usk River Authority
at the head of Ebbw Vale to evaluate the contribution to local groundwater and spring
supplies arising from approximately 1.4 m.g.d. of water flowing through a disused
mine. The knowledge was required to assess the effect of abstraction of this water
as a high quality industrial source.

Tritiated water (50 Ci) was added to the flow in the mine over a period of
29 hours so that the tracer in the flowing water was initially at such a low concentration
that after a further dilution by a factor of two it would conform with the drinking water
tolerance recommended by the International Commission on Radiological Protection
(ICRP) (see Sect. 2.4.). Samples were taken from three mine adits or seepages which
discharged small flows into the head waters of the Ebbw Fach, some $1\frac{1}{2}$ km from the
injection site. Samples were also taken from two pumped groundwater supplies used for
industrial purposes in Ebbw Vale and from springs 4 km east, supplying domestic water.
Other springs and streams in the district were also regularly sampled at distances up to

7 km from the injection site.

Tracer was first located two days after injection at a minor mine adit. It issued from a second mine adit one day later and was also found in the river head waters. After eleven days, low concentrations were observed in the pumped groundwater supplies used in the valley steelworks. Peak concentrations occurred at the seepage points between four and eight days after injection and in the groundwater after 23 days. These peak concentrations were all less than the ICRP recommended drinking water tolerance. By analyzing the tritium concentration and the flow at the observation points, it was found that all the injected tracer had been accounted for after about 100 days.

The groundwater flow was rapid and complex in this extensively mined area. Tritium proved a satisfactory tracer. Fluorescein had been used in an earlier investigation and had not been located, probably due to loss in the disturbed strata.

2.4. Safety considerations

The radioactive tracer is in a concentrated form only before injection into the system under investigation. With simple precautions, such as lead shielding and handling tongs, there is no hazard to a trained operator. When using tritium it is only necessary to take precautions against ingestion or inhalation.

It is essential that the use of radioactive tracers should be planned to minimize the radiation dose to any person exposed to the radioactivity. With this objective, investigations use the minimum amount of tracer compatible with the required sensitivity of detection and use the shortest possible half-life isotope. Tracer is injected into the system at a large dilution or over a long period so that highly concentrated material cannot occur in, say, a local groundwater supply.

The recommendations of the International Commission on Radiological Protection provide a guide to the maximum permissible concentration of radioactive isotopes in drinking water. These are summarized for the relevant isotopes in a Guide to the Safe Handling of Radioisotopes in Hydrology (IAEA, Vienna, 1966). It is usually possible to carry out an investigation at a considerably lower concentration than that recommended by the ICRP and the short duration of most investigations provides an additional safety factor.

3.   NATURAL TRITIUM TECHNIQUES

3.1.   Tritium in rainfall

Tritium is an isotope of hydrogen which emits low-energy β radiation with a half-life of 12.26 years.   It occurs naturally in the atmosphere where it is incorporated in the water molecule and enters the hydrological cycle principally as rain.

The original concept of tritium measurements in groundwater was to 'date' the water using the fact that rainfall contained 5 to 10 TU[*] and this would reduce with a half-life of 12.26 years.   This is analogous to the use of Carbon-14 in archaeological dating.

This approach has been changed by the production of tritium during thermonuclear testing and the tritium concentration in rainfall has increased to mask the truly natural level.   Tritium concentration rose to a peak value of 600 TU in 1954, a second peak of 460 TU in 1958 and a major peak of 3300 TU in 1963.   These peaks occurred during early summer when tritium is transferred from the stratosphere to the troposphere and the tritium concentration oscillates annually with the lowest values in winter.   The mean annual values of tritium weighted for rainfall and allowing for the tritium decay to 1972 are of more interest in current groundwater studies and these show values of 38 TU in 1954, 109 TU in 1958 and 707 TU in 1963/64 since when they have declined to approximately 110 TU from 1968 to 1972.   The present values for 1972 are expected to show a peak of about 200 TU in the summer and a low of 30 TU in winter.

With this complex input of tritium over a period of 18 years quantitative interpretation of groundwater tritium measurements is difficult.   By taking a series of measurements over a period of time, interpretation may be made easier and less ambiguous.   Nevertheless, valuable information can often be obtained on the sources and interconnections of groundwater supplies.   In addition, much tritium 'data' may be stored in the unsaturated zone of an aquifer and the measurement of natural tritium provides a valuable additional technique for groundwater and pollution investigations.

3.2.   Tritium techniques in the unsaturated zone

Surface or rainfall water can travel downwards through the unsaturated zone of an aquifer either by intergranular seepage so that the incoming water displaces the water stored in the zone or by crack or fissure transport where downward movement is more rapid and relatively independent of the water held within the body of the strata.   By measuring the tritium content of the water at different depths below surface, these two transport mechanisms can be separated.

[*1] Tritium Unit (TU) occurs when there is 1 tritium atom to 10^{18} hydrogen atoms and produces 7.2 disintegrations of tritium per minute per litre of water.

Tritium measurements in chalk groundwater where the unsaturated zone was greater than about 15 m thick consistently showed very low tritium values in the groundwater. A joint project was undertaken with the Water Resources Board to investigate this unexpected result. A core was obtained by percussion drilling using no water lubrication and samples of the chalk were taken at 50 cm intervals. Water was abstracted by vacuum distillation and the tritium content of each sample was measured.

Fig. 116 shows the mean tritium value of rainfall, weighted for the volume of rain and corrected for half-life decay to the sampling time in the chalk bore. Fig. 117 shows the tritium distribution in the core. The similarity of the shape is evident, the prominent peak due to high tritium values in 1963/64 being located at approximately 4 m depth and the 1954 tritium input being at about 11 m. The results show that water movement down through the chalk is predominantly by intergranular flow with a smaller contribution due to crack or fissure flow as shown by the lower tritium values at depths below 12 m. This does not agree with the classical concept of fissure flow in the chalk but is not inconsistent with the observed water level increase in the water table after heavy rainfall. Except near the surface, the unsaturated zone is at field capacity and heavy rainfall produces a displacement of the water and causes a discharge to the water table. From the aspect of pollution, it is important to note that this investigation shows that this is not the same water which had fallen as rainfall immediately prior to the rise in the water table.

The tritium profile technique is applicable to other aquifers and is being used to examine downward movement of water in clay. Drilling to obtain cores or samples in harder formations in the absence of drilling fluid is proving a practical problem.

The technique has obvious application to pollution or potential pollution studies. Although it is limited to areas where the water table is deep (below 10 or 15 m) and would not be applicable where water movement is predominantly by fissures (eg. Carboniferous limestone), there are still very many areas in which it could be used.

In the event of the selection of a major tipping site, tritium data in the unsaturated zone are 'storing' information relating to nearly 20 years of water movement below the site, providing the water table is sufficiently deep. This information would be complementary to any tracer studies which were undertaken since such investigations would necessarily only be able to examine fairly short-term water movement.

Tritium moves with the velocity of the groundwater and as such is transported more rapidly than pollutants which may be subject to delay due to viscosity, adsorption or ion exchange. This should be taken into consideration when considering specific pollutants which are subject to loss.

3.3. Tritium in groundwater supplies

A single measurement of tritium in a groundwater supply can provide valuable information about the origin of the water. Absence of tritium would show that the supply consisted entirely of pre-1954 water and thus was safe from short or medium term pollution effects from the surface. High tritium (60 to 100 TU) would show it to be primarily recent rainfall and therefore potentially susceptible to pollution from the surface.

The interpretation of this type of data must be done with a full consideration of the hydrogeology of the area to reduce ambiguity. A high tritium value can arise from current spring and summer rainfall or from a mixture of the previous few years rainfall. Intermediate values indicate mixtures of old and recent water. By measuring the tritium content over a period of time, it can often be established whether the recent component is current rainfall or is an average of the last few years of rain. This could be of some importance for interpretation of the pollution risk.

Many hundreds of tritium measurements of groundwater have been made and these can already be used to establish a general pattern for a number of types of aquifer. In general, water from deep sources in chalk or sandstone shows low or zero values of tritium and most confined aquifers are near zero even a short distance from the recharge area. Limestones are an exception and may be high even when confined conditions have applied for a distance of 10 km. However, general measurements are not always applicable to specific sources and several individual boreholes within a low tritium area have been found to contain tritium arising from faulty casing, local leakage or induced river recharge.

Information on the origin of water containing no tritium can be obtained by measuring the Carbon-14 content of the bicarbonate dissolved in the water. This technique is capable of measuring the 'age' of waters back to about 25 000 years and may prove useful in separating connate waters from recharged groundwater and in examining water movement where the transport rate is very low.

4. CONCLUSIONS

Radioisotope techniques involving the use of added radioactive tracer or the study of the distribution of environmental radioisotopes continue to be used increasingly to provide a better understanding of groundwater behaviour. They add to and complement the many techniques already available to the geohydrologist and in particular, natural tritium studies can provide unique information relating to water movement and to the potential movement of water-borne pollutants.

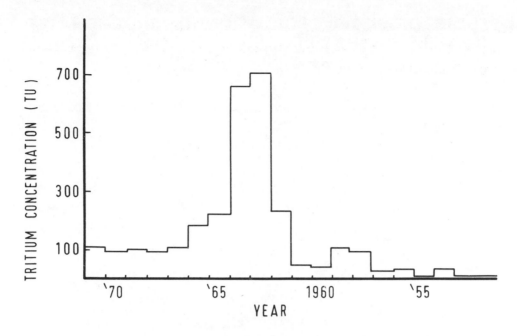

FIG. 116.　TRITIUM IN RAINFALL (UK)
(Weighted means corrected to Jan. 1972)

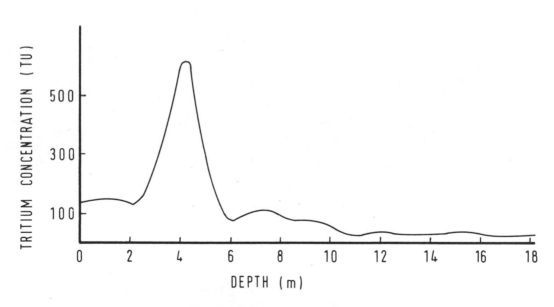

FIG. 117.　CHALK PROFILE
(Tritium content of water in core, Oct. 1968)

LOCAL STUDIES OF MISCIBLE POLLUTIONS OF GROUNDWATER: THE SINGLE WELL PULSE TECHNIQUE

by J.J. Fried, P.C. Leveque, D. Poitrinal and J. Severac
Ecole des Mines de Paris, Fontainebleau, France
Laboratoire de Radiogéologie du Bassin d'Aquitaine — Faculté des Sciences de Bordeaux, France

1. INTRODUCTION

The propagation of pollutants in soils which may be considered as stratified depends on local parameters. The problem is to determine these parameters in each stratum at relatively low cost. In order to reach this aim, a method has been developed at the Laboratoire de Radiogéologie du Bassin d'Aquitaine and at the Centre d'Informatique Géologique. It consists briefly in the following steps:

A well, completely penetrating the aquifer is screened throughout its entire depth. It is filled with water labelled by a radio-tracer at constant concentration. This labelled water is pushed back by water labelled at the well, then by fresh water and then pumped back into the injection well. At each level, tracer concentration is recorded with respect to time. The measurements have been used in two ways:

i) Concentration variations at a given level with respect to time and at a given time with respect to depth yield relative permeabilities of various strata (1)(2)(3).

ii) Concentration variations at a given level with respect to time allow a quantitative evaluation of the longitudinal dispersion coefficient at each level, using mathematical models and the dispersion scheme (4)(5).

2. EXPERIMENTAL STUDIES

2.1. Description of the method (shown in Fig. 118)

2.1.1. Background to the investigation.

The radioactive tracer is injected into the aquifer from a screened piezometer. The radioactivity is measured along the whole depth of a screened pipe. At each point, the activity is a function of:

i) the concentration of the tracer in the piezometer itself;

ii) the volume of the aquifer invaded by tracer around the point. The cloud of tracer is distorted according to the hydrodynamical properties of each stratum: it penetrates easily transmissive layers; its extension is far smaller in semi-pervious strata;

iii) the volumic radiation emitted by the invaded aquifer: in addition
to the tracer concentration, this radiation depends on the effective
porosity and on the density of the solid matrix.

The characteristics of the probe and the mathematical simulation show that
the effects pointed out in i) and iii) are widely predominant.

2.1.2. The tracer.
Different radioactive tracers have been tried:

- Br 82   half-life = 36 hours   peak of energy = 0.78 - 1.47 MeV
- I 131   8.05 days   0.36 - 0.64 MeV
- Cr 51   27.8 days   0.32   MeV

In practice I 131 or Br 82 are generally used.

The required activity is very low: some μCi.   These isotopes are
respectively in form of sodium bromide and sodium iodide solution contained in a medical
flask.   The solution is taken out from the flask using two hypodermic needles by means
of a water flow.   The labelled water is mixed with about 4 m^3 of water (for a well 30 m
deep) by feed back pumping.   During this mixing, the carrier is added.

The apparatus needed for such a process is very simple and safe.   It is
especially suitable for field work.

Such a tracer has some very interesting properties:

i)   It does not present any danger for the field staff or the environment
because of its short half-life and its low activity.

ii)   It has a good detectability.

iii)   It does not modify the water density.

2.1.3. Gauging apparatus.
The probe measures activities with solid medium scintillation process.   The
impulse sent by the probe through the supporting cable is received by an integrator and a
counting scale.   A recorder controlled by the moving down of the probe draws the log
(Fig. 119).

The properties of the probe have been investigated with two physical models:

i)   The resolutive power (see Appendix)

ii)   The size of the volume of influence: this investigation has been
      carried out on a cubic metre container divided in cylindrical
      coaxial sectors.   It was possible in each one to inject a
      radioactive tracer.   A concentric cloud around the piezometer
      is not observed further than one metre away at best (Fig. 120).

2.1.4. The piezometer.

Its diameter must take into account the diameter of the probe such as to reduce
the effect described in (i) of Sect. 2.1.1.).   It must be screened (> 20%) on the whole
length of the investigated aquifer.   The screened pipe is covered with nylon linen the
mesh of which is about 0.5 to 1 mm.   A gravel pack must not be present since this would
act as a screen between the injected aquifer and the probe.

2.1.5. Description of the experiment.

i)   First step
      Filling up the piezometric column with tracer by pumping at the head
      of the borehole and injecting at its base.

ii)   Second step
      When the activity of the water pumped out is constant, pumping is stopped.
      The injection of the tracer begins.   In the case of an unconfined aquifer,
      the elevation of the water table must be small for two reasons:

            - in order to insure safety, and
            - in order not to inject tracer above the investigated aquifer.

iii) Third step
      Injection of unlabelled water.

iv)   Fourth step
      Pumping back into the injection borehole.   The injection of unlabelled
      water pushes the cloud of labelled water further from the well;   then,
      during the pumping back, it is possible to investigate the return of
      the radioactivity in each stratum.

In each step, the flow-rate must be low enough: then it is possible to consider that the time necessary for a log is very short in regard to the evolution of the system "traced aquifer". Logs are made as often as possible during the whole test.

Such tests have been carried out about fifty times for several purposes; e.g. relationships between lakes and rivers, industrial settlements, dam site investigations, earth dam surveys, etc.

2.2. A qualitative evaluation of permeability

An experiment with Br 82 is given as an example. Logs in Figs. 121 and 122 have been drawn considering the radioactive decay of the isotope. This test has been carried out in the alluvial aquifer of Gironde in south west France.

From these logs there are indications of two strata characterized by;

- peaks of activity during the injection,
- nodes during injection of non-active water, and
- the fact that during the pumping, these strata send back more activity
  than others.

Then it is evident that the transmissivity calculated by a classical pumping test is mostly due to these strata.

Fig. 123 shows the evolution of activity for various permeable layers during the pumping back. It may be seen that in most cases the recovery rate of activity is about 85 to 95%.

3. MATHEMATICAL MODELS

The mathematical models are based upon the dispersion equation, which is assumed to be the general diffusion equation:

$$\text{div} \left(\overline{K} \rho \, \text{grad} \, \frac{c}{\rho} \right) \; - \; \text{div} \; (uc) \; = \; \frac{\partial c}{\partial t} \qquad \dots\dots\dots\dots 1$$

where $\overline{K}$ is the dispersion tensor, c the pollutant concentration, u the pore velocity and ρ the mixture density.

Of course, this equation is not used as it stands but is simplified by as many assumptions as the problem allows (Sect. 3.1).

3.1. Description of the models

The simplifying equations, justified by the types of soil where the method is applied and by its experimental setting, are the following.

The medium is made up of homogeneous, horizontal, independent strata. Only small quantities of radioactive tracer are used, which implies that the density and viscosity of the mixture do not vary and are equal to those of fresh water. The coordinate system is chosen to coincide with the principal axes of the dispersion tensor, which is then put in its diagonal form.

The velocity regime is that of dynamic dispersion (5) and the dispersion coefficients K_i can be written as:

$$K_i = \alpha_i \left| u \right|$$

where u is the pore velocity and α_i the intrinsic dispersion coefficient (which has the dimension of a length).

It is intended to measure α_L, the longitudinal dispersion intrinsic coefficient. The horizontal and vertical transverse dispersion are assumed to be negligible. The model is then;

i) $\operatorname{div} \left(\alpha \left| u \right| \operatorname{grad} c \right) - \operatorname{div} (uc) = \dfrac{\partial c}{\partial t}$

ii) $\dfrac{\partial^2 h}{\partial r^2} + \dfrac{1}{r} \dfrac{\partial h}{\partial r} = \dfrac{S}{T} \dfrac{\partial h}{\partial t}$

iii) $u = u_n - \dfrac{k \rho g}{\mu \phi} \operatorname{grad} h$ 2

where h is the piezometric head during the injection or the pumping and u_n the natural aquifer velocity. μ is the dynamic viscosity, ϕ the porosity, k the permeability coefficient, T the transmissivity and S the storage coefficient. Convenient boundary and initial conditions are added.

Two cases are considered: either the aquifer velocity (u_n) may be neglected with respect to the imposed velocity, or u_n is taken into account in a) and b) respectively.

a) u_n is neglected: dispersion is purely radial and cylindrical coordinates are used to write the dispersion equation as:

$$\frac{1}{r}\frac{\partial}{\partial r}\left(\alpha_L\, r\, |u|\, \frac{\partial c}{\partial r}\right) - \frac{1}{r}\frac{\partial c}{\partial r} = \frac{\partial c}{\partial t} \qquad \ldots\ldots\ldots\ldots 3$$

b) u_n cannot be neglected: the principal directions of the dispersion tensor are respectively tangential to the streamlines and the equipotential lines and Equation 2-i) reduces to:

$$u^2\frac{\partial}{\partial \phi}\left(\alpha\, |u|\, \frac{\partial c}{\partial \phi}\right) - u^2\frac{\partial c}{\partial \phi} = \frac{\partial c}{\partial t} \qquad \ldots\ldots\ldots\ldots 4$$

in the ϕ - ψ coordinates system, where ϕ = constant are equipotential lines and ψ = constant are streamlines.

It should be noted that in most of the experiments we performed, u_n could be neglected and the axially symmetrical case was taken.

3.2. Determination of the longitudinal dispersion coefficients

Two methods may be used to determine the dispersion coefficients:

- a semi-analytical formula derived from Equation 3.
- direct simulation and curve fitting.

3.2.1. Semi-analytical formula (6).

Case a) above is assumed when u_n is neglected. It has been shown that the velocity steady state is obtained after a relatively short lapse of time (a few minutes of injection or pumping), thus the velocity is taken as

$$u = \frac{A}{r}$$

with $A = 2\,\pi bQ\,\phi$ (where b is the aquifer width, Q the injection rate and ϕ the porosity). Velocities are oriented outward from the well (a minus sign indicates pumping). α is deduced from Equation 3 as:

$$\alpha = \frac{r\frac{\partial c}{\partial t} - A\frac{\partial c}{\partial r}}{A\frac{\partial^2 c}{\partial r^2}} \qquad \ldots\ldots\ldots\ldots 5$$

$\frac{\partial c}{\partial t}$ is obtained from experimentation and Equation 5 yields:

$$\alpha = \dfrac{r \dfrac{c_{n+2}(1) - c_{n+1}(1)}{dt} - \dfrac{c_{n+3}(1) - c_{n+2}(1)}{r_3 - r_2} A}{\dfrac{A}{r_3 - r_1} \left| \dfrac{c_{n+3}(1) - c_{n+2}(1)}{r_3 - r_2} - \dfrac{c_{n+2}(1) - c_{n+1}(1)}{r_2 - r_1} \right|} \quad \dots\dots\dots\dots 6$$

where $c_n(i)$ is the concentration at time t_n at a point i lying at the distance r_i from the well axis; r_1 is the well radius.

Equation 6 has been obtained from a discretization of 5, assuming that the transition zone variation during a relatively short time step dt may be neglected with respect to its width. Then, the tracer moves without dispersion and, under this assumption, the concentration at the well at time t+dt is the concentration at time t at a point r_1+dr where dr is given by the formula:

$$dr = \int_t^{t+dt} u \; dt$$

Note that the probe 'sees' radioactivity at some distance from the well. Thus, as developed previously, the deconvolution method could be used to determine the concentration gradients directly. Work along these lines is being carried out.

3.2.2. Simulation.

The finite difference form of Equation 2, reduced to a mono-dimensional system, has been used. The scheme adopted is totally implicit, with backward differences in the velocity direction for Equation 2 (i). Direct resolution of the linear system of discretized equations has always been used.

The computed solutions are fitted to the experimental solutions by curve fitting. It has been shown (7) that it is possible to obtain the parameters (α, u) in one experiment by such a fitting.

As an example, one curve fitting obtained during an experiment conducted in the Rhone Valley, near Lyon is given (Fig. 124). It represents the concentration variations at the well versus time at a given level. During this experiment, the dispersion coefficients were measured at nearly all levels, every 25 centimetres. For heterogeneous media, values of α of the order of 60 cm and for more homogeneous parts of the aquifer, values of the order of 10 cm, were found.

394.

This experiment is part of a project to build predictive models for accidental pollution near the pumping wells of an urban area.

4. CONCLUSION

The single well pulse technique has proved a good way to localize zones which are most sensitive to pollution in an aquifer, i. e. where permeability and dispersion are highest. It gives quantitative evaluation of the dispersion coefficient in each stratum and thus allows the derivation of a predictive model for the most pollutable strata.

From an experimental point of view, this method is interesting as most of the radioactive tracer is pumped back; furthermore, if the lifetime of the radioactive elements is short, the aquifer will not be polluted at all by the experiment. Besides, this is a good argument to use artificial radioactive tracers in hydrology.

APPENDIX
INVESTIGATION OF THE RESOLUTIVE POWER OF THE PROBE

The measurement is made with a loss of information which characterizes the resolutive power of the probe.   A deconvolution process is used to avoid this effect. The following assumption was made

$$s(j) \ = \sum (e(j-i) \, \phi(i))$$

where s(j) is the activity measured with the apparatus at depth j, e is the distribution of tracer along the piezometer, ϕ is the kernel function characteristic of the probe working with a given isotope.

First step: with an experimental model, the response of the probe for one stratum invaded by tracer is measured.   Using the deconvolution numerical process, calculations were made as shown in Fig. 125, considering that the activity 1 is the spatial concentration of tracer in a sandy stratum (the sand of the stratum is calibrated at 2 to 3 mm).

Second step: it is now possible by deconvolution of s by ϕ to find the distribution of the tracer in an invaded borehole.

Fig. 126 shows the result in a case where measurements were made every 10 cm.   On this figure, for e, 1000 is the spatial concentration as previously stated.

REFERENCES

1. DEGOT, B., and others

    Two applications of BR 82 in underground hydrodynamics.
    Symposium on Radioisotopes in hydrology, Tokyo, 1963.

2. LEVEQUE, P. C.

    The use of natural and artificial radioactive tracers in underground hydrology.
    Houille Blanche, 1969, (8), pp. 833-

3. LEVEQUE, P. C., and others

    Hydrodynamic criteria to guarantee perenniality for the surface geology storage of radioactive waste.
    71st Asian Conference, Tokyo, 1971.
    Ass. Int. des Hydrog, 1248-III-71-S.

4. FRIED, J. J.

    Miscible pollutions of groundwater: a study in methodology.
    Proceedings of the Intern. Symp. on Water Quality Modelling, Ottawa, Canada.

5. FRIED, J. J., and COMBARNOUS, M. A.

    Dispersion in porous media.
    Advances in Hydroscience, 1971, 7, pp 169-282.

6. FRIED, J. J.

    A mathematical model for the single well pulse technique.
    Symp. on Water Res., Bangalore, India, 1971.

7. FRIED, J. J.

    Doctoral thesis, Bordeaux 1972.

8. EMSELLEM, Y., and others

    Deconvolution and automatic identification of parameters in hydrology.
    International Symposium on Mathematical Models in Hydrology,
    Warsaw, 1971.

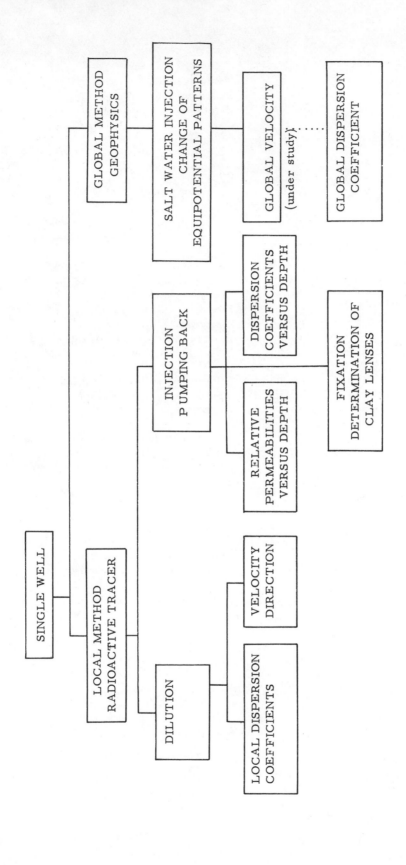

FIG. 118.   THE SINGLE WELL POSSIBILITIES

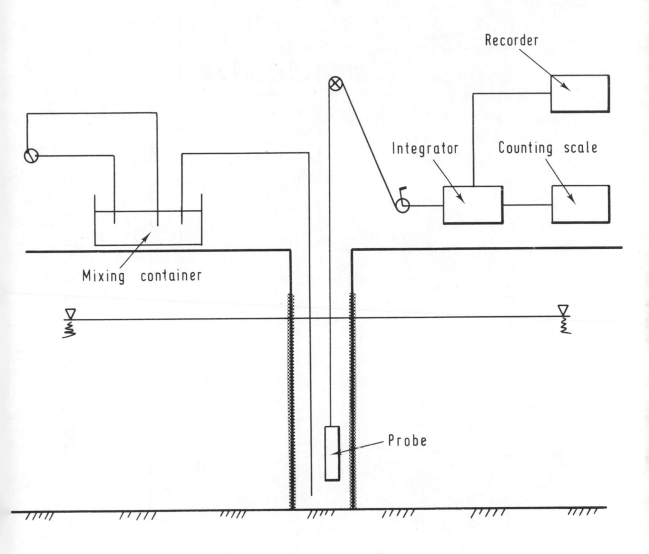

FIG. 119.   INJECTION IN THE SINGLE WELL

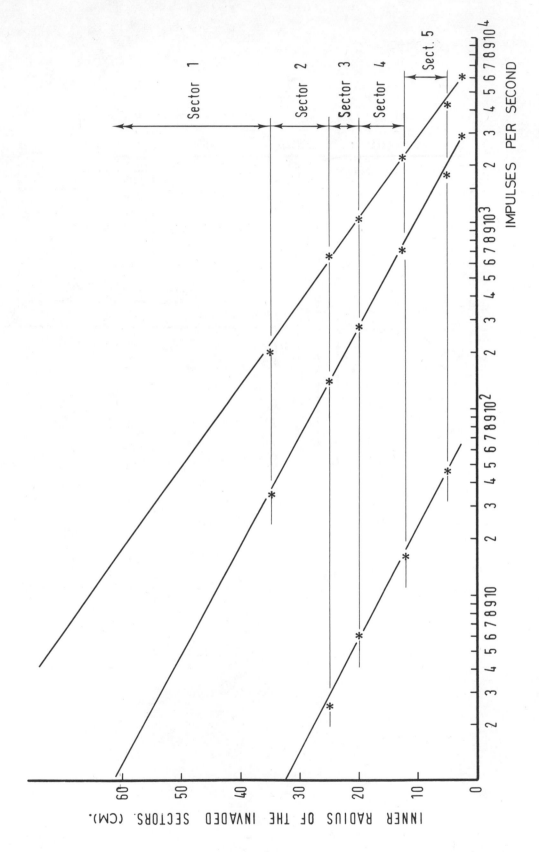

FIG. 120.   INVESTIGATION OF VOLUME OF   INFLUENCE OF THE PROBE.   ACTIVITIES IN INVADED SECTOR ARE SUCCESSIVELY: 1; 1 and 2; 1, 2 and 3; 1, 2, 3, 4; 1, 2, 3, 4 and 5.   THE CONCENTRATION OF THE TRACER AND THE LEVEL OF ENERGY OF MEASUREMENT ARE THE SAME AS THOSE USED ON THE FIELD

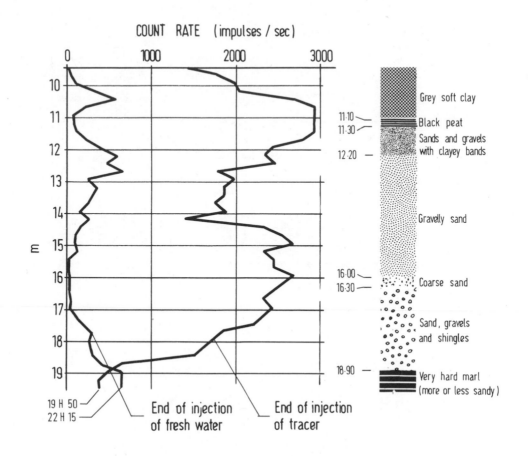

COUNT RATE (impulses / sec)

0   1000   2000   3000

11·10 — Grey soft clay
11·30 — Black peat
Sands and gravels with clayey bands
12·20 —

Gravelly sand

16·00 — Coarse sand
16·30 —

Sand, gravels and shingles

18·90 — Very hard marl (more or less sandy)

19 H 50 —
22 H 15 —

End of injection of fresh water   End of injection of tracer

FIG. 121.   Br^{82} INJECTION AT WELL 6, BORDEAUX-NORD PROJECT

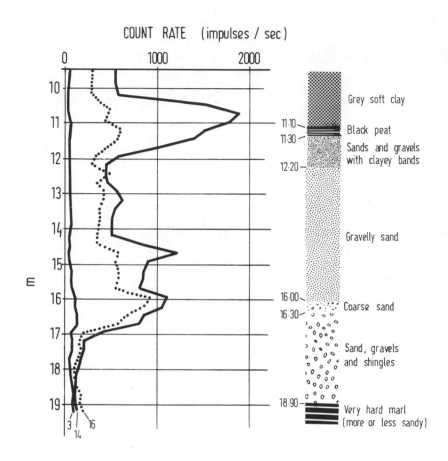

COUNT RATE (impulses / sec)

Grey soft clay

11·10 — Black peat
11·30 — Sands and gravels
with clayey bands
12·20 —

Gravelly sand

16·00 — Coarse sand
16·30 —

Sand, gravels
and shingles

18·90 — Very hard marl
(more or less sandy)

FIG. 122.   MOVEMENT OF WATER PUMPED AFTER INJECTION OF Br[82] AT WELL 6, BORDEAUX-NORD PROJECT

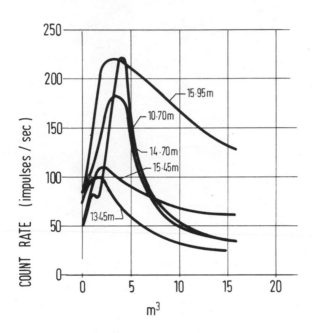

FIG. 123. ACTIVITY v. PUMPED VOLUME, FOR
VARIOUS PERMEABLE LAYERS, AT
WELL 6, BORDEAUX-NORD PROJECT

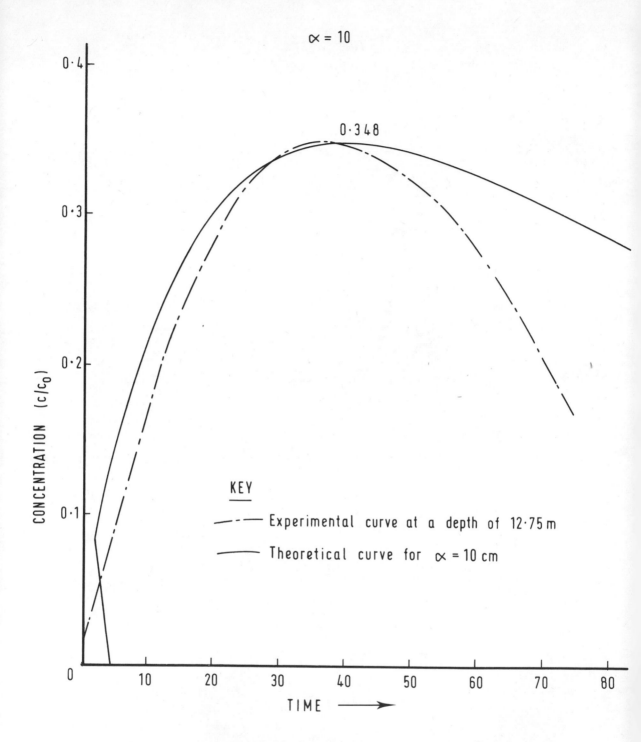

FIG. 124. CONCENTRATION VARIATIONS AT THE WELL VERSUS TIME AT A GIVEN LEVEL

404.

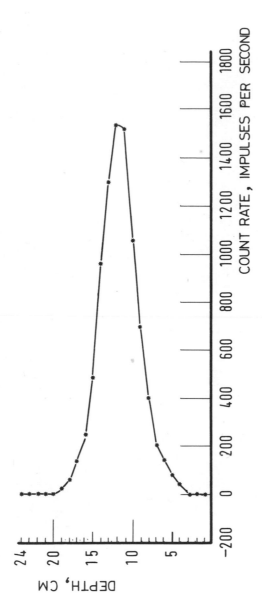

FIG. 125.   IMPULSE RESPONSE OF THE PROBE

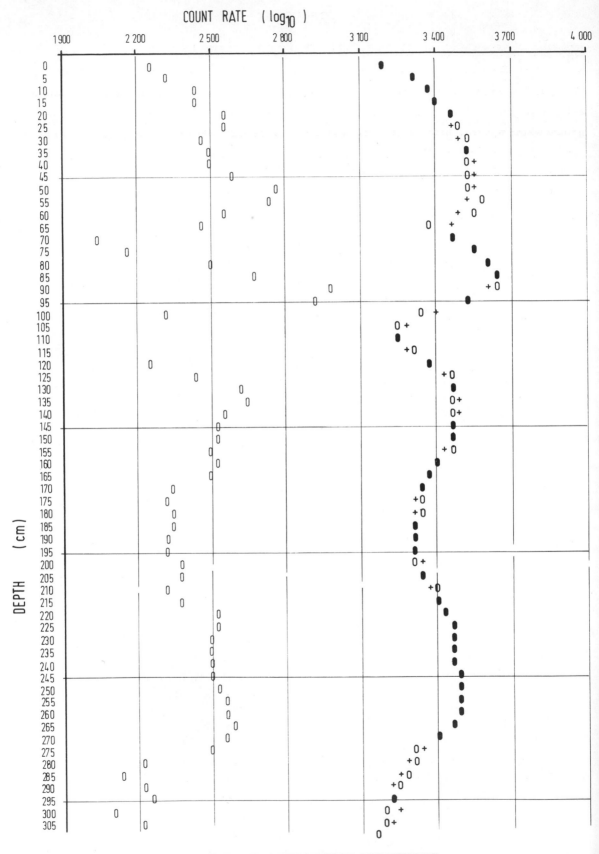

FIG. 126. THE DECONVOLUTION FUNCTIONS

406.

DISCUSSION ON TRACING THE MOVEMENT OF GROUNDWATER

1.   D. B. Smith remarked that the wide range of chemical forms of isotopes enabled the hydrologist to choose those least prone to adsorption within aquifers, namely, chelated complexes and anions.   He asserted that tritium stood alone as the perfect tracer in the groundwater sphere, being part of the water molecule HTO where $T = {}^{3}H$: thus it was the only tracer for which a nil retrieval was interpretable.   Any loss of tritium tracer by exchange of tritiated water with exchangeable water of clay minerals was not an unrecoverable loss, but served to disperse the tracer arrival.   Its main disadvantage lay in the need for assay of its weak activity in the laboratory.

2.   Bomb tritium peaks in rainfall for 1954 (50 T. U.), 1958 (150 T. U.) and 1964 (850 T. U.) were a diagnostic feature in the percolation regime, for which Smith produced a supplementary diagram to his paper (Fig. 127, p.408).

3.   In discussing the case history mentioned above, several speakers commented on the probable duality of flow regime in some chalk strata: a relatively rapid percolation through a system of fine cracks superimposed on a much slower migration through the uncracked but microporous material was hypothesized.   Laboratory tests on undisturbed core samples were envisaged.

4.   As an engineer specializing in groundwater problems, Dr. R. Herbert remarked that there is already a well-established body of knowledge relating to flow in unsaturated media.   Generally the medium receiving infiltration changes its hydraulic conductivity (via the water/voids ratio) so as to accommodate the flow to a great extent.   This was so for steady state conditions; a brief input however could lead to a progressively attenuated column of polluted water and he was inclined to suppose that such could be the explanation of failure to trace some suspected pollutions away from their origin.

5.   Others warned that geology could throw up complexities, such as stratigraphic breaks and preferntial flow zones which were liable to impose a high load of exploratory work such as drilling and sampling from boreholes, in which context the relatively inexpensive radioactive tracer work has to be set.

6.   Tracing the connection between sinks and resurgences in karstic limestones was the subject of case histories contributed by D. I. Smith and T. C. Atkinson (p409-413 of this volume).   These authors employed artificially coloured Lycopodium spores and, more recently, the fluorescent dyestuff, Rhodamine WT.

7.   A note by H. Hirsch and others (p. 414-416) instances non-radioactive chemical tracers that have been used in groundwater work.

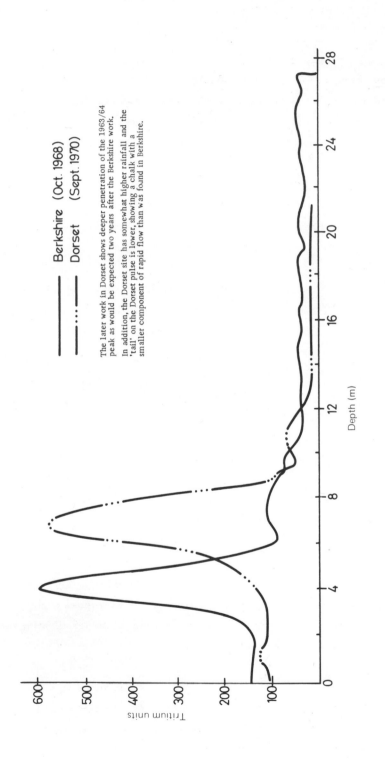

FIG. 127.   A COMPARISON OF TRITIUM PROFILES IN CHALK

Berkshire (Oct. 1968)
Dorset   (Sept. 1970)

The later work in Dorset shows deeper penetration of the 1963/64 peak as would be expected two years after the Berkshire work. In addition, the Dorset site has somewhat higher rainfall and the 'tail' on the Dorset pulse is lower, showing a chalk with a smaller component of rapid flow than was found in Berkshire.

Depth (m)

Tritium units

UNDERGROUND FLOW RATES IN CAVERNOUS AND FISSURED LIMESTONES

by D.I. Smith and T.C. Atkinson
University of Bristol, England

Introduction

This contribution is based upon work undertaken over a period of about ten years by members of the Department of Geography at the University of Bristol on underground water tracing in limestone terrains. Two techniques have been used for the tracing: one involving the use of Lycopodium spores and the other dyes. Lycopodium spores are small cellulose particles with a diameter of 25 microns. As they occur naturally in some streams it is usual for flow tracing to use dyed spores, in up to five different colours. Several kilograms of spores comprising many millions of individual particles can be introduced into each stream sink. The spores have a slightly higher specific gravity than water but remain in suspension in turbulent flow (and tend to move with the maximum velocity thread Ed.). At resurgencies a portion of the flow is filtered through fine mesh plankton nets from which sediment samples are recovered and examined under the microscope for the presence of spores. A detailed account of the Lycopodium method can be found in Drew and Smith (1).

In the second method a fluorescent dye is introduced into the groundwater circulation. Various methods have been employed for detecting the presence of the dye in water from springs but the most successful is the use of fluorometer. The instrument used (Turner Instruments Inc.) has the facility of continuous operation and recording in the field. The dye can successfully be detected in dilutions as small as one part in 10^{10}. A full description of the method is given by Brown and Ford (2).

Results

The results given here are all for experiments in which the tracer was introduced into a stream sink or swallet. There have been two areas of intensive work by Bristol University researchers, namely the Mendip Hills in Somerset (Atkinson (3)) and the Maroon Town area of northern Jamaica (Smith (4)), although other areas have also been investigated. In the Mendip Hills (Fig. 128, p.412) 43 stream sinks have been traced and the catchment areas of 11 major springs defined, as in Figs. 129 and 130 p.412 and 413.

The underground drainage divides differ from the surface watersheds and the prior preparation of a map of subterranean catchments is of prime importance in dealing with groundwater pollution. This is illustrated in the case study described by the Bristol Waterworks Company (p. 374-376) which describes pollution in the Ashwick area seen as the most easterly swallet system in Fig. 128, p. 412.

TABLE 39

FLOW RATES IN KARST FORMATIONS

Fig. 130, p. 413, is a more detailed map of the eastern Mendip catchments. The resurgence at St. Dunstan's Well (marked as V on Fig. 130) consists of two separate springs at the same altitude and only a few metres apart. Each of the springs is fed from different stream sinks and some of the flow lines cross one another without mixing. From this it can be inferred that the flow from each sink to the spring occurs in separate and discrete conduits. Similar patterns of crossing flow lines have been discovered in all of the other areas studied.

Velocities of underground travel (based upon straight line distances between sink and spring) are of the order of kilometres per day; further details are shown in the table. These velocities are faster than most quoted figures for normal groundwater movement by a factor of 1000 to 10 000 times. It can be inferred that the form of the fluid flow is turbulent in order to accommodate these velocities.

Flow rates (km/day)

	Mean flow velocity	Standard deviation	Number of tracer experiments
White Limestone Jamaica	3.45	4.05	40
Carboniferous Limestone Central Mendip Hills	7.36	5.91	23
Eastern Mendip	6.00	1.68	16

The percentage of spring flow composed of water from stream sinks varies between 0 and 40% at different times and for different springs. In the Mendip Hills it is normally between 1 and 10%. The remaining spring discharge is composed of water which has percolated directly into the limestone and this has been less thoroughly investigated. Preliminary tracer tests on this percolation water suggests that, while it moves at velocities less than that of water from stream sinks, values of hundreds of metres per day are not unusual. It is also evident from field studies that much of the percolation water moves in fissures and conduits similar to those of stream sinks although on a smaller scale.

Implications for groundwater pollution

The consequences of these results are clear. The underground drainage of massive, fissured or cavernous limestones such as the Carboniferous Limestone of upland Britain is largely confined to conduits which afford no natural filtration to the groundwater. Thus, any pollutant introduced into the limestone, especially below the level of the soil, is likely to re-appear at a spring within a matter of hours or days and to have undergone relatively little dilution. This applies not only to stream sinks but also to percolation water and especially to closed depressions which are foci of natural drainage. The practice of waste disposal or refuse tipping in cavernous limestones of this type is potentially extremely dangerous.

Conduit flow is not exclusively confined to cavernous limestones and in at least two cases local stream drainage sinks into fissured Chalk to flow to springs several kilometres distant at velocities comparable with those described above. It is possible that the velocity of groundwater associated with closed depressions in the Chalk is also rapid.

Acknowledgements

This work has been supported by the following bodies: the Bristol Avon River Authority, Bristol Waterworks Company, the Government of Jamaica, the Natural Environmental Research Council, the Somerset River Authority and the United Nations Food and Agricultural Organization.

References

1.   DREW, D. P. and SMITH, D. I.   Techniques for the tracing of subterranean drainage.
Brit. Geomorph. Res. Gp., 1969, Tech. Bull no. 2., 36 p.

2.   BROWN, M. C. and FORD, D. C.   Quantitative tracer methods for investigation of karst hydrologic systems.
Trans. Cave Res. Gp. G. B., 1971, 13, (1), 37-51.

3.   ATKINSON, T. C.   The dangers of pollution of limestone aquifers.
Proc. Univ. Bristol Speleo. Soc., 1971, 12, (3), 281-290.

4.   SMITH, D. I. ed.   Limestone geomorphology - a study from Jamaica.
J. Brit. Speleo. Ass., 1969, 6, (43/44), 85-166.

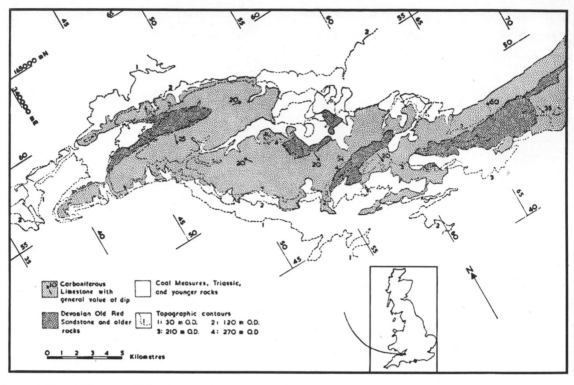

FIG. 128.   GEOLOGY AND TOPOGRAPHY OF THE MENDIP HILLS

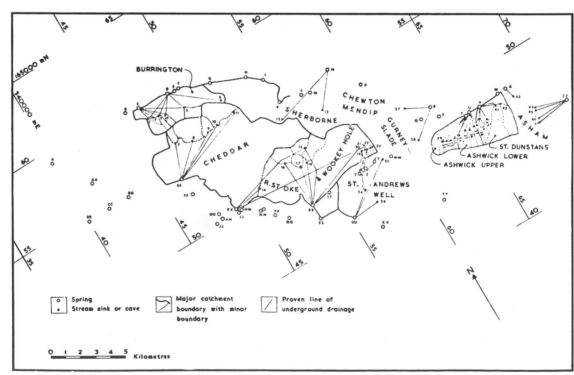

FIG. 129.   CATCHMENTS AND UNDERGROUND FLOW PATTERNS
IN THE MENDIP HILLS

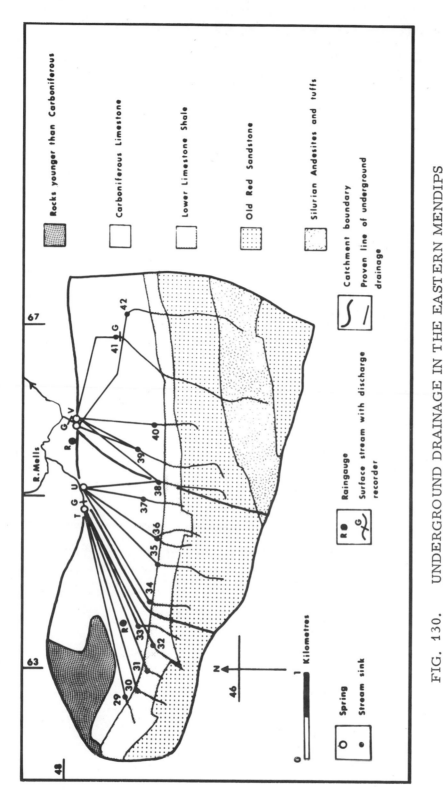

FIG. 130.   UNDERGROUND DRAINAGE IN THE EASTERN MENDIPS

413.

SOME CHEMICAL TRACERS OF GROUNDWATER FLOW

by H. Hirsch, E. Hall and J.A. Cole
The Water Research Association, Medmenham, Bucks, England

A distinction should be drawn between (a) the tracing of a groundwater connection, and (b) the more quantitative uses of tracers for determination of groundwater velocities and dispersion.   Both purposes would be served best by a tracer conforming to the criteria below quoted from Kaufman and Orlob (1):

1.   A satisfactory tracer should be susceptible to quantitative determination in very low concentration.

2.   It should be entirely absent from the injected water or present only at low concentrations in the displaced water.

3.   It must not react with the injected water or displaced water to form a precipitate.

4.   It must not be absorbed by the porous medium.

5.   It must be cheap and readily available.

But criteria 1 and 4 could be relaxed somewhat for application (a), in that only quantitative presence or absence of tracer may suffice and because reversible adsorption does not halt the tracer, only delaying it.

Kaufman and Orlob (1) cite the classical Slichter technique, involving the injection at a borehole of sodium chloride in solution and its detection by electrical conductance changes in and around observation boreholes, as one of the most common methods of groundwater velocity measurement.   However the technique is limited to relatively permeable materials and to distances of 10 to 100 m.   For more general tracing of groundwater, they instance other chemicals including the following.

TABLE 40

SOME CHEMICAL TRACERS

Tracer	Notes	Commonly occurring conc.	Limit of detection	Maximum desirable limit
NH_4^+	Presence is consequent on pollution of the aquifer	0	0.01 mg/litre	0.1 mg/litre
Li^+		0.005	0.005	7 mg/litre
$B(OH)_4^-$	Produced from ash tips	<0.1 mg/litre	0.1 mg/litre	<2 mg/litre B for irrigation
$Cr_2O_7^{--}$	*	0	0.01 mg/litre Cr	0.05 mg/litre Cr
NO_3^-	Derived from sewage effluent	5 mg/litre N	0.06 mg/litre	10 mg/litre N
$CH Br_3$	Odorous above desirable limit	0	10^{-5} mg/litre	0.01 mg/litre
Dextrose	*	0	-	-
Detergents	Occur in sewage effluent	0.03 mg/litre	0.03 mg/litre	0.3 mg/litre

*These two tracers are not recommended since they are both unstable and therefore limited in their application. As with many heavy metals, chromium compounds are toxic, the limit quoted relating to the WHO drinking water recommendations.

Further chemicals and dyes for studying groundwater movement, not mentioned in (1), are tabulated below. The dyes in this table are cited in the UNESCO Groundwater Studies handbook (2), which details ranges of quantities to use according to the strata. (In fact, the ranges average around 20 g of dye per 10 m of flow path, which appears an excessive dose: we suggest starting work with only one tenth this quantity).

TABLE 41

FURTHER CHEMICAL TRACERS

Tracer	Notes	Limit of detection	Max. desirable limit
Fluorescein Eosin Erythrosine Congo Red	Fluorescent dyes for use in alkaline waters pH $\geqslant$ 7. Fluorescein (sodium salt) has strong pH dependency on fluorescence. Liable to adsorption in clayey media.	~0.02 mg/litre	2.0 mg/litre
Ponceaux Red Methylene blue Aniline blue	Coloured dyes for use in acidic waters pH < 7.		
K^+	Commonly present in leachates from tips	0.5 mg/litre	5.0 mg/litre
I^-	Uncommon in aquifers and not prone to adsorption. Convenient to analyze using selective ion electrode.	0.01 mg/litre	1.0 mg/litre

References

1. KAUFMAN, W.J. and ORLOB, G.T.

    Measuring ground water movement with radioactive and chemical tracers.
    Journal of Am. Wat. Works Association, 1956, Vol. 48, No. 5.

2. BROWN, R.H. and others

    Ground-Water Studies.
    UNESCO, Paris, 1972, Sect. 7.4, 3-7.

Extra References

a. KNUTSSON, G.

    Tracers for ground water investigations.
    In 'Ground Water Problems' -
    Proceedings of the International Symposium, Stockholm, 1966, pp. 123 - 152 (Pergamon Press 1968; edited by E. Eriksson, Y. Gustaffson and K. Nillson).

b. SCHMOTZER, J.K., JESTER, W.A., and PARIZEK, R.R.

    Ground water tracing, with post-sampling activation analysis.
    J. Hydrolog., 1973, 20 (3), pp. 217 - 237.

DEEP WELL INJECTION OF LIQUID WASTE [*]

by J.S. Mokha [**]
University of Tulsa, Oklahoma, USA

1. ## INTRODUCTION

   Unprecedented industrial growth has enormously increased the production of highly toxic and obnoxious wastes which are difficult or impossible to treat economically by the traditional methods. The increase in population and changing land uses and practices are also taxing the ability of existing disposal facilities, surface waters, groundwaters and the atmosphere, to handle the growing pollution loads. The human environment is a closed system, so however reluctantly, the disposal of industrial and household wastes often consists of making a choice between polluting one or more of its facets. In this regard, the injection of dangerous effluents into vertically isolated deep formations has special significance, because if properly implemented, it removes from the human environment a possible source of nuisance.

2. ## DEEP-WELL CONCEPT

   The storage capacity of some geological formations is a natural asset just like the deposits of gold, oil, or any other useful mineral. This characteristic has been utilized for many years, primarily for the disposal of oil-field brines brought to the surface during the production of oil and gas. In recent years, injection wells have been used more commonly for the subsurface storage of natural gas and permanent underground disposal of various industrial wastes (including radioactive wastes) in suitable geological formations. There still exist great possibilities of using the storage capacity of different formations to store toxic and unwanted fluids and so it is imperative that this concept is well understood, investigated, and where conditions permit, exploited.

   General subsurface disposal or storage contemplates pumping liquid waste through wells into a permeable formation having adequate porosity to accept large waste volumes at low injection pressure (Fig.131). Wastes pumped down a well into a disposal formation advance due to the injection head, displacing a nearly equal volume of the formation fluid. Because the volumes are so nearly equal, a knowledge of the fluid content per unit volume of reservoir rock gives first approximation of the volume which will be occupied by the injected fluid. The

---

[*] This paper is based on a report written when the author was on the staff of The Water Research Association

[**] Now with Universal Oil Products Inc., St. Paul, Minnesota, U.S.A.

efficiency with which the injected fluid displaces the formation liquid increases with homogeneity of the waste and absolute permeability of the formation (1). Other factors which affect the efficiency are the relative viscosities of the two liquids and their mutual miscibility. Higher viscocity of the injected waste with regard to formation fluid and their better mutual miscibility would increase the displacement efficiency.

If the accepting horizon is horizontal, homogeneous and isotropic, the waste front would progress as an ever increasing disc whose centre is the well bore. But the presence of a natural hydrodynamic gradient in the disposal zone will cause the injected waste to be distributed asymmetrically about the well bore and the advancing front will spread at a higher rate downstream than sideways and upstream (2)(3)(4).

It should be stressed that there are no empty pores in subsurface formations below the unsaturated zone; all of them are occupied by natural formation fluids, for example, brine and oil and gases like ethane and methane. Withdrawal of any of these fluids does not completely empty the formation pore space; it only temporarily reduces the local formation pressure until other fluids can move in to replace those withdrawn (5)(6). Space for injecting waste is created by the injection pressure and can be attributed to the following factors, acting singly or in combination: (6)(7):

a) compressibility of the formation fluids,

b) elasticity of the formation rocks and

c) hydrodynamic connection between the disposal formation and a sea.

Liquids are only slightly compressible but the presence of dissolved gases in formation liquids can increase their compressibility to a certain extent. Compressibility factors of 5×10^{-5} to 10^{-4} per atmosphere are normal for most formation fluids (1)(8).

All rocks are to some extent pliable and subject to deformations by stress changes applied over large areas. It is also generally accepted that rocks are not only supported by the physical properties of those formations lying below but by the pressures of fluids contained in them (6)(9). Oil-field experience has shown that compressibility of rocks is almost totally due to a decrease in porosity of the pores (1). There are many examples of surface subsidence resulting from groundwater withdrawals (6). If the withdrawal of a fluid from a formation can cause surface subsidence, it is argued that the inverse might also be true and, more pore space in rocks can be created by injecting liquid waste into formations under pressure. The compressibility factor for most of the rocks ranges between 1.10^{-5} to 2.10^{-5} per atmosphere, but considering, the enormous volume of rock in which pressure changes occur, the extra space thus created could be substantial.

Some land-based formations outcrop under adjacent seas. In such cases, when injected material enters a disposal stratum, an equal volume of formation liquid seeps out into the sea, and consequently there is little pressure build-up in the formation. Such formations can store vast volumes of liquid waste. But this explanation may not be acceptable where inland basins are concerned.

It must be emphasized that the storage space in deep formation is restricted because injection pressures cannot be increased beyond certain reasonable limits due to the danger of fracturing the confining layers and other technological limitations. Consequently, only very concentrated, highly toxic and relatively untreatable fluids should be considered for injection.

While planning a deep-well disposal programme it should be kept in view that under less than ideal conditions, subsurface disposal can and does produce contamination problems which are difficult to discover and harder to rectify than the surface disposal methods (10). Therefore, great emphasis should be laid on the geological suitability of the disposal formation and well sites, and on the design, operation and monitoring of the injection wells.

3. GEOLOGICAL REQUIREMENTS AND SITE SELECTION

Local geology is of foremost importance because definite geological conditions are necessary for a successful deep-well disposal system. Two of the most vital conditions are the following:

1. The disposal formation should have sufficient porosity, permeability and areal extent to accept wastes at safe and practical injection pressures.

2. The disposal formation should be vertically below the level of fresh-water and should be confined vertically by impermeable rocks in order to prevent the pollution of fresh water aquifers and interference with any other natural resources.

The injection formation normally would be either sandstone, limestone or dolomite, although other rocks can have sufficient porosity and permeability under suitable conditions. Rocks with solution or fracture porosity are generally preferred to those with intergranular porosity, because the former have comparatively bigger flow channels and there is less likelihood of their being plugged by suspended solids in the waste liquid.

The amount of liquid which can be injected into a formation depends upon its areal extent, effective thickness, porosity, permeability and injection pressure and it can be calculated using various methods (8).

All those rocks which have very slight permeability could serve as confining layers (aquicludes). Unfractured shales, clay, slate, anbydrite, gypsum, dense siltstones are some of the good aquicludes. The thickness of the confining layer should be considerable because thin beds are susceptible to hydraulic fracturing (a method of breaking down formations hydraulically with a sand-carrying fluid) which is often done to develop wells or to increase their intake capacity. The safe thickness of confining stratum depends upon the nature of its rocks, depth of its burial, the expected injection pressure and the natural stresses in the area. The basic function of the confining layers is to contain the wastes injected in the disposal formation and consequently, all necessary precautions must be taken to prevent vertical fractures in the confining layers.

It has been demonstrated (5) that vertical fractures induced in petroleum reservoirs do not penetrate adjacent soft formations which have a high Poisson ratio relative to the reservoir rock. Fractures will propagate, however, through adjacent formations that are hard and brittle and have a low Poisson ratio as exemplified by some dense lime-stones and quartzites. At depths of more than 3000 ft and in regions of horizontal natural stresses, hydraulically caused fractures are also vertical (11).

The minimum depth of burial can be considered to be the depth at which a confined saline water-bearing zone is present; it may range from a few hundred to several thousand feet.

The minimum salinity of water in the disposal formation should be specified keeping in view the future demands of water and the technological developments in the treatment of waters. In some areas (for example, Illinois, USA) groundwater with less than 5000 mg/litre of total dissolved minerals is considered usable or fresh and is protected from waste disposal (12). Future demineralization technology will probably bring about revision of the acceptable total dissolved minerals figure to about 10 000 mg/litre.

The geometry of the injection formation is also important. A thick lens of highly permeable sandstone may not be very satisfactory for waste injection if it is small and surrounded by impermeable beds, because pressure build-up would be rapid compared to that in a similar formation of large areal extent. However, in some cases it is desirable to inject wastes into a known geological structural feature, because under favourable circumstances it would ensure the confinement of the waste within a specified area and might also allow its recovery at a later date if so desired. Wastes of high specific gravity can be safely stored in confined synclines.

Other considerations in the determination of site suitability are the following (2),

a) The presence of abnormally high natural fluid pressure and temperature in the potential injection zone that may make injection difficult or uneconomical.

b) Local earthquakes that can cause movement along faults and damage to the subsurface well facilities.

c) The presence of abandoned, improperly plugged wells that penetrate the injection zone and provide a means for escape of injected waste to groundwater aquifers or to the surface.

d) The mineralogy of the injection zone and chemistry of interstitial (native) waters, which may determine the injectability of a specific water.

Preliminary information to evaluate a deep-well disposal site can be obtained by studying detailed geological maps of the area concerned, but sometimes it will be necessary to undertake further investigations to determine the geologic and hydrogeologic conditions and assess the probable effects of waste injection. The site testing requirements are governed by the sensitivity of the area to possible damage and might include regional study, test drilling, geophysical logging, coring and core analysis, formation fluid sampling, formation pressure measurement, and pumping or injection tests (13). The petroleum industry has developed valuable testing procedures for evaluating subsurface conditions which are given in Table 39.

TABLE 42

SUBSURFACE EXPLORATION

Information desired	Methods available for evaluation
1. Thickness and character-istics of stratum and its location.	Drilling logs, core samples, electric logs, radioactive logs, sonic logs, caliper logs and photographic logs.
2. Porosity.	Core samples, electric logs, radioactive logs, sonic logs.
3. Permeability.	Core samples, drill stem tests, pumping (injection) tests, electric logs, micro logs.
4. Natural formation pressure.	Drill stem test, submerged manometers.
5. Formation fluid.	Drill stem test, core samples, electric logs, radioactive logs.
6. Formation temperature.	Temperature log.
7. Amount of flow in various strata.	Flow meter logs.

4. SUITABILITY OF WASTE FOR INJECTION

The injectability of a particular waste depends upon the physical and chemical characteristics of the waste, the disposal formation, and the formation fluids. This is because physical and chemical interactions between the waste and the formation minerals or fluids can cause plugging of the formation pores and consequently decrease the injectivity index or intake capacity per unit of pressure differential at the injection horizon. Plugging is normally caused by the following

a) Suspended solids and gases initially present in the waste liquids.

b) Precipitation reaction between disposable liquids and formation rocks or fluids.

c) Production of gaseous reaction products.

d) Development of reaction coatings on formation minerals.

e) Dispersion of clay minerals as a result of ion exchange or salinity reduction in formation fluids (14)(7)(2).

If organic matter is present in the waste it might promote the growth of bacteria, algae or mould on the formation rocks, which could also decrease permeability (15)(16).

Normally, two tests are conducted to determine the compatibility of waste material with formation fluids and minerals. The first is to determine if a precipitate forms from a mixture of formation fluid and waste liquid at a temperature equivalent to that of the disposal formation. The duration of such a test can be from some hours to many days depending upon the types of reactions expected. In the second test, waste liquid is passed through cores taken from the disposal formation, to determine any change in permeability to the waste.

Wastes that are not initially injectable can be treated to make them so. Such treatment problems are identical to those found by petroleum companies operating a water flood and can be resolved by some of the conventional processes employed in industrial water conditioning, such as chemical flocculation, sedimentation, filtration and chlorination (15), which remove suspended solids, dissolved iron and manganese, and entrained air. In some cases, where it is not possible to make waste compatible due to economic or technological considerations, a front of non-reactive liquid might be injected ahead of the waste liquid to form a buffer between the waste and the formation fluid (7)(17).

5. WELL DESIGN AND DEVELOPMENT

The basic design requirements of a disposal well are similar to those of oil wells, injection wells for oil-field flooding and water wells. They involve selecting a drilling system, casing programme and completion procedures on the basis of a detailed study of the geological profile of the area (depth of burial, thickness and physical character-istics of various formations; zones of lost circulation, heaving shales, abnormal pressure, and high temperature; and dense zones where drilling is abnormally slow), technological and economical considerations.

Two systems of well drilling are normally employed - cable-tool and rotary. The cable-tool system is widely used for drilling shallow wells (up to 2000 ft deep), disposal formations where casing is set above the formation, and the zones of lost circulation. However, there are many areas where the use of cable-tools is not feasible because of high pressure, extreme depths and soft formations. The rotary system is more universal and can be used to penetrate any hazardous formation by intelligently employing the correct drilling regime (drilling fluid, type of bit, weight on bit and rotary speed). The plugging of the disposal formation can also be minimized by using mud solutions which can build viscosity, hold it for certain amount of time and then can revert back to a viscosity about the same as water (18). With a few exceptions, the cost and time of drilling is always less when the rotary system is employed (19). Rotary drilling also provides better subsurface core samples (14).

Once the drilling system is decided the next step is to plan the construction of the well. By 'construction' is meant the number, diameters and lengths of various casings which are to be set in a well and the depth to which cement is to be placed. The following points deserve consideration:

a) It is vital that steps be included to prevent mixing of fluids from different penetrated formations and to prevent the intrusion of the injected waste into any formation other than the disposal horizon.

b) The possibility of using the chosen drilling system or systems.

c) The possibility of achieving the planned drilling speed.

d) Type of completion method to be used, whether open-hole, perforated casing, screened open-hole, screen-perforated casing or gravel packing.

e) The injectivity index (intake capacity per unit of pressure differential at the injection horizon) should be the maximum possible.

f) Convenience of using different methods of well stimulation.

g) Prospects of deepening well at a later date.

h) The possibility of using tools to retrieve lost pieces of equipment and repairing wells in case of accident.

From an analysis of all the above mentioned factors a casing bit-size programme is formulated. The first consideration here is the number of strings of casing which will be required. One or more of the following is required in every well,

a) Conductor casing.
b) Surface casing.
c) Intermediate casings.
d) Injection casing.

Conductor casing has the largest diameter and is required only where the surface soils are so incompetent that the washing and eroding action of the drilling mud would create a large cavity at the surface. The principal function of the surface casing is to protect freshwater aquifers. If the well is unusually deep or severe drilling problems such as abnormal pressure formations, heavy formations or lost circulation zones are encountered, it may be necessary to set one or more intermediate strings of casing to seal off the zones causing trouble. The final installation required is the pipe through which liquid waste is to be injected. The lengths of various casings depend upon the lithology and hydrogeological properties of various formations underlying the well site.

When the casing requirements have been determined, the size of each string must be fixed, beginning with the injection casing. To minimize costs it is desirable to use the smallest diameter of casing possible commensurate with the above considerations. Sufficient clearance (about 1 in.) should be allowed around the outside of a casing to provide a satisfactory thickness of the cement bond between the casing and the wall of the hole, and also to allow free passage of the casing into the hole.

Normally, lengths of steel pipe either screwed or welded together to form a continuous string, are used as casing. To prevent the corrosion of the injection casing by the liquid waste, its inside is coated with asphalt, plastic or some other protective substance. Where very corrosive wastes are to be disposed of, the use of an inner tube (Fig. 132) is generally preferred, since this can be readily replaced while corroded casing cannot. The inner tube can be made of a corrosion-resistant alloy, or of plastic or fibre-glass. Alternatively steel tubing can be plastic, cement or epoxy lined (18)(7).

The annular space between the inner tube and the injection casing is isolated from the disposal formation, and the liquid waste is injected by mechanical or fluid seals (14) (6)(20). In the first case, a packer is set between the inner tubing and the injection casing (Fig. 133) and the annulus between them is filled with inhibited water, hydrazine, treated drilling mud, oil or any other suitable fluid to assure that the packer will not become unseated and to balance the internal pressure in the injection tubing (14)(21). The fluid in the annulus can be recycled to prevent a stagnation area forming. Pressure in the annular space is normally monitored to detect any leak in the inner tubing or the packer. In the case of a fluid seal (Fig. 132), a non-corrosive liquid of lighter density than the waste is filled into the annulus. Due to the density difference, the non-corrosive liquid floats over the injected waste and their interface achieves the same effect as the mechanical seal.

In disposal wells, great care should be taken in cementing the annulus between various casings and the wall of the hole in the annular space between two adjacent casings. The cementing is done to protect casings from external corrosion by sub-surface waters, to increase casing strength, to prevent mixing of the fluids contained in formations behind a casing and to prevent waste being injected into formations other than the disposal zone. As a safety factor, all casings should be cemented throughout their entire length. Where tests have shown that the more economical Portland cement deteriorates due to the action of subsurface fluids or waste stream, special types of cements can often be employed. Acidic fluids are normally harmful to Portland cement type materials and mineral acids are more hostile than the weak organic acids (21). Slagle K.A. and Stogner J.M. (21) have suggested the use of siliceous additives, polyvinyl acetate latexes or certain other types of resins into the Portland cement slurry to increase its resistance to wastes containing acetic acid, ferric chloride and some other acids. Gypsum cement containing water soluble resin has proved durable in hydrochloric acid disposal and Portland cement/resin slurry has shown good resistance to sulphuric acid. Low water slurries when prepared with the proper type of Portland cement also offer satisfactory resistance to the majority of waste fluids (21).

Disposal well development includes those completion steps which aim to;

a)   remove the damage done to the accepting formation by the plugging
        action of mud solutions and cement slurry during drilling and
        grouting;

b)   enhance the acceptance rate of the formation by opening and
        creating new pores and fissures; and

c)   stabilize the formation around the well bore.

Some of the common methods of well development involve circulating with clean fresh water, mechanically scratching the well bore, swabbing, acidifying, hydraulic fracturing and shootings wells with explosives (7)(23)(6)(14)(20)(24)(11).

6.   SURFACE EQUIPMENT

Surface equipment varies considerably with the waste being handled and the necessary treatment prior to waste injection. Basically, the equipment consists of single or multiple stage centrifugal pumps, waste collecting, storage and treating facilities.

426.

7. INJECTION MONITORING

Deep-well disposal operations should be monitored to obtain performance data, detect failure of well casing or cement, assure safety of groundwater and other resources and to determine the movement of injected waste. The following parameters are normally monitored (14)(2).

1. Rate of injection, injection pressure and composition of injected material.

2. Pressure and composition of the fluid in the annular space between the injection tubing and well casing.

3. Pressure and water composition in observation wells.

Information about pressure and rate of waste injection is vital for evaluating well performance. A constantly decreasing injectivity index would indicate gradual plugging of the formation which may necessitate the redevelopment of the well. Monitoring of pressure and composition of the fluid in the annular space would indicate failure of well casing, cementing, injection tubing or unseating of the mechanical packer if that is being used.

Observation wells, which terminate in the confining layer or freshwater aquifers just above it, can be useful in detecting leakage from the disposal horizon and should be carefully monitored for pressure and water quality. In Florida, the Chemstrand Company, which disposes of effluents from nylon manufacture in a confirmed formation, uses two observation wells (25). One of them is drilled to the top of aquiclude at a distance of 100 ft from the injection well and is used to evaluate the effectiveness of the aquiclude in confining the wastes. The second observation well is at a distance of 1300 ft from the injection well and it fully penetrates the disposal horizon. Its primary use would be to evaluate the hydraulic characteristics of the disposal horizon and to calculate the future injection pressures and rate of waste movement.

8. ECONOMICS

The deciding factor for selecting the deep-well injection method as a means of waste disposal and pollution abatement is its economic attractiveness (26)(27)(21). The economics of such a project depend upon the following cost factors (14)(7)(2)(6)(26).

1. Initial investigation to determine the possibility of locating a suitable disposal formation.

2.   Well construction, completion and testing.

3.   Surface equipment.

4.   Waste treatment prior to injection.

5.   Operation and maintenance of the injection well and other equipment.

9.   INJECTION SYSTEMS IN USE

The case histories of many injection systems in use provide abundant information about the design operation and maintenance of such projects. A wide variety of wastes is being injected through deep wells. In the United States the Dow Chemical Company has been using this disposal method for the last two decades and is discharging saturated sodium chloride solution; concentrated calcium - magnesium liquors; waste liquids containing phenols, chlorophenols, biphenols, methocel and weak caustic wash waters from petrochemical process (14)(20). Other companies are injecting organic and inorganic acids (18)(27), hardwood pulping liquors (26), nylon manufacture effluent (22)(25), radioactive and solid wastes (21) and many other waste liquids (2).

Many new injection wells are being drilled and planned in the United States. ARMCO Steel Corporation, Ohio, propose to inject spent acid waste through a 3000 ft deep well into a sand formation 274 ft thick (28). Chemetron's Pigment Plant, Michigan, would be using another mile-deep well to inject acid waste into a sandstone formation, while their existing disposal well is already disposing of 150 000 g.p.d. of the waste (29). In Tuscaloosa, Alabama, Reichlord Chemicals Inc., is planning to drill a well 5500 ft deep as a means of disposing of their waste water (30).

10. CONCLUSIONS

Deep well injection is one of the economical and technically feasible means of waste disposal. It is enjoying increased acceptance and utilization by those industries faced with severe pollution control problems, and should play an increasingly important role in the future. Each particular waste disposal system has its own distinct problems but the experience already available should be useful in their solution. Geological investigations, well-planning and its implementation, deserve careful and serious consideration to eliminate the possibility of waste intrusion into formations other than the disposal formation.

REFERENCES

1. MURAVYOB, I.,
   ANDRIASOV, R., and others

   Development and exploitation of oil and gas fields; trans. from Russian. Moscow, Peace Publishers.

2. WARNER, D.L.

   Subsurface disposal of liquid industrial wastes by deep-well injection. American Association of Petroleum Geologists, Tulsa, Memoir 10, 1968, pp. 11-20.

3. STERNAU, R., SCHWARZ, J., and others

   Radioisotope tracers in large-scale recharge studies of ground water. Vienna, International Atomic Energy Agency, Proceedings of Symposium on Isotopes in Hydrology, 1967.

4. BROWN, R.H.

   Hydrologic factors pertinent to ground water contamination. Cincinnati, R.A. Taft Sanitary Engineering Center, Technical Report no. WG1-5, 1961, pp. 7-16.

5. GALLEY, J.E.

   Economics and industrial potential of geologic basins and reservoir strata. American Association of Petroleum Geologists, Tulsa, Memoir 10, 1968.

6. TALBOT, J.S.

   Some basic factors in the consideration and installation of deep well disposal systems. Wat. Sewage Wks, 29 Nov. 1968, pp. R213-219.

7. WARNER, D.L.

   Deep-well injection of liquid waste - a review of existing knowledge and an evaluation of research needs. U.S. Public Health Service, Division of Water Supply and Pollution Control. Cincinnati, Ohio, 1965, p. 55.

8. VAN EVERDINGEN, A.F.

   Fluid mechanics of deep-well disposals. American Association of Petroleum Geologists, Memoir 10, Tulsa, 1968, pp. 32-42.

9. JOHNSON DRILLERS JOURNAL

   Johnson Drillers Journal. VOP Johnson Division, St. Paul, Minnesota, 1968, 40, (5).

10. CLEARY, E.J., and WARNER, D.L.

    Some considerations in underground waste water disposal. J. Am. Wat. Wks. Ass., 1970, 62, (8), pp. 489-498.

11. CRAFT, B.C., HOLDEN, W.R., and GRAVES, E.D. Jr.

    Well design: drilling and production. Englewood Cliffs, N.J., Prentice-Hall Inc., 1962, p. 571.

12. BERGSTROM, R. E.

Feasibility criteria for subsurface waste disposal in Illinois.
Ground Wat., 1968, 6, (5), pp. 5-9.

13. DAVIS, Stanley N., and De WIEST, R. G. M.

Hydrogeology.
New York, Wiley, 1966, pp. 260-317.

14. WALKER, W. R., and STEWART, R. C.

Deep-well disposal of wastes.
Proc. Am. Soc. civ. Engrs, J. sanit. Engng Div., 1968, 94, pp. 945-968.

15. KAUFMAN, W. J., EWING, B. B., and others

Disposal of radioactive wastes into deep geologic formation.
J. Wat. Pollut. Control Fed., 1969, 33, pp. 73-84.

16. CALIFORNIA STATE WATER POLLUTION CONTROL BOARD

Report on the investigation of travel of pollution.
Sacramento, Calif., The Board, publ. no. 11, 1957.

17. SCOTT, J., and CRAWFORD, P. B.

Oil recovery in 1970's: key to survival.
Petrol. Engr., 1970, 42, (7), pp. 41-54.

18. MARSH, J. H.

Design of waste disposal wells.
Ground Wat., 1968, 6, (2), pp. 4-8.

19. McCRAY, A. W., and COLE, F. W.

Oil well drilling technology.
Norman, Oklahoma University, 1967, p. 492.

20. QUERIO, C. W., and POWERS, T. J.

Deep well disposal of industrial waste.
J. Wat. Pollut. Control Fed., 1962, 34, pp. 136-144.

21. SLAGLE, K. A., and STOGNER, J. M.

Oil fields yield new deep-well disposal technique.
Wat. Sewage Wks, 1969, 116, (6).

22. DEAN, B. T.

Design and operation of a deep well disposal system.
J. Wat. Pollut. Control Fed., 1965, 37, pp. 245-254.

23. GIBSON, U. P., and SINGER, R. D.

Small wells manual.
Washington, D. C., Dept. of State, Agency for Int. Dev., 1969, p. 156.

24. JOHNSON, Edward E.

Ground water and wells.
St. Paul, Minnesota, Johnson, 1966, p. 440.

25. BARRACLOUGH, J. T.

Waste injection into a deep limestone in north western Florida.
Ground Wat., 1966, 4, (1).

26. BROWN, R. W., and SPALDING, C. W.

Deep-well disposal of spent hardwood pulping liquors.
J. Wat. Pollut. Control Fed., 1966, 38, pp. 1916-1924.

27. VEIR, B.B.

Deep well disposal pays off at Celanese chemical plant.
Wat. Sewage Wks, 1969, <u>116</u>, (5),
pp. IW/21-IW/24.

28. GROUND WATER

News Notes.
Ground Wat., 1968, <u>6</u>, (6).

29. GROUND WATER

News Notes.
Ground Wat., 1969, <u>7</u>, (6).

30. GROUND WATER

News Notes.
Ground Wat., 1970, <u>8</u>, (2).

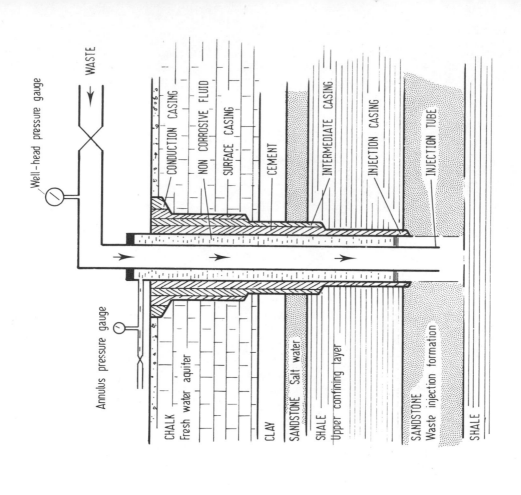

FIG. 132. DISPOSAL WELL WITH FLUID SEAL

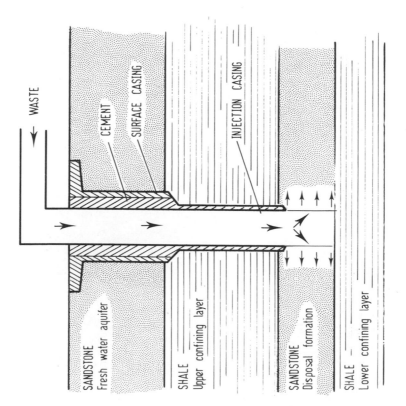

FIG. 131. TYPICAL DISPOSAL WELL

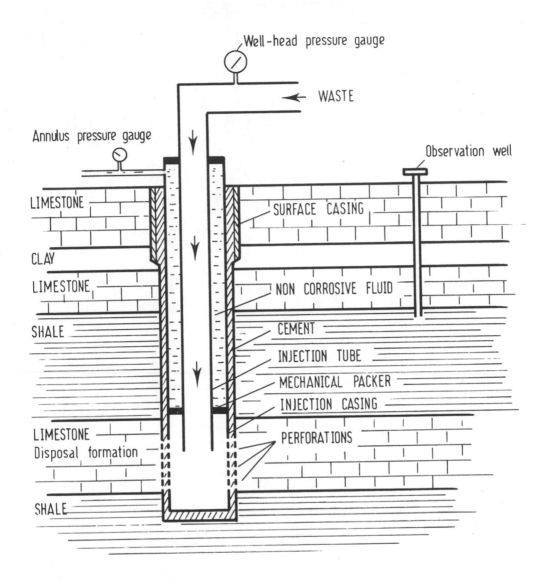

FIG. 133. DISPOSAL WELL WITH MECHANICAL SEAL

BIBLIOGRAPHY

1.   Industrial wastes forum: subsurface disposal.
Sewage Ind. Wastes, 1953, <u>25</u>, (6), pp 715 - 720.

2. TALBOT, J.S., <u>and</u> BEARDON, P.   The deep well method of industrial waste disposal.
Chem. Eng. Progress, 1964, <u>60</u>, (1), pp 49 - 52.

3. ADINOFF, J.   Disposal of organic chemical wastes to underground formations.
Purdue, Proceedings of Ninth Conference on Industrial Waste, 1954.
Purdue University, 1955, Ext. Ser. no. 87, pp 32-38.

4. CECIL, L.K.   Underground disposal of process waste water.
Ind. Engng Chem., 1950, <u>42</u>, pp 549-599.

5.   Spent pulping liquors to be discharged underground.
Chem. Engng News, 1963, 2 Sept., pp 128-129.

6. HENKEL, H.O.   Surface and underground disposal of chemical wastes at Victoria, Texas.
Sewage Ind. Wastes, 1953, <u>25</u>, (6), pp 1044-1049.

7. HENKEL, H.O.   Deep-well disposal of chemical wastes.
Chem Eng. Progress, 1955, <u>51</u>, (12), pp 551-554.

8. HOLLAND, H.R., <u>and</u> CLARK, F.R.   A disposal well for spent sulphuric acid from alkylating iso-butane and butylenes.
Paper presented at Nineteenth Industrial Waste Conference, Purdue, 1964.

9. HUNDLEY, C.L., <u>and</u> MATULIS, J.T.   Deep well disposal.
Purdue, Proceedings of Seventeenth Conference on Industrial Waste, 1961.
Purdue University, 1962, Ext. Ser. no. 112, pp 175 - 180.

10. HUNDLEY, C.L., <u>and</u> MATULIS, J.T.   Deep well disposal of industrial waste by FMC Corporation.
Ind. Wat. Wastes, 1962, Sept/Oct., pp 128-131.

11. LANSING, A.C., <u>and</u> HEWETT, P.S.   Disposal of phenolic waste to underground formations.
Proceedings of Ninth Industrial Waste Conference, Purdue, 1954.
Purdue University, 1955, Ext. Ser. no. 87, pp 184 - 194

12. MACLEOD, I.C.   Disposal of spent caustic and phenolic water in deep wells.
Proceedings of Eighth Industrial Waste Conference, Ontario, 1961, pp 49-50.

13. MECHEM, O.E., <u>and</u> CARRETT, J.H.   Deep injection disposal well for liquid toxic waste.
Proc. Am. Soc. Civ. Engrs. J. Construct. Div., 1963, pp 111-121.

14. PARADISO, S. J. Disposal of fine chemical wastes.
Proceedings of Tenth Industrial Waste Conference, Purdue
Purdue University, 1956, Ext. Ser. no. 89, pp 49 - 60

15. de LAGUNA, W. D., and Disposal of radioactive waste by hydraulic fracturing.
HOUSER, B. L. U. S. Atomic Energy Commission, Report no. ORNL-
2994, 1960, pp 128-136

16. CECIL, L. K. Injection water treating problems.
World Oil, 1950, 131, (3), pp 176 - 180

17. Problems in the disposal of radioactive waste in deep
wells.
Dallas, Texas. Am. Petrol. Inst. Div. Prod., 1958, 27p.

18. THEIS, C. V. Geologic and hydrologic factors in ground disposal of
waste.
U. S. Atomic Energy Commission, Report no. WASH-275,
1954, pp 261 - 283.

19. BAKER, W. M. Waste disposal well completion and maintenance.
Ind. Wat. Wastes, 1963, Nov/Dec., pp 43 - 47.

20. ROEDDER, E. Problems in the disposal of acid-aluminium high level
radioactive waste solutions by injection into deep-
lying permeable formations.
Washington, U. S. G. P. O., US Geological Survey Bull.
no. 1088, 65p, 1959.

21. STAHL, C. D. Compatibility of interstitial and injected waters.
Producers Monthly, 1962, 26, (11), pp 14 - 15.

22. LAIRD, R. W., and Incompatible waters can plug oil sands.
COGBILL, A. F. World Oil, 1958, 146, (6), pp 188 - 190.

23. BERNARD, G. B. Effect of relations between interstitial and injected
waters on permeability of rocks.
Producers Monthly, 1955, 20, (2), pp 26 - 32.

24. HARLEMAN, D. R. F., and Longtitudinal and lateral dispersion in an isotropic
RUMER, R. R., Jr. porous medium.
J. Fluid Mech., 1963, 16, (3).

25. SELM, R. F., and Deep well disposal of industrial wastes.
HULSE, B. T. Chem. Eng. Progress, 1960, 56, (5), pp 138 - 144.

26. DONALDSON, E. C. Subsurface disposal of industrial wastes in the
United States.
U. S. Bureau Mines Inf. Circ. 8212, 1964, 34 p.

27. BATZ, M. E. Deep well disposal of nylon waste water.
Chem. Eng. Progress, 1964, 60, (10), pp 85 - 88.

28. SHELDRICK, M. G. Deep well disposal: are safeguards being ignored?
Chem. Engng, 1969, 7 April, pp 74-76, 78.

29. ROSE, J.L.   Injection of treated waste water into aquifers.
                        Wat. Wastes Engng, 1968, 5, (10).

30. LUM,DANIEL   Preliminary report on geohydrologic exploration for
                        deep well disposal of effluent,
                        Waimanalo Sewage Treatment Plant, Waimanalo, Oahu.
                        Honolulu, Hawii Division of Water and Land
                        Development Circular no. C54, 1969.

31. WARNER, D.L.   Deep-well disposal of industrial wastes.
                        Chem. Engng, 1965, 4 Jan., pp 73 - 78

32. GRAVES, B.S.   Underground disposal of industrial wastes in Louisiana.
                        Paper presented at Soc. of Mining Engineers, 1964.

33. POWERS, T.J., and   Fundamentals of deep well disposal of wastes.
    QUERIO, C.W.   Proceedings of Ohio Water Pollution Control
                        Conference,
                        Cleveland, Ohio, 1961, pp 11 - 12.

TESTING AQUITARDS FOR LEAKAGE [*]

by P.A. Witherspoon
University of California, Berkeley, California, USA

I now want to take your attention deep underground. We will not consider the flow in the unsaturated regions. Fig. 134, p. 440, shows a multiple aquifer system - the unshaded layers are aquifers, the stippled layer represents low permeability beds, we call them not aquicludes but aquitards. You'll have chosen one of these aquifers, the lower one, and you intend to inject into it, and your problem is how much fluid will move across the aquitard and perhaps reach the upper bed, aquifer 2.

It can be shown that there's a way of getting a steady state solution for this problem if you will assume that these beds go out for many, many miles. And this solution is shown on the graphs in Fig. 135. Here on the vertical axis we have simply plotted the percent of injected fluid across that upper aquitard, shown in the little diagram above, as a function of distance away from the injection point. We assume radial movement and isotropic homogeneous beds. And the factor that can be used to describe this system is the B factor introduced by Hantush and Jacob many years ago and defined in Fig. 134, where the k's are hydraulic conductivities.

If you take B as low as 100, you would have a very leaky situation, so you find on that first curve on the left, that within the distance of about 1000 ft, all of the injected liquid at the steady state would be migrating across the aquitard up into aquifer 2. On the other hand, if you chose the system where B was 10^5, the next to the last curve on the right, the leakage within 1 mile of the point of injection would be so low that I cannot plot it on this particular graph, i.e. far far less than 1%. Within the first 10 miles it's a little bit less than 20% and so forth. This kind of curve then might be useful provided your sub-surface conditions match the conditions used in this analytical solution. One could use this approach in comparing two different kinds of situation. If the B factor is larger in one case than in another, the larger B factor offers the safer situation. Now this should be conservative as the transient behaviour has not been considered, it takes time to reach the steady state, and you see from this sort of approach you can calculate if you can assume plug type flow - simple convective motion, you could calculate how far it would take to get out radially around an injection well. This system says all of the motion, if there's any leakage, goes upwards but there's no reason why leakage cannot be down. So this I think is a conservative approach, provided the mathematical conditions can be found to be reasonably approximated in the field.

Well, now someone's going to ask immediately "How do I get k_1'?" How do I

[*] Because of their importance, Professor Witherspoon's remarks on procedures for testing aquitards are reproduced here on p. 437-441 from the verbatim record, together with diagrams shown as slides.

measure the permeability of an extremely low conductivity bed adjacent to an aquifer of relatively much higher permeability? And from some of the same theory we've developed what we call the ratio method. That simply means that if before injecting I pump water out of aquifer 1 and observe the behaviour of the draw-down in aquitard 1, by a well that stops in the aquitard and allowed to reach a static fluid level. Water is pumped out of aquifer 1 and changes in head are observed within the aquitard. The ratio of draw-down in the aquitard to the draw-down in the aquifer immediately below it, and at the same radial position, is related to k_1'.

Fig. 136, p. 441, shows some field results. Here is a situation where the pumping well operating at 100 g.p.m. was situated 50 ft away from what I will call an aquitard observation well. In this instance, we were trying to test below the aquifer. The case happens to be in conjunction with a gas storage project and if you follow developments in the United States you'll know that we've had a great amount of difficulty. We inject natural gas into an aquifer and are liable to find it coming out of a water well a few days, a few weeks, a few months later. So, some time ago, it occurred to us that we ought to be very careful where we inject natural gas. You can't imagine the problems that we have when the farmer goes out to his water well house and strikes a match to light his pipe and finds the building disappears before his very eyes! The well was opened in the aquitard 30 ft below the base of the upper aquifers; the storage zone was deeper down. The well operated for over 100 days - a very long test, meaning money involved here. Draw-downs within both the aquifer and the aquitard are shown here - the solid line being calculated at 50 ft. The aquitard well, you'll notice, doesn't begin to respond until about 40 days. The water level fell but not until about 40 days had elapsed and, if we go through the calculation of this method I've described, we get a hydraulic conductivity for that aquitard of about 10^{-10}. We also cored the same section and got exactly the same result.

Fig. 137, p. 441, shows an opposite effect where we were attempting to test a carbonate cap rock above a sandstone aquifer. The behaviour of the aquifer well, which you notice is at a position 880 ft away from the pumping well, departed from the Theis solution in about 100 minutes. And if you know the theory of leaky aquifers you'll recognise that that alone was sufficient evidence of a leaky cap rock. We've verified it by putting in an aquitard well at 1144 ft away. It also began to respond in 100 minutes, as contrasted with the previous case of 40 days. Well, the limestone beds above were cored and if you believe their matrix permeabilities, they were of the same order as the other case, 10^{-10} cm/sec etc. But if you go to the analysis the result averages 1 cm/sec - the difference is enormous because the limestone beds are badly fractured.

The observation wells that I mentioned are cased down to the point where you desire to make your observation; cemented on the outside, if necessary you perforate through that plus the cement; it's a typical oil field completion. And then one has to wait sometimes a few weeks because of the very low permeability of the system to reach the static level. We can show from theory that the delay required to fill up that well in no way affects the ability of that system to respond because the primary thing is the time lag - how long does it take for a transient pulse to reach the well? We've done this now from about 1000 ft down to 4000 ft in the United States.

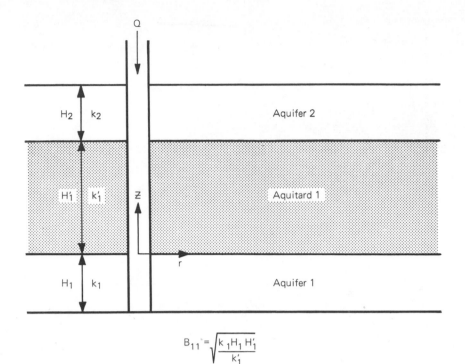

$$B_{11} = \sqrt{\frac{k_1 H_1 H_1'}{k_1'}}$$

$k_2 \gg k_1$

FIG. 134.   SCHEMATIC DIAGRAM OF TWO-AQUIFER SYSTEM

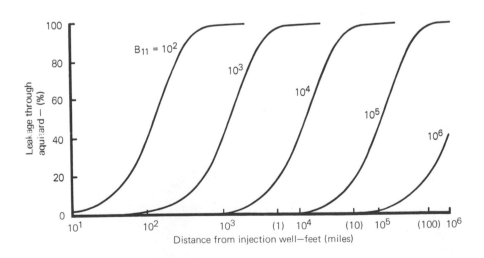

FIG. 135.   VARIATION OF AQUITARD LEAKAGE WITH DISTANCE
FROM INJECTION WELL

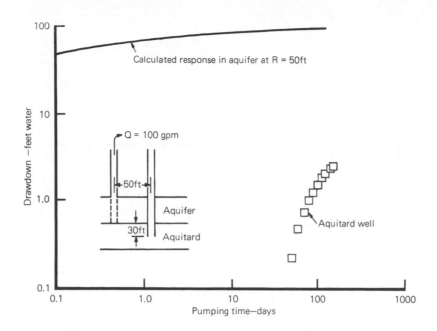

FIG. 136.   RESULTS OF PUMPING TEST WITH SLIGHTLY LEAKY AQUIFER

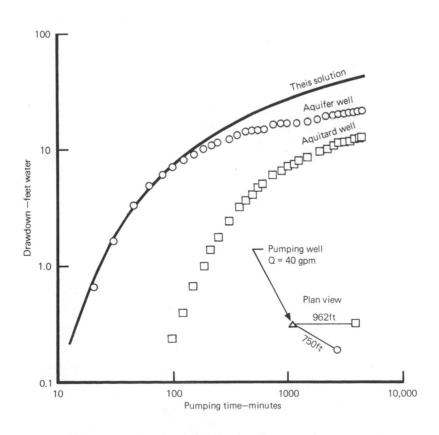

FIG. 137.   RESULTS OF PUMPING TEST WITH LEAKY AQUIFER

DISCUSSION ON DEEP INJECTION OF LIQUID WASTES

1. It was pointed out by a geologist that

    (i) one should distinguish waste disposal and brine disposal wells;

    (ii) an injection well should operate at pressures far below those at which hydraulic fracturing occurs.

2. The cost of typical fully-equipped injection wells was in the order of £40 000 to £100 000. The term 'disposal wells' should be discouraged where 'disposal' only meant indefinite storage.

3. Injection wells were declared to be of immense current concern to the Environmental Protection Agency in the USA, who were wary of accepting further incursions into aquifers having future worth. Two-thirds of current groundwater pollution court cases in the USA arose from deep well disposal; new Federal legislation was on its way soon. US Geological Survey Water Supply Paper No. 2020 is a useful annotated bibliography on deep well waste disposal.[+]

4. Professor Witherspoon answered a query as to the type of observation well needed in an aquifer at risk from injected waste by saying that a 'typical oilfield completion' would be routine. He also mentioned the fluid injection/earthquake phenomenon, citing the Denver example and research under way into earthquake modification.

5. Another discussant put the question, 'if waste disposal is increasing, what will happen when the 'sinks' are full?'. It was thought essential that priorities for disposal needs should be declared.

6. In the author's absence, the suggestion was cited from J.S. Mokha's May 1972 report that, as underground waste disposal was limited in possibilities, it might have to be reserved for the most noxious wastes. This viewpoint was countered by a chemist who foresaw the danger of putting highly undesirable products into strata whose safety was never known to a high degree of certainty.

[+]RIMA, D.R., CHASE, E.B. and MYERS, B.M.
Subsurface waste disposal by means of wells - A selective annotated bibliography. 1971, 305 p.

UNCLOGGING OF BOREHOLES

by G.R.S. Stow
George Stow & Co. Ltd., Henley-on-Thames, Oxon, England

Boreholes may require unclogging initially as part of the construction procedure or after a time due to mechanical or chemical clogging which may be independent of the corrosion resistance of the lining or screens. Acidization unclogging methods may be used to improve greatly the yields of Limestone aquifers.

1. INITIAL UNCLOGGING

Use of the mud flush drilling system for production boreholes is rarely permitted in the UK except for comparatively deep boreholes, so that de-mudding techniques using dispersants such as Magcophos (sodium tetraphosphate) to thin and break down the Bentonite mud are not enlarged upon, but reference is made to 'Mud Engineering' by Dresser Industries Inc. It is difficult to de-mud completely a borehole in which an in situ sand/gravel filter pack is placed and in these circumstances it is better to use a fine slotted taper wire screen or a pre-packed screen without an in situ pack. The use of 'Revert' or other short gel life drilling mud can be advantageous.

Removal of fines in sands/gravels by swabbing, jetting, back-washing and surge pumping through the screens is practised to unclog the formation around the screen to increase the specific yield of the borehole. However, this is not practicable if a thick in situ pack is employed or if formation sands are relatively uniform. In some cases, for shallow boreholes, a temporary mild steel tube with large slot openings is placed and surge pumped to remove fines before installing the permanent screens with or without an in situ pack.

Back-washing and surge pumping is often effective in cleaning out joints and fissures in stable formations to improve the specific yield.

Inter-bedded sands and clays can result in slotted tubes, screens or packs becoming partially choked with clay which may be removed by jetting and/or surge pumping after the introduction of Calgon (Sodium Hexametaphosphate) to give about a 4% solution in the borehole.

Swabbing is effected by a plunger reciprocated in the borehole and removing the material drawn in by air lift or bailing.

Jetting usually involves two or more horizontal high pressure water jets with the nozzles close to the wall of the screen. This tool is rotated and lifted slowly from the bottom of the screen.

Back-washing and surge pumping involve pumping down at a high rate, stopping the pump and allowing the water to rise to near rest level or, better still above rest level, by introducing water from a circulating tank; this process is repeated many times. Air lift pumping and blowing can be used in the same way.

Slurry, particularly from cable drilling in Marls, Clays and Chalk, may plaster borehole walls and penetrate joints and fissures. This is normally cleared on pumping but gives rise to a cloudy effluent for some hours on test pumping. Chalk slurry is difficult to settle out and if the water has to be discharged under conditions where the milkiness is objectionable, the slurry can be very much reduced by the introduction of a relatively small dosage of hydrochloric acid with an inhibitor such as Corinex in the borehole.

Although it is not strictly unclogging, it is very common practice to acidize boreholes in Limestones when constructed, or at a later date, to increase the yield by enlarging joints and fissures. As examples, a 30 in. x 300 ft borehole in the Oolitic Limestone near Malmesbury gave 51 g.p.h. per foot drawdown before treatment with 15 tons of hydrochloric acid and 486 g.p.h. per foot drawdown after treatment. A 24 in. x 400 ft borehole in the Chalk at Andover gave a yield of 165 g.p.h. per foot drawdown before treatment with 20 tons of hydrochloric acid and 5670 g.p.h. per foot drawdown after treatment. Bulletin 49 of the State of Illinois Department of Registration and Education quoted a number of cases where acid treatment of Dolomite wells increased the specific yields by several hundred per cent using rather similar quantities of acid.

For best results, the borehole should be lined with cemented casing to below the water table. The rapid injection of acid under pressure, which is maintained by the CO_2 generated, causes the acid to flow outwards from the borehole and to enlarge fissures and joints for a considerable distance. If the acid is placed in an unsealed borehole or the gas can be dissipated into dry fissures, the acid merely attacks the walls of the borehole and is less effective.

A substantial flange is welded to the top of the casing to carry a bolted counter flange. PVC piping is flanged through the counter flange which is also fitted with a blow off connection. Valves and pressure gauges are fitted to the injection pipework and blow off pipework. The acid with inhibitor is injected at a pressure of up to 689.5 kN/m^2 (100 p.s.i.) by a rubber-lined pump through PVC piping to points just above the best fissures indicated by resistivity logs.

The hydrochloric acid strength is usually about 32% and the quantity used about 50 to 100% of the volume of water in the borehole. One ton of acid would react with 0.44 tons of Chalk and produce about 0.19 tons of CO_2.

The injection time is generally under one hour and the reaction with the Limestone is usually complete less than one hour later, although considerable gas pressure (CO_2) may remain in the borehole for some hours and have to be blown off.

Boreholes have been clearance pumped and used for supply within 24 hours of acidification when the chlorides have fallen below 200 mg/litre. In a typical case, after an injection of 32 tons of acid and starting clearance pumping 3 hours later at $0.052 \text{ m}^3/\text{sec}$ (1 m.g.d.) after 2 hours pumping, the chlorides were 2000 mg/litre, after 23 hours pumping 200 mg/litre, after 79 hours pumping 30 mg/litre and after 240 hours pumping 14.5 mg/litre. It is usual to aerate the discharge over land or into a lagoon for a few hours until the CO_2 is greatly reduced before discharging into a water course. Grassland does not appear to be affected adversely by chlorides over a short period.

If there are no useful aquifers near the bottom of the borehole, the clearance pumping should be with the pump suction close to the bottom of the borehole otherwise high chlorides would remain in this section. Precautions are necessary to ensure that no dangerous accumulations of CO_2 gas can occur in working spaces and that means of washing off and neutralizing acids are to hand. Polythene sheets to cover the acid pump and pipework and face masks are desirable.

2. LATER UNCLOGGING

A borehole may become clogged by encrustation or the products of corrosion on the borehole linings or screens. The likely causes are (a) deposition of calcium carbonate due to release of CO_2 from the water by reason of high groundwater temperature, reduction in pressure on drawdown and to a small extent to entry velocity into the borehole, (b) precipitation of iron and possibly manganese compounds from the water, particularly if oxygen is introduced by drawdown in a relatively shallow borehole, and (c) corrosion attack on unsuitable lining or screen tubes which may block the slots with say iron oxide. Corrosion may be due to chemical or electrolytic causes.

Deposition of carbonates is not a problem in the UK but precipitation of iron compounds is common, particularly for boreholes in the ferruginous beds of the Lower Greensand and the Wealden Series. The introduction of oxygen is caused by cascading water into the borehole on pumping and by the rise and fall of the water table below the top of the aquifer. A shallow borehole into the Lower Greensand near Silsoe in Bedfordshire became so choked with iron compounds that it had to be unclogged at about yearly intervals.

Encrustation may also take place in the strata close to the borehole walls and in any filter pack around the screens.

The usual treatment is the introduction of hydrochloric acid, and after some hours, contact swabbing or surge pumping followed by initial clearance pumping, then Calgon injection, followed by further swabbing or surge pumping and final clearance pumping. Iron bacteria may occur in ferruginous aquifers and produce a clogging oxide slime which may be broken down by heavy chlorine injection to give about 100 mg/litre followed by swabbing and surge pumping. In some cases, it is necessary also to remove any filter pack and/or the screens and replace the screens after cleaning if they are otherwise suitable. Encrustation on the screens and slotted tubes is also removed by scraping and jetting.

Where possible boreholes should be sited down dip to be deeper and preferably where the aquifers may be confined by overlying clays and possibly of larger diameter to reduce the drawdown and entry velocity. This procedure would reduce the risk of calcium carbonate deposition by reducing the release of CO_2 and minimizing the introduction of oxygen. The writer cannot recall trouble with precipitation of iron compounds in boreholes over 500 ft deep where pumping water levels are at a considerable height above the top of the aquifer.

It has been suggested that continual substantial abstraction from a ferruginous aquifer could stimulate recharge of surface waters with oxygen and CO_2 and at some time cause chokage of the aquifer with iron compounds. On the other hand, continuous pumping of iron in solution would remove iron from the strata.

Corrosion of lining tubes and borehole screens can be avoided by the use of suitable materials such as stainless steel, bronze, plastics and suitably coated mild steel. In some cases, special consideration should be given to the materials used in pumps and rising mains.

Entry velocity of water into borehole screens is usually so low that there is no problem with erosion, but failures have occurred, particularly where high velocity water from a fissure in a Sandstone has impinged on the filter pack around the screen, causing movement of the pack and, in effect, sand blasting holes through the screen. Such Sandstone might be relatively thin in a sandy aquifer and require screening. One solution may be to use a preformed pack and screen against such a Sandstone aquifer.

Borehole screens and slotted linings may become partly choked in time with clays, silts or fine sands and these may be cleared as outlined for initial unclogging of boreholes in sand and gravels. However, it may be necessary to remove and replace the screens and repack the boreholes.

There may be sand flow into boreholes from soft fissured Sandstones such as the Bunter. This sand is mainly derived from fissures and joints due to erosion by water flow and it is difficult to screen out as the area of fissures and water yielding joints exposed to any screen is relatively small. A sand retaining screen to exclude sands of say 52/150 B.S. Sieve could cause serious resistance to flow. Where slotted tubes surrounded by a fine gravel pack have been employed, chokage of the pack with fines has occurred and caused reduction of yield. The yield has been restored by removal of the filter pack. In cases of very soft Sandstone it may be advisable to place a very coarse gravel pack (in no way sand retaining) to support the borehole walls against coarsely slotted lining tubes. It is suggested that the best way of dealing with this problem is to spread the abstraction between several boreholes and reduce the velocity from joints and fissures so that sand inflow is negligible.

3. SHOOTING WELLS

The use of explosive charges, usually mass gelignite charges or smaller and aimed charges of R.D.X. T.N.T. is not often effective in improving borehole yields. There have been cases where such charges have reduced well yields by stemming Marls or Clays into joints or fissures. The purpose is to connect a borehole with nearby joints or fissures or enlarge those encountered in a borehole drilled in hard rocks. The joints and fissures can be located by an electrical resistivity borehole log and shaped R.D.X. T.N.T. charges fired at these places. Such charges can make a circular slot around the borehole and penetrate some 2 ft or make cylindrical holes 3 to 5 ft in depth.

Directional charges are used to connect at depth adjacent boreholes.

GENERAL REFERENCE

STOW, G.R.S.   Modern waterwell drilling techniques in use
                                      in the United Kingdom.
                                      Proc. I.C.E., 1962, 23, 1-14.

WELL SCREENS; SOME NOTES ON MATERIALS

by E. Elliott *
Stainless Steel Development Association, London, England

The author of this paper is not an expert on groundwater but he has been asked to provide some information about the choice of material for well screens with particular reference to experience in the United Kingdom.

The two principal causes of reduction in flow rate through well screens are the growth of incrustations and corrosion of the material from which the screen is made; it is with the latter phenomenon that these notes are concerned. A number of materials may be satisfactory in resisting corrosive attack from soil water, and those principally used in the past have been copper alloys, aluminium, plain carbon steel, stainless steel, plastics and glass reinforced plastics. Of these, unprotected plain carbon steel and aluminium are suitable only in carefully selected waters of low corrosivity. Plain carbon steel may be successful if thoroughly protected, for example with efficient and continuous rubber or nylon coatings; this type of product is used for perforated screens with large slot openings surrounded by bonded-on sand filter packs. Correctly chosen copper alloys are satisfactory as regards resistance to corrosion, for example phosphor bronze. In general, however, considerations of cost and strength have led to the choice of stainless steel, and it is with this material that this paper is principally concerned. Stainless steel is used mainly in two ways, namely as a fine wire wound screen or as a machine or press slotted tube surrounded by an in situ sand filter pack.

The corrosion of metals is now well understood, and is an electro-chemical phenomenon. Due to slight and inevitable variations across the surface of the metal, certain areas become cathodic and others anodic in the presence of an electrolyte, and the anodes dissolve while the cathodes are protected. There are many methods of preventing corrosion of susceptible metals, for example coatings which obviate contact between the metal and the electrolyte or are themselves corroded sacrificially. Stainless steels are inherently resistant to corrosion, by the mechanism described below.

There are many types of stainless steel; the material is defined in this country as a corrosion resisting steel containing not less than 9% of chromium. It is the chromium in stainless steels which is principally responsible for their high resistance to corrosion. As is nowadays well known, this durability is due to the presence on the surface of the material of a naturally occurring and highly protective oxide film

* now with British Steel Corporation, Special Steels Division, Sheffield.

containing oxides of chromium (the "passive" film) which is invisible, inert and very tightly adherent; it also possesses the valuable property, in contrast to applied protective films, that it is self-repairing when damaged by abrasion, by reaction with atmospheric or other sources of oxygen. Stainless steel is thus corrosion resisting throughout its thickness and its durability is unaffected by accidental damage.

For the application under consideration, choice of stainless steel is limited to the group classified as austenitic, a term stemming from the metallurgical structure of these steels which are easily recognisable because of their non-magnetic nature. Such steels most commonly contain between 17 and 19% chromium and between 7 and 12% nickel, although there are steels outside these limits. As has been mentioned the chromium provides the corrosion resistance while the nickel improves mechanical properties, formability and weldability and gives some added durability in certain corrodants. The most versatile of these steels, and one that has been used for well screens made from plate, is 304S15 in B.S. 1449 "Specification for Steel Plate Sheet and Strip. Part 4. Stainless and heat resisting plate, sheet and strip" and its composition is given in Table 40. The mechanical properties of this steel in the fully softened condition are as follows:

0.2% Proof Stress tons/sq. in. min.	Tensile Strength tons/sq. in. min.	Elongation % min.
13.5	33	40

It will be noted that the strength is high but so also is the ductility, so that the material may readily be formed from plate into tubular shape. Strain hardening on cold working is quite rapid so that the initial 0.2% proof stress is quickly augmented. This also means that the hardness of the material in the vicinity of pressed slots is high and so is the abrasion resistance.

This material is eminently suitable for fusion welding, and this is usually carried out in plate thicknesses either by inert gas shielded methods or by metal arc welding using coated electrodes. To ensure the best resistance to corrosion crevices should be avoided in design since they can in the presence of severe corrodants act as sites for differential oxygenation and corrosive attack. Similarly, welding scale should be removed as this can hydrate and give the appearance of corrosion; it is, however, understood that for well screens this may be accepted.

A well known phenomenon that can be associated with welded austenitic stainless steels is that of sensitization to intercrystalline corrosion, which also has the popular appellation of weld decay. This is caused when material is heated in the range 500 to 800 $^{\circ}$C, and particularly around 650 $^{\circ}$C, when chromium carbides migrate to the grain boundaries of steel leaving adjacent areas deficient in chromium, and therefore of lower

resistance to corrosion. When the material is subsequently exposed to a suitable
corrodant this penetrates along the grain boundaries, resulting in loss of mechanical
properties. It will be noted that 304S15 has a maximum carbon content of 0.06%,
and since sensitization depends upon carbon level and time at temperature, this steel
is not liable to be sensitized unless welded in considerable thickness. For greater
thicknesses, three methods of avoiding sensitization are possible, namely heat
treatment of the weldment, the use of a material of extra low carbon content, or the
addition of a stabilizing element. The first method is not really applicable to well
screens, as it involves heating to about 1050 $^{\circ}$C and rapid cooling, in order to take the
precipated chromium carbides back into solution in the austenite. Steel 304S12 is
similar to 304S15 but has a maximum carbon content of 0.03%, and may therefore be
safely welded in all thicknesses. The third procedure is to choose a stabilized steel
and this may contain either titanium or niobium, both elements with which carbon
combines in preference to chromium, thereby preventing the loss of chromium adjacent
to the grain boundaries described above. The compositions of two of these steels,
321S12 and 347S17, are included in Table 40.

The general corrosion resistance of all the steels described is very similar;
they will resist corrosion by soil water of all normal types, whether hard or soft, and
containing for example carbon dioxide or hydrogen sulphide. It is understood that in
some countries and under certain situations additional corrodants have to be borne in
mind, and a few notes are therefore included about these.

In considering such aggressive corrodants, it must be mentioned that when
necessary another austenitic stainless steel may be chosen with considerably higher
corrosion resistance than that of those listed above. This is a steel of similar
composition but containing an addition of molybdenum to the extent of about $2\frac{1}{2}$%, which
in addition to considerably enhancing general resistance to corrosion is particularly
effective in improving resistance to pitting attack. This steel is available additionally
in low carbon and stabilized varieties and the composition of these steels, namely
316S12, 316S16 and 320S17 is given in Table 41.

Carbon dioxide, oxygen and hydrogen sulphide in solution are without effect on
stainless steel and 304 is quite suitable for service with them. This is also true of
chlorides in solution in water in the proportions normally encountered with potable
waters. With sea water, however, it is necessary to choose the molybdenum-
containing steel of 316 type, if pitting corrosion is to be avoided.

Organic compounds are in general without effect on stainless steel, and 304 will be satisfactory. Of the mineral acids, 304 will resist nitric acid, but with sulphuric acid 316 should be chosen, as it may be expected to be satisfactory except in high concentrations. Hydrochloric acid attacks all stainless steels, but in dilute solutions 316 will give good service.

A further aspect of the corrosion hazard which must be considered is that of bimetallic corrosion. It is well known that where two unlike metals or alloys are in contact in the presence of an electrolyte corrosion of one of them may be accelerated due to the contact, and electromotive series such as that given in Table 42 are frequently published. It will be noted that austenitic stainless steel is noble towards all metals with which it is likely to be in contact in the application concerned, but the effect on other metals must also be considered. If good metallurgical contact is involved, as for example if stainless steel is welded to plain carbon steel, acceleration of attack on the plain carbon steel will be caused. With mechanical joints, the effect may be expected to be less marked, but the danger still exists. It must be borne in mind that the areas of anode and cathode are important. Thus if a large area of stainless steel is in contact with a small area of plain carbon steel the rate of attack on the plain carbon steel will be much greater than if the relative areas are reversed. As an example, stainless steel bolts in a large plain carbon steel structure represent no particular hazard, but the other way round trouble may be expected in the presence of a corrodant.

Since one of the advantages of stainless steel is good strength, mention must be made of recently developed steels which provide even greater improvements in this respect; they are collectively termed high proof stress stainless steels, and may offer advantages for well screens, since thinner and therefore cheaper materials may be used. As has been mentioned, austenitic stainless steels strain harden rapidly and therefore material in sheet thicknesses may be used in the cold-rolled condition giving 0.2% proof stress as high as 60 tons/sq. in. with good ductility. In plate thicknesses, cold-rolling is not feasible, and strengths are enhanced by "warm working" or alloying with nitrogen; typical properties given in Table 43. No reduction in corrosion resistance results from either of these treatments.

In summary, it may be stated that correctly chosen and fabricated stainless steel will provide well screens and other components in which there is no danger of corrosive attack, and therefore of blocking of orifices due to corrosion product.

451.

TABLE 43

AUSTENTITIC STAINLESS STEEL PLATE;
COMPOSITION'S TO BS.1449

Steel	C	Si	Mn	Ni	Cr	Ti	Nb
304S12	0.03	0.20/1.00	0.50/2.00	9.0/12.0	17.5/19.0	-	-
304S15	0.06	0.20/1.00	0.50/2.00	8.0/11.0	17.5/19.0	-	-
321S12	0.08	0.20/1.00	0.50/2.00	9.0/12.0	17.0/19.0	5C/0.70	-
347S17	0.08	0.20/1.00	0.50/2.00	9.0/12.0	17.0/19.0	-	10C/1.00

TABLE 44

AUSTENTITIC STAINLESS STEEL PLATE
(MOLYBDENUM STEELS): COMPOSITIONS TO BS.1449

Steel	C	Si	Mn	Ni	Cr	Mo	Ti
316S12	0.03	0.20/1.00	0.50/2.00	11.0/14.0	16.5/18.5	2.25/3.00	-
316S16	0.07	0.20/1.00	0.50/2.00	10.0/13.0	16.5/18.5	2.25/3.00	-
320S17	0.08	0.20/1.00	0.50/2.00	11.0/14.0	16.5/18.5	2.25/3.00	4C/0.60

453.

TABLE 45

ELECTROMOTIVE SERIES

Least noble end

Zinc

Aluminium

Plain carbon steel

Cast iron

Brass

Copper

Phosphor bronze

Austenitic stainless steel

Molybdenum - containing austenitic stainless steel

Gold

Most noble end

TABLE 46

HIGH PROOF STRESS STAINLESS STEEL PLATE:
TYPICAL MAXIMUM MECHANICAL PROPERTIES

Steel type	High proof stress type	0.2% Proof stress ton/sq.in.	Tensile strength tons/sq.in.	Elongation %
304	Not highproof stress	13.5	33	40
304	Warm worked	26	40	20
	Nitrogen strengthened	19	38	35
321	Warm worked	26	40	20
316	Warm worked	26	40	20
	Nitrogen strengthened	20.5	40	35

DISCUSSION ON BOREHOLE TECHNOLOGY

1.   Elliott's paper was introduced as a topic of relevance to the well engineer having to act in difficult conditions of water quality, such as where the pH or chlorides of a water rendered mild steels prone to corrosion attack.

2.   Polluted waters were not referred to other than a mention of acid drainage near mines, but the paper had implications for monitor wells in landfill also.

3.   A further cause of groundwater acidity was instanced by a Dutch speaker: when groundwater is heavily pumped from a sulphide-bearing horizon, oxidation of the sulphide can ensue in the cone of depression.

4.   Mr. Elliott also mentioned the usual UK practice which was to employ the stabilized stainless steels for borehole work, for ready weldability of the cold sheet.   When great sheet thicknesses were employed, warm working and nitrogen alloying were adopted.

5.   Questioned on the site identification of stainless well screens and casings, Mr. Elliott said that the British Standard grading was the best guarantee and that spot tests on site would hardly serve a useful purpose.   Another speaker suggested spectrographic determinations at £10 per sample would be worthwhile in any case of doubt.

6.   Mr. Stow commented that, at the time of writing his paper, data on the clearance of the products of acid treatment from the vicinity of a borehole were rarely set down as a matter of record.   But in response to a request from the organizers of the Conference, he had ascertained the following for clearance pumping from a borehole near Cheam, Surrey.   The hole was 24 in. diameter, bored in chalk to 550 ft depth.   Fig. 138 shows the effect of 4 days treatment with 30 tons of H Cl on chlorides at two pumped boreholes in use, each about 1000 ft from the acidified borehole.   These both appear to remove most of the chlorides reaching their catchments, even before the clearance pumping proper commenced.   The effect of the latter upon water sampled from the acidised hole is dramatic, with chlorides and CO_2 dissolved showing an interesting parallelism of behaviour.

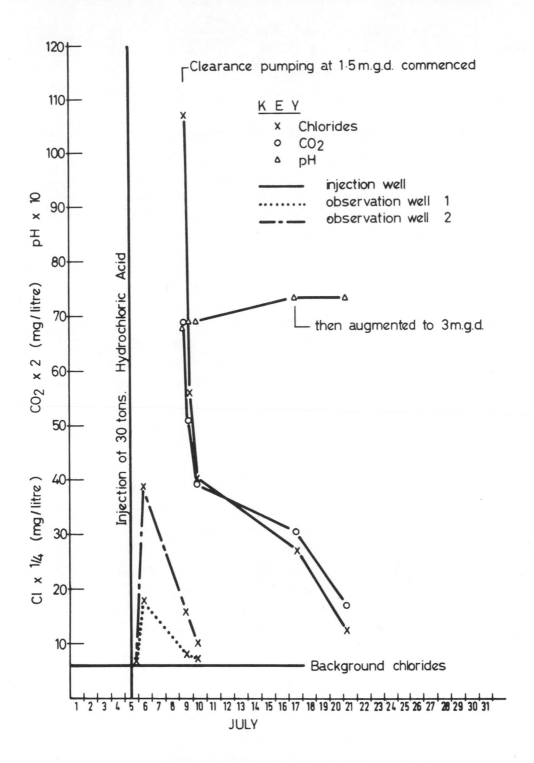

FIG. 138.   LEVELS OF CHLORIDES, CO2 AND pH FOLLOWING
ACIDIFICATION OF A BOREHOLE

457.

A LOOK TO THE FUTURE

<u>Mr. D.A. Gray (Chairman)</u>

I propose just to mention three features which have struck me during this conference and which I think do concern everyone here.

The first is the existence of problems on two time scales: namely the pressing day-to-day problems for which the disposers of wastes will require solutions, and the longer term research problems which will have to be resolved before many of the questions which the papers have thrown up are adequately resolved. I mention just the nitrate one in passing.

The second feature is that there is very considerable knowledge spread around in the numerous organizations and with numerous individuals. The problem is, of course, to get this co-ordinated in terms of our existing framework. They are spread so widely that persons trained in particular disciplines have found it difficult in the past to talk effectively to those in alternative disciplines. One has known that there has been work going on in other areas, but the financing and structure of the research is such that it is not being co-ordinated and adequately organized.

This really brings me to the third point, which is that on numerous occasions speakers have referred to the need for more finance in groundwater pollution research affairs. This is undoubtedly true, but it is clear that at an <u>ad hoc</u> level there is already quite significant expenditure going on. The research programmes which should be undertaken should not, in my view, be mounted at the expense of the finance allocated on this day-to-day <u>ad hoc</u> solution, because we have problems on the two time scales. None the less the objectives for the long-term research programme do need defining on a disciplinary basis, and defining very firmly.

I won't dwell upon these points, despite the temptation to illustrate the further principles, but now pass over to our two speakers for this session. They are going to produce a discussion on two aspects; Dr. Packham is to speak first on investigation and identification, and then Mr. Cole will speak on cures.

<u>Dr. R.F. Packham</u>

Looking into the future is a very fascinating pastime which we all indulge in from time to time. When it falls to one's lot to crystal-gaze in public at a conference such as this, it does give rise to some mixed feelings. On the credit side one does not have to prepare a paper in advance and one has also a good excuse for keeping reasonably quiet during the proceedings. On the other side, when the moment comes to declare oneself there is the problem of making fairly innocuous statements or very positive predictions which might cause one to be branded as a genius in the years to come, but more likely, the opposite.

The first prediction, however, which I will make unhesitatingly in relation to groundwater pollution or indeed almost any other topic, is that things are going to get much worse. By that I mean that life will become more complicated and more expensive. It is necessary of course to specify for whom things are going to be more difficult. It is quite clear that the future holds problems for the polluter whether he is responsible for the disposal of industrial or municipal waste. Public opinion is strongly against him, and it is on that account rather than on the basis of any technical judgement that legislation has been, and will continue to be enacted to further protect us from his actions. At this conference, particular interest has been centred on our new Deposit of Poisonous Waste Act, 1972. Although this has only been in force for a few months, we have heard that it is leading already to important management decisions in some sectors of private industry resulting in a significant reduction both in the quantity and the toxicity of the materials being disposed of. I think this is a very encouraging sign and although a number of years will pass before we will be able to judge the full effect of this Act, it does seem reasonable to expect that as a result of it, a growing hazard to groundwater supplies and indeed to other parts of our environment, will be significantly diminished. It was said yesterday that one result of the notification scheme required by the Act was that Local Authorities and River Authorities were being made aware for the first time of the extent and nature of the tipping going on in their areas. This is an excellent thing and there can be no doubt that in some areas situations will be revealed that will require detailed investigation and rectification. In another five years time there will be many more case histories to be added to those recorded at this conference. In addition to this, of course, we can look forward to the results of some new investigations.

Although in some areas we may expect improvements in the type and quantity of waste to be disposed of, waste can never be entirely eliminated. It has been suggested that in some cases treatment and incineration are alternatives to dumping. This is certainly a challenging area for research and development. In other cases, liquid wastes currently being tipped may be disposed of to the Local Authority, although this may necessitate greater vigilance at the sewage works. It is obvious, however, that there will always be large quantities of waste to be disposed of by dumping and at this conference we have had papers showing what can be done to minimize the risk of groundwater pollution from a variety of sources including waste tips. The most important considerations in the design and operation of tipping sites have been listed by

Hunter Blair in his paper, and I think that Professor Dracos's contribution gives an outstanding example of the development and implementation of a reliable scheme for combating the risk of pollution from a shunting yard and oil refinery.   This kind of approach will be adopted more widely in the future.   Its main feature is that groundwater pollution hazards are considered at the design stage.

It has been suggested that most of the science and the technology needed to prevent groundwater pollution is available now, and, as is the case with so many other problems, that of groundwater pollution is simply the problem of finding money to take the measures that are necessary.   One delegate went further and suggested that people were not prepared even to pay the cost of finding out whether groundwater pollution was a problem. Have we, indeed, too much faith in the traditional purity of our underground water supplies. The worst feature of groundwater pollution is that often, by the time it is detectable at the point of abstraction, it is already too late to do much about it.

It seems likely that with increased knowledge of the potential sources of groundwater contamination in a given area, there will be a marked increase in the extent of water quality monitoring for specific contaminants, both at the point of abstraction and from boreholes located near to the potential contamination source.   There was much talk about the cost of this operation yesterday, though it seemed that the sum of money involved for each sampling borehole sunk was not much more than that required for, say, a pH meter.   The problem of sampling from the unsaturated zone has been referred to and this is probably a major limitation of this approach and certainly an area for further research. Whether or not such sampling comes into routine use there is clearly a need for more detailed studies of the type described by Dr. Exler and others.   This is the only way of truly assessing the magnitude of any hazard.   As with nearly all problems we must have data and we hope that this conference will have stimulated interest in this type of study, if necessary on a smaller scale.   Someone said that the old proverb 'out of sight out of mind' was applicable to the present situation in this country and I am inclined to agree with this.   On the other hand, we must not get so obsessed with pollution hunting that we make unreasonable demands.   This is a danger when highly sensitive analytical techniques such as gas-liquid chromatography come within the grasp of everyone.   However, in my opinion, this conference has established a clear-cut need for more quality monitoring and more data collection of all types.

There are some fundamental problems that will have to be cleared up before long, and high on the priority list, I think, is the question of nitrate in water supplies.   There still seems to be a major difference of opinion here.   On the one hand we have the water undertakings who have noticed that increased nitrate levels and increased fertilizer use, if they don't go hand in hand, certainly follow parallel paths. On the other hand, we have the agriculturalists who have asked us if we really believe that farming is so lucrative that the farmer can afford to throw large quantities of nitrate into the water supply. Could the answer be a discrepancy between what is desirable and what is actually obtainable in practice, bearing in mind the nature of both people and the English climate. I hope that someone can resolve this problem as there are certainly some areas where action is called for to prevent any further increase in nitrate levels.

Mr. J.A. Cole

This conference has been gratifyingly lively both here and in the discussions which led up to it. I have never at a previous occasion had to write down so many triangle symbols for 'warning' and so many Rs in a circle for 'research need' and Ds in a circle meaning 'development need' as on the 25 pages of notes taken down in these three days. It has really revealed a massive list of future actions to consider. I think it is significant that this conference has had about 30% of the time given to basic scientific discussion, another 30% given to pollution prevention and 30% reporting the detection and occurrence of pollution: the 'cure and clean-up' aspects only amount to 10% of our efforts here. In this 10% our emphasis appeared to be on petroleum products: e.g. we have had some good discussion on pipeline emergency procedures and some excellent guidance from Professor Dracos and Dr. Sidenvall on how to cope with various types of spillage.* Let us now see how this expertise on remedial measures can be knit together, to cater for all types of pollutant - inorganic or organic - solid or liquid.

Our Chairman to this session (Mr. Gray) and my colleague, Dr. Packham, have already uttered some useful guidelines on preventive measures. It is on guidelines for cures that I would like to focus now, because there is a gross dearth of ready guidance in this matter. The contractor is wondering what he is allowed to do. The industrialist is wondering whether a new Act is going suddenly to leave him with ten filled containers with nobody to take their contents. Then the waterworks always have had the problem of the totally unexpected pollution, usually nothing to do with landfill, happening right out of the blue. Granted one has to admit that we have often a lot of indirect evidence, where there has been something suspected but the situation hasn't been fully documented; this is just the sort of thing we will have to pursue further before we are able to prescribe all the cures. But at least let us have a check-list for symptoms. We ought also to have a check-list of urgent remedial actions. We ought to have a code of practice on the more routine sort of monitoring. We know that circumstances alter cases, but I refuse

---

*Post-conference note: the UK Atomic Energy Authority operate a Hazard Control Unit, based on Harwell.

to admit that as reason for vagueness of action and think we ought to name what the circumstances are. We ought to be in a position to say that where there is a toxic substance which might reach the land, at a landfill or as an injection of some sort, it has to be looked at geologically in such and such a way.

I doubt whether there is a textbook that has orientated itself yet towards groundwater pollution. Perhaps this conference's proceedings form an embryo textbook, but they are also a contribution to the present day information explosion! Because groundwater pollution is a complex subject, none of us is going to rush away saying we know it all. But we already know some important things to pursue as I have indicated already. This brings me to the question of what organizations can cope with groundwater pollution. That is quite beyond my scientific remit - but the analogy is close to that of a fire brigade. I have heard some amusing gentlemen here refer to the 'karst police' coping with fissured limestones and the like. I think we require more than the karst police, I think we need the 'fissured-rock fire brigade' and the 'intergranular ambulance'. And should these be available as a dial 999 service?

Generally, if a spill occurs near the ground level you can actually dig it up (it becomes a problem for someone else then!), but if it is within the well, you have to bail the contaminant out and flush out the well with water, maybe using some detergent. If a contaminant gets into the aquifer, and has not got stuck to the rock matrix, you can suck it out with scavenger wells. Most of these are situations where there are some urgent actions which can be taken and we ought to decide more explicitly what those should be. It's especially necessary to know in advance what cleansing or emulsifying agents can safely be used, without compounding the pollution problem. I don't think this is the place to say where such recommendations will be collated but I can assure you that we at the Water Research Association intend to collate a lot of the material of this working conference.

You have worked hard at it, there is a good lot of material and we intend to go on working at it. You are welcome to come back to us and say 'what was said there?' or indeed, 'what opinion have you come to on some of this material?' Even though we are not the experts in every case, we do know to whom else one can go. There is some wonderful documentation in the form of Bibliographies prepared by Mr. J. L. Robinson of our library and by Mr. Stringer of Herts County Council; we will try and get these to you. Written contributions will be accepted from people who came in with a question but didn't get a chance to put it. Well, let us make good use of all this!

<u>Mr. D.A. Gray (Chairman)</u>

I'd like now to invite Professor David Todd from the University of California at Berkeley to address us and we will conclude with a word from Mr. G.M. Swales of the Research Advisory Committee of the Water Research Association.

<u>Professor D.K. Todd</u>

I shall keep my remarks very brief because our time is running out and the schedule has been very assiduously followed for the last three days. From a personal standpoint, first of all, this conference to me has been a very great pleasure and extremely worthwhile. It's somewhat like dining at a three-star restaurant in France, as they say, it is well worth the trip. Early in 1973, our College of Engineering, together with the University Extension, will be organizing a conference on groundwater pollution in San Francisco. The conference here at Reading, of course, has for me been a very nice antecedent because it has exposed me to many points of view, provided ideas and topical material that will be invaluable in organizing our own conference.

We will be discussing essentially all of the topics that have been discussed and presented here. The emphasis, however, will be less on chalk and chalk problems and more on unconsolidated alluvium. We will not, I should emphasise, discuss tips but we will discuss dumps and landfills. In fact, I must make a confession when I first looked at the conference papers and ran across this word 'tips' which is completely unknown in the United States, I assumed it was a typographical error and it should have been 'pits'.

But in addition to the wide scope covered here, we will add others, such as the problem of sea-water intrusion, which is one of our major groundwater pollution problems in the United States. Also, the problem of the movement of radio-active waste underground, which is going to become increasingly important, and the problems involving water reclamation using artificial groundwater recharge, which is now an operational matter, will be included. I can only hope that we will generate as much interest in our conference as has been shown here these last three days.

I think the Water Research Association is to be commended for its recognition of the importance of groundwater pollution, for its leadership in organizing this conference, and for its encouraging presentation of all aspects of the complex subject, including the hydraulic, geologic, biological and economic, and for presenting viewpoints of government, water supply organizations, and private industry. Speaking, if I may, on behalf of all of the delegates from outside of the United Kingdom, I would like to express appreciation for the organization of this conference to Dr. Allen, Director of the

Water Research Association, and to Mr. Cole and his colleagues who have made this conference so successful.   Thank you.

Mr. G. M. Swales

I don't think we can close here today, without the Association itself, paying a tribute to a number of people who have taken part in this Conference.   I would first like to thank, on behalf of the Association, the various Chairmen we have had for our sessions.   These are all busy men in their own field, all experts; time to them is an important thing.   We thank them very much for all they have done.   I think we would like to thank the authors and the contributors for the papers.   They again are invaluable.   And last, on behalf of the Association, and the Committee I would like to thank those members of our own staff who have put in all the work in organizing, running and testing everything.   I think, with that Sir, I would only add one further thing.   That the Association feels that it is doing a useful job in bringing together all the various disciplines which are required to discuss, fairly fully, a particular topic.   Thank you.

Mr. D. A. Gray (Chairman)

Thank you; That is the final contribution and I now declare this conference closed.

BIBLIOGRAPHY ON GROUNDWATER POLLUTION

Compiled by J. A. Cole, E. S. Hall, A. Hunter Blair and J. L. Robinson,
The Water Research Association

This is a collection of references, mostly for the period 1950 to 1973, classified as follows:

The selection of references was primarily based on journals and books in the Association's library and therefore can be said to be representative of the European and North American literature on this topic. Substantial help was given by Dr. K. H. Schmidt of the Dortmunder Stadtwerke, who supplied 118 references retrieved by the keyword method from a computer-stored list: many of these references are included.

The references do not necessarily include all items cited by authors in their papers to the Groundwater Pollution Conference. In fact, the intention has been to supplement the papers, rather than match their ground exactly.

References appear in alphabetical order of author within each section.

Additional material, in the form of annotated references, supplied by Mr. J. L. Stringer of Hertfordshire County Council, appears at the end of this volume, pp. 500-516.

1. BIBLIOGRAPHIES AND SYMPOSIA

AMERICAN WATER RESOURCES
ASSOCIATION

Proc. National Symposium on Groundwater
Hydrology, San Francisco, November 1967.
Urbana, Ill, The Association, 1967.
365 p. including papers on the following:

Chemical Geohydrology by J.D. Hem
107-112.

Aquifers and Models by C.V. Theis, 138-148
(with discussion of dispersion).

Analysis and predictive methods for
groundwater flow in large heterogeneous
systems by R.W. Nelson and D.B. Cearlock,
301-318.

BOYER, J.F. and GLEASON, V.E.

Coal and coal mine drainage (Literature
review).
J. Wat. Poll. Contr. Fed., 1972, **44**, (6),
1088-1093.

CHERRY, J.A. et al

Hydrogeological Factors on Shallow
Subsurface radioactive-waste management
in Canada.
Proceedings International Conference on
Land for Waste Management, Ottawa, Canada,
October 1973.

INTERNATIONAL ASSOCIATION
OF HYDROLOGICAL SCIENCES
with AMERICAN ASSOCIATION
OF PETROLEUM GEOLOGISTS
and US GEOLOGICAL SURVEY

Underground waste management and artificial
recharge.
International Symposium, New Orleans,
August 1973, 2 vols.

International Association
of Hydrogeologists

Symposium
"Grundwasser und Umwelt"
(Groundwater and the environment)
Essen, W.Germany, May 1973.

INTERNATIONAL WATER SUPPLY
ASSOCIATION

Pollution of surface and underground waters.
Subject No. 2 at Internat. Water Supply
Association 5th Congress and Exhibition,
Berlin, May-June 1961, 92 p.

JOHNSON, A.I.

Report on groundwater (Literature review)
Bull. Int. Assoc. Hydrol. Sciences, 1972,
17, 44-51.

LEEDEN, F. van der (ed)

Ground Water: a selected bibliography.
Water Information Center, Port Washington,
N.Y., 1971, 116 p.

STONE, R.V.

'The way we do it'
Symposium. Groundwater contamination,
US Dept. Health Educ. and Welfare Public
Health Service, Washington, DC, 1964.

UNITED STATES
Dept. of Health, Education and
Welfare

Groundwater contamination.
Proceedings of Symposium, 1961.
The Robert A. Taft Sanitary Engng Center,
Techn. Report W61-5, 218 p.

UNITED STATES
Environmental Protection Agency

Subsurface Water Pollution, a selective
annotated bibliography (1968-1971).
Part 1 - Subsurface waste injection,
Part 2 - Saline water intrusion,
Part 3 - Percolation from surface sources.
E.P.A., 1973.

UNITED STATES
Environmental Protection Agency

Polluted Groundwater: some causes effects,
controls and monitoring.
EPA 600/4-73-001b, July 1973.

UNIVERSITY OF CALIFORNIA
Water Resources Engineering
Educational Series

WREES Program X, Groundwater Pollution:
Sect. 1 Causes of groundwater pollution.
     2 Chemical pollution.
     3 Biological pollution.
     4 The California pollution situation.
     5 Agricultural pollution sources and
       their control.
     6 Seawater intrusion and its control.
     7 Disposal wells.
     8 Radioactive pollution.
     9 Solid waste disposal effects.
    10 Prevention and detection of pollution.
    11 Wastewater reclamation effects.
    12 Administrative and legal considerations.
The College of Engineering, University of
California, Berkeley, 1973, 263 p.

UNIVERSITY OF VERMONT

Sanitary landfill leachate travel in various
soil media - a bibliography.
Technical Information Center, College of
Technology, Burlington, Vermont,
December 1971.

VARIOUS AUTHORS

Symposium on pollution of groundwater.
Proc. Soc. Wat. Treat. Exam., 1962, 11,
pp. 11-49, 75-105.

YOUNG, R.H.F.

Effects of water pollution on groundwater
(Literature review)
J. Wat. Poll. Contr. Fed., 1972, 44, (6),
1208-1211.

2. ADMINISTRATIVE AND GENERAL

AIRAKSINEN, K.

On the provisions concerning waste water in
our new water law (Finnish, English summary)
Vesitalous, 1961, 2, (1), 24-26.

BOLTON, P.

Prevention of water source contamination.
J. Amer. Wat. Wks Ass., 1961, 53, (10),
1243-1250.

BORNEFF, J.

Occurrence and evaluation of carcinogens in
water. (in German)
Gas-und-Wasserfach, (Wasser/Abwasser),
1967, 108, (38), 1072-1076.

BORNEFF, J.

Harmful substances in water: origin,
significance and removal.
Zbl. Bakt, Parsit., Infektionskr. Hyg.
1971, 155, (3), 220-230.

BRONZES, P.

Study of the metabolization of pollutant
products.
Water Research, 1972, 6, (4/5), 457-464.

CALLAHAN, J.T.

The role of the US Geological Survey in waste
disposal monitoring.
Ground Water, 1972, 10, (3), 6-9.

CARTER, G.

Pollution of underground water.
Water and Water Engineering, 1958, 62,
244-246.

DEPARTMENT OF THE ENVIRONMENT	Refuse disposal. Report of the working party on refuse disposal. London, HMSO, 1971, 199 p.
DEUTSCH, M.	Ground water contamination and legal controls in Michigan. USGS Water Supply Paper 1691. U.S.G.P.O., Washington, D.C., 1963, 86 p.
DIXON, M.P.	Taste and odour evaluation methods. Proc. Soc. Wat. Treat. Exam., 1970, 19, (2), 161-168.
DIXON, R.D. and DIGIALLONARDO, A.	A survey of methods used to detect leaks in underground installations. Water and Sewage Works, 1972, 119, (6), 78-81.
ELDRIDGE, E.F.	Irrigation as a source of water pollution. J. Wat. Poll. Contr. Fed., 1963, 35, 614-625.
ELIASSEN, R. and CUMMINGS, R.H.	Analysis of waterborne outbreaks 1938-1945. J. Am. Wat. Wks Ass., 1948, 40, 509-528.
FLYNN, J.M.	Impact of suburban growth on ground water quality in Suffolk County, New York. Cincinnati, Robert A. Taft Sanit. Engng Center Tech. Refs. W61-5. Ground Water Contamination, Proc. of Symposium, 1961, 71-82.
FRANK, W.H.	Even groundwater is threatened. (in German) Wasser-Bedrohtes Lebenselement. Montana-Verlag, Zurich, 1964, 145-147.
GELDREICH, E.E.	Waterborne pathogens in "Water pollution microbiology", ed. R. Mitchell. New York, Wiley, 1972, 207-241.
GELHAR, L.W., and others	Gravitational and dispersive mixing in aquifers. Proc. ASCE, 1972, 98, (HY12), 2135-2153.
GERMAN SOCIETY OF GAS AND WATER ENGINEERS (DVGW)	Guidelines for protected areas to safeguard drinking water supplies (in German). Working Paper W101, DVGW, Frankfurt/Main, 1972, 16 p.
GOLWER, A. and MATTHESS, G.	Effect of pollution by wastes on ground water yield. Deutsche Gewasserlamdliche mitteilungen, 1969, 51-55.
GREAT BRITAIN - STATUTES	Deposit of poisonous wastes Act. London, HMSO, 1972, 7 p.
GROVER, H. EMRICH	Guidelines for sanitary landfills. Ground Water and Percolation. Compost Science J. of Waste Recycling, 1972, 13, (3), 12.

GRUNE, W.N.   Natural radioactivity in ground water.
                                   Part 1.
                                   Wat. and Sewage Wks., 1961, 108, (11),
                                   409-411.

HACKETT, J.E.   Groundwater-contamination in an urban
                                   environment.
                                   Ground Water, 1965, 3, (3), 27-30.

HEALTH, MINISTRY OF   Report No. 31.
                                   London, HMSO, 1925.

HEALTH, MINISTRY OF   Report of Chief Medical Officer.
                                   London, HMSO, 1937, 159.

HEALTH, MINISTRY OF   Safeguards to be adopted in the day-to-day
                                   administration of water undertakings.
                                   London, HMSO, 1939, Memorandum 221,
                                   Revised 1948, 8p.

HEALTH, MINISTRY OF   Report on the outbreak of enteric fever in
                                   the Malton Urban District.   Report on Public
                                   Health and Medical Subjects No. 69.
                                   London, HMSO, 1933.

HENKEL, H.O.   Surface and underground disposal of chemical
                                   wastes at Victoria, Texas.
                                   Sewage Ind. Wastes, 1953, 25, (6), 1044-1049.

HOUSING AND LOCAL GOVERNMENT   Taken for granted; report of the working
MINISTRY OF   party on sewage disposal.
                                   London, HMSO, 1970, 65 p.

HOUSING AND LOCAL GOVERNMENT   Safeguards to be adopted in the operation and
MINISTRY OF   management of waterworks.
                                   London, HMSO, 1967, 11 p.

HUGHES, G.M. and   Scientific and administrative criteria for
CARTWRIGHT, K.   shallow waste disposal.
                                   Civ. Engng (ASCE), 1972, 42, (3), 70-73.

JACKO, R.   The quality of the groundwater in the Danube
                                   Plain (in Czech, German summary).
                                   Bratislava, Vyskumny Ustav Vodohospodarsky
                                   (VUV), Works and Studies No. 17, 1962,
                                   127 p.

JORDAN, D.G.   Groundwater contamination in Indiana.
                                   J. Am. Wat. Wks Assn, 1962, 54, 1213-1220.

KUMPF, W.   Contamination of surface and groundwater
                                   (in German).
                                   Gas-und-Wasserfach, (Wasser/Abwasser),
                                   1961, 102, (42), 1137-1146.

LANCE, L.C.   Nitrogen removal by soil mechanisms.
                                   J. Wat. Poll. Contr. Fed., 1972, 44, (7),
                                   1352-1361.

LeGRAND, H. E.

Management aspects of groundwater contamination.
J. Wat. Poll. Contr. Fed., 1964, 36, 1133-1145.

MacLEAN, R. D.

Pollution of underground waters and the recharging of supplies.   Part A, some public health aspects.
Proc. R. Gov. Health (meeting 19.7.61.), 1961, 81, 289-291.

METROPOLITAN WATER BOARD

40th report on the results of bact. chem. and biol. examination of London waters for years 1961-1962, (4), 35-36.

PETERS, Helen J.

Groundwater management.
Mainly concerns protection of water quality.
Water Resources Bulletin, 1972, 8, (1), 188-198.

PETERS, Helen J.

Groundwater management.
Water Resources Bulletin, 1972, 8, (1), 188-197.

POPEL, F.

Water Management in Town Developments (in German).
Schriften Reihe Vereinigung Deutscher Gewässerschutz, 1957, 2, 62-79.

PRATT, P. F.

Nitrate in the unsaturated zone under agricultural lands.
United States, Environmental Protection Agency Report No. 16060DOE04/72, 1972.

PREUL, H. C.

Underground Pollution Analysis and Control.
Water Research, 6, 1972, 1141-1154.

ROBERTS, F. W.

Contamination of rivers and other water supply sources.
Municipal Engineering, London, 1966, 143, 678, 683 and 685.

ROSEN, A. A. and others

Recent developments in the chemistry of odour in water; the cause of earthy/musty odours.
Proc. Soc. Wat. Treat. Exam., 1970, 19, (2), 106-119.

SANZOT, E.

Pollution of water and protection of reserves.
Bulletin Centre, Belge. et Document Eaux, 1958, 40, 155-158.

SCHINZEL, A.

Pollution of ground water by waste materials (in German).
Wass. und Abwass., 1957, 2, 192-217.

SCHROEYERS, G. A.

Water Pollution: Development and remedies (in French).
Trib. CEBEDEAU, 1963, 16, 384-388.

SIMMONDS, M.A.

Notes on pollution of ground water supplies
in Queensland, Australia.
Proc. Soc. Wat. Treat. Exam., 1962, 11,
76-83.

SWENSON, H.A.

The Montebello incident.
Proc. Soc. Wat. Treat. Exam., 1962, 11,
(1), 84-88.

TAYLOR, E.W.

The pollution of surface and underground
waters.
J. Brit. Wat. Wks Ass., 1960, 42, 582-603.

UNITED NATIONS FOOD AND
AGRICULTURAL ORGANIZATION

Groundwater legislation in Europe.
Rome 1964.
F.A.O. Legislative Series No. 5, 175 p.

Delacourt - The Public Health Aspect of the
Struggle against Pollution of
Waters.
Paper presented to meeting of Committee
of Public Administration of Union of Western
Europe, 1959, 14 p.

UNITED STATES Department of
Health, Education and Welfare

Manual of individual water supply systems.
Public Health Service Publ. No. 24,
USGPO, 1962, 121 p.

VECCHIOLI, J.

Experimental injection of tertiary-treated
sewage in a deep well at Bay Park, Long
Island, N.Y., A summary of early results.
New England W.W. Ass. J., 1972, 86, (2),
87-103.

VROCHINSKII, K.K. and others

Pesticide level in the water of reservoirs and
sources of water supply (in Russian).
Gig Naselennykh Mest 1969, 8, 112-114.
CA 1970, 73, No. 97780.

WALKER, T.R.

Groundwater contamination in the Rocky
Mountains Arsenal area, Denver, Colorado.
Geological Society of America Bulletin,
1961, 72, 489-494.

WALKER, W.H.

Illinois groundwater pollution.
J. Am. Wat. Wks Ass., 1969, 61, (1), 31-40.

WALTON, G.

Problems in groundwater pollution.
Cincinnati, Robert A. Taft Sanit. Engng
Center, 1962, 7 p.

WARRICK, L.F. and TULLY, E.J.

Pollution of abandoned well causes 'Fond du
Lac' epidemic.
Engng News Record, 1930 (March 6th),
104-410-1.

WATER POLLUTION CONTROL
FEDERATION

The review of the literature of 1966 on waste
waters and water pollution control.
J. Wat. Pol. Cont. Fed., 1967, 39, 689-749,
867-945, 1049-1154.

WHITE, D.E., HEM, J.D. and
WARING, G.A.

Chemical composition of subsurface waters.
Washington, USGPO, USGS Professional Paper
440-f, 1963, 67 p.

WINQUIST, G.

Groundwater in Swedish eskers.
Kuugl. Tekn Hogskolaus haull nr. 61, 1953.

YORKSHIRE OUSE AND HULL
RIVER AUTHORITY

Survey of water resources.
Leeds, The Authority, 1969, 3 vols.

ZOETMAN, B.C.J. and others

Oil pollution and drinking water odour.
H_2O, 1971 August 5th, (16), 367-371.

*   *   *

Considerations preliminary to development
of a waste pesticide disposal system.
J. Environmental Health, 1971, 13, (1), 55-57.

Pollution of ground water.
J. Amer. Wat. Wks Ass., 1957, 49, (4),
392-396.

Food and waterborne disease outbreaks.
Public Health Report Washington DC, 1952,
67, 1089-1095.

Ground water deserves sanitary protection.
Johnson National Drillers Journal, 1959, 31,
(6), 4.

3. GROUNDWATER FLOW AND POLLUTION DISPERSION

ATKINSON, T.C.

The danger of pollution of limestone aquifers
with special reference to the Mendip Hills,
Somerset.
Proc. Univ. Bristol Speleological Soc.,
1971, 12, 281-290.

BAARS, J.K.

Travel of pollution and purification en route
in sandy soil.
Bull. World Health Org., 1957, 16, 727-747.

BACK, W. and HANSHAW, B.B.

Geochemical interpretations of groundwater
flow systems.
Water Resources Bull., 1971, 7, (5),
1008-1016.

BEAR, J.

Some experiments in dispersion.
J. Geophys. Res., 1961, 66, (8), 2455-2467.

BEAR, J., and others

Physical principles of water percolation and
seepage.
Paris, UNESCO, 1968, 465 p.

BEREND, J. E.

An analytical approach to the clogging effect of suspended matter.
Bull. Int. Ass. Sci. Hydrol., 1967, 12, (2), 43-55.

BERNARD, G. B.

Effect of relations between interstitial and injected wastes on permeability of rocks.
Producers Monthly, 1955, 20, (2), 26-32.

BRENNER, H.

The diffusion model of longitudinal mixing in beds of finite length.   Numerical values.
Chem. Eng. Sci., 1962, 17, 229-243.

BROWN, R. H.

Hydrologic factors pertinent to ground water contamination.
Cincinnati, Robert A. Taft Sanit. Engng Center Tech. Report No. WG1-5, 1961, 7-16.

BUTLER, R. G., ORLOB, G. T. and McGAUHEY, P. H.

Underground movement of bacterial and chemical pollutants.
J. Am. Wat. Wks Ass., 1954, 46, 97-111.

CALIFORNIA STATE WATER POLLUTION CONTROL BOARD

Report on the investigation of travel of pollution.
Sacramento, Calif., The Board, Publ. No. 11, 1954, 218 p.

CEARLOCK, D. B.

A systems approach to the management of the Hanford groundwater basin.
Ground Water, 1972, 10, (1), 88-97 (discussion 97-98).

CHILDS, E. C.

Introduction to the physical basis of soil water phenomena.
New York, Wiley, 1969, 255-261.

COLE, J. A.

The flow of groundwater (use of tracing chemicals to measure velocity and dispersion in saturated flow).
University of Cambridge, MSc dissertation, June 1957, 199 p.

Da COSTA, R. H. and BENNETT, R. R.

The pattern of flow in the vicinity of a recharging and discharging pair of wells in an aquifer having a real parallel flow.
Inte. Ass. Sci. Hydrol., Athens, Symposium 1961, Subterranean Water, Publ. 56.

DAGAN, G.

Perturbation solutions of the dispersion equation in porous mediums.
Water Resources Res. 1971, 7, 135-142.

DEGOT, B. and others

Two applications of ^{82}Br in underground hydrodynamics.
Proc. of Symposium on Radioisotopes in hydrology. Tokyo 1963, pub. Int. Atomic Energy Agency, Vienna, 1963, 321-346.

DREW, D.P. and SMITH, D.I.

Techniques for tracing subterranean drainage. Brit. Geomorph. Research Group, Technical Bulletin No. 2, Nottingham Univ., Geog. Dept., 1969, 36 p.

DUTT, G.

Quality of percolating waters: No. 1 Development of a computer program for calculating the ionic composition of percolating waters. Davis, Univ. of California, Dept. of Irrigation, Water Resources Center, Contribution No. 50, 1962, 35 p.

ELRICK, D.E.

Dispersion and reaction in unsaturated soils; applications to tracers. Bull. Int. Ass. Sci. Hydrol., 1969, **14**, (2), 49-60.

FILMER, R.W. and COREY, A.T.

Transport of virus-sized particles in porous media. Colorado State University, Sanit. Engng Papers No. 1, June 1966.

FOURNELLE, H.J., and others

Experimental groundwater pollution at Anchorage, Alaska. US Public Health Report, 1957, **72**, 203.

FRIED, J.J.

Miscible pollutions of groundwater - a study in methodology. Proc. Internat. Symp. on Mathematical Modelling Techniques in Water Resources Systems. Environment Canada, Ottawa, 1972, **2**, 362-371.

FRIED, J.J.

A mathematical model for the single well pulse technique. Symposium on Water Research, 1971, Bangalore, India.

FRIED, J.J. and COMBARNOUS, M.A.

Dispersion in porous media. Advances in Hydroscience, ed. Ven Te Chow, New York Academic Press, 1971, **7**, 169-284.

FRIED, J.J. and UNGEMACH, P.

A dispersion model for quantitative study of groundwater pollution by salt. Water Research, 1971, **5**, (7), 491-496.

FRIED, J.J. and UNGEMACH, P.

In-situ determination of the coefficient of longitudinal dispersion in a natural porous medium (in French). Comptes Rendus Acad. Sci. Paris Series A, 1971, **272**, 1327-1329.

HARLEMAN, D.R.F. and RUMER, R.R. Jr.

Longitudinal & lateral dispersion in an isotropic porous medium. J. Fluid Mech., 1963, **16**, (3), 385-394.

474.

HAZZAA, I.B.

Determination of the velocity and direction of groundwater flow by radioisotopes. Bull. Int. Ass. Sci. Hydrol., 1970, 15, (2), 11-16.

HAZZAA, I.B., and others

Investigation of groundwater radial flow using radioactive tracers. Bull. Int. Ass. Sci. Hydrol., 1967, 12, (3), 55-59.

HAZZAA, I.B., and others

Dispersion of radioactive tracers in a model basin study aquifer. Bull. Int. Ass. Sci. Hydrol., 1967, 12, (4), 75-85.

HENRY, H.R.

The effects of temperature and density gradients upon the movement of contaminants in saturated aquifers. Int. Ass. Sci. Hydrol., Publ. 78, General Assembly Berne, 1967, 54-65.

HERZIG, J.P., LECLERC, D.M., and Le GOFF, P.

Flow of suspensions through porous media-application to deep filtration. Ind. Eng. Chem., 1970, 62, 8-35.

HILLEL, D.

Soil and water. Physical principles and processes. New York, Academic Press, 1971, 288 p.

KAUFMAN, W. and ORLOB, G.T.

Measuring groundwater movement with radioactive and chemical tracers. J. Am. Wat. Wks Ass., 1956, 48, 559-572.

LAU, L.K., KAUFMAN, W.J. and TODD, D.K.

Dispersion of water tracer in radial laminar flow through homogeneous porous media. Progr. Rep. No. 5, Univ. of Calif., Berkeley, 1959, 93 p.

LeGRAND, H.E.

Patterns of contaminated zones of water in the ground. Water Resources Research, 1965, 1, 83-95.

LeGRAND, H.E.

Monitoring of changes in quality of ground water. Ground Water, 1968, 6, (3), 14-18.

LEVEQUE, P.C.

The use of natural and artificial radioactive tracers in underground hydrology. Houille Blanche, 1969, (8), 833-848.

LAIRD, R.W. and COGBILL, A.F.

Incompatible waters can plug oil sands. World Oil, 1958, 146, (6), 188-190.

MANDEL, S. and WEINBURGER, Z.

Analysis of a network model for dispersive flow. J. Hydrol., 1972, 16, (2), 147-157.

MATHIU, R. P. and GREWAL, N. S.

Underground travel of pollutants.
6th Conference of Int. Ass. Wat. Polln. Res.,
Jerusalem, 18-23 June 1972, Paper No. B31.

MERCADO, A.

The spreading pattern of injected water in a
permeably stratified aquifer.
IASH Symposium of Haifa, 1967, Pub. No. 72,
23-36.

MERCADO, A. and BEAR, J.

Mixing of labelled water by injection and
pumping at same well.
UWSS Tech. Rep. No. 24, Tahal Water
Planning for Israel Ltd., 1963, P. N. 284, 28 p.

MIKELS, F. C. and KLAER, F. H.

The application of groundwater hydraulics
to the development of water supplies by
induced infiltration.
Gentbrugge, Int. Ass. Sci. Hydrol., 1956,
Publ. No. 4, 232-242.

MILDE, G. and MOLLWEIDE, A. V.

Hydrological factor in groundwater
contamination.
Wasserwirtschaft-Wassertechnik., 1970, 20,
(7), 234-237.

NELSON, R. W. and CEARLOCK, D. B.

Analysis and predictive methods for ground-
water flow in large heterogeneous systems.
Proc. Nat. Symp. on Ground Water Hydrology
(American Water Resources Assoc.), San
Francisco, 1967, 301-318.

RIFAI, M. N. E. and others

Dispersion phenomena in laminar flow
through porous media.
Progr. Rep. No. 3, Univ. of Calif., Berkeley,
July 1956, 157 p.

ROBECK, G. G.

Groundwater contamination studies at the
Sanitary Engineering Center.
Cincinnati, Robert A. Taft Sanit. Engng
Center, Tech. Report W61-5, 193-197.

SCHOELLER, M.

The geochemistry of the water table of the
lower sands of the Aquitane in relation to the
flow and nature of the aquifer (in French).
IASH (Berne Symposium), 1968, Publ. 78,
7-11.

SHEIDEGGER, A. E.

The physics of flow through porous media.
Toronto, Univ. of Toronto Press, 1957, 236 p.

SMITH, D.B., WEARN, P.L., RICHARDS, T.H. and ROWE, P.C. Water movement in the unsaturated zone of high and low permeability strata using natural tritium.
IAEA Symposium on Use of Isotopes in Hydrology, 1970, 917 p.

STAHL, C.D. Compatibility of interstitial and injected waters.
Producers Monthly, 1962, 26, (11), 14-15.

STERNAU, R., SCHWARZ, J. and others Radioisotope tracers in large-scale recharge studies of groundwater.
Vienna, International Atomic Energy Agency.
Proceedings of Symposium on Isotopes in Hydrology, 1967, 708 p.

THEIS, C.V. Aquifers and models.
Proc. Nat. Symp. on Ground Water Hydrology (American Water Resources Assoc.)
San Francisco, 1967, 139-148.

VERRUIJT, A. Steady dispersion across an interface in a porous medium.
J. Hydrol., 1971, 14, 337-347.

WITHERSPOON, P.A. and NEUMAN, S.P. Hydrodynamics of fluid injection.
Paper presented at Symposium on Underground Waste Management and Environmental Implications, Houston, Texas, December 1971.

WOODING, R.A. Perturbation analysis of the equation for the transport of dissolved solids through porous media. Part I.
J. Hydrol., 1972, 16, 1-15 (later parts in press).

YOUNGS, E.G. Redistribution of moisture in porous materials after infiltration.
Soil Sci., 1958, 86, 117-125; 202-207.

YOUNGS, E.G. Two and three dimensional infiltration seepage from irrigation channels and infiltrometer rings.
J. Hydrol., 1972, 15, 301-315.

YOUNGS, E.G. Some aspects of water movement through soils.
Society of Chemical Industry Monogram, 1970, 37, 107-119.

ZILLIOX, L. and MUNTZER, P. Physical model of the mechanism of groundwater pollution by miscible liquids (brines) and immiscible ones (hydrocarbons) (in French).
La Houille Blanche, 1971, 26, (8), 723-730.

4. ARTIFICIAL RECHARGE

BIZE, J., and others

Artificial recharge of underground waters (in French).
Masson et Cie, Paris, 1972, 200 p.

GOTAS, H.B.

Field investigation of waste water reclamation in relation to groundwater pollution.
California State Water Pollution Control Board Publ. No. 6, 1953.

HUNTER BLAIR, A.

The present position of artificial groundwater recharge research in the United Kingdom.
2nd National Conference on Artificial Recharge, Gottwaldov, Czechoslovakia, 1972.

MATLOCK, W.G.

Sewage effluent recharge in an ephemeral channel.
Wat. Sewage Wks, 1966, 113, 224-229.

SCALF, M.R., and others

Movement of DDT and nitrates during groundwater recharge.
Water Resources Research, 1969, 5, (5), 1041-1052.

SIGNOR, D.C., and others

Annotated bibliography on Artificial Recharge of Ground Water, 1955-1967.
USGS Wat. Sup. Paper 1990, 1970, 141 p.
International Association of Scientific Hydrology. Artificial recharge and management of aquifers.
Symposium of Haifa, 1967, IASH Publ. No. 72, 7-289.

SUTER, M.

The Peoria recharge pit. Its development and results.
Proc. Am. Soc. Civ. Engrs J. Irrig. Drain Div., 1956, 82 (IR3), 1102-1117.

TODD, D.K.

Annotated Bibliography on Artificial Recharge of Ground Water through 1954.
USGS Wat. Sup. Paper 1477, 1959.

WATER RESEARCH ASSOCIATION

Proceedings Artificial Groundwater Recharge Conference, Reading, 1970.
Medmenham, The Association, 1972, 2 vols.

5. UNDERGROUND WASTE DISPOSAL

ADINOFF, J.

Disposal of organic chemical wastes to underground formations.
Proceedings Ninth Conf. on Industrial Wastes, Purdue, 1954.
Purdue Univ. 1955, Ext. Ser. No. 87, 32-38.

AMERICAN PETROLEUM INSTITUTE

Problems in the disposal of radioactive waste in deep wells.
Dallas, Texas, Am. Petrol. Inst. Div. Prod., 1958, 27 p.

AMERICAN WATER WORKS ASSOCIATION TASK GROUP E4-C	Control of underground waste disposal. J. Am. Wat. Wks Assoc., 1952, 44, 685-689.
BAKER, W. M.	Waste disposal well completion and maintenance. Wat. Wastes, 1963, (Nov/Dec), 43-47.
BARRACLOUGH, J. T.	Waste injection into a deep limestone in north-western Florida. Ground Water, 1966, 4, (1), 22-24.
BATZ, M. E.	Deep well disposal of nylon waste water. Chem. Engng Progress, 1964, 60, (10), 85-88.
BERGSTROM, R. E.	Feasibility criteria for subsurface waste disposal in Illinois. Ground Water, 1968, 6, (5), 5-9.
BROWN, R. W. and SPALDING, C. W.	Deep-well disposal of spent hardwood pulping liquors. J. Wat. Poll. Contr. Fed., 1966, 38, 1916-1924.
BUCHANAN, F.	Disposal of drainage water from coal mines into the chalk in Kent. Symp. on Pollution of Ground Water, Paper 12 in Soc. of Wat. Treat. Exam., 1962, 11, 101-105.
CECIL, L. K.	Underground disposal of process waste water. Industrial Engng Chem., 1950, 42, 594-599.
CLEARY, E. J. and WARNER, D. L.	Some considerations on underground waste-water disposal. J. Am. Wat. Wks Ass., 1970, 62, (8), 489-498.
DEAN, B. T.	Design and operation of a deep-well disposal system. J. Wat. Poll. Contr. Fed., 1965, 37, 245-254.
De LAGUNA, W. D. and HOUSER, B. L.	Disposal of radioactive waste by hydraulic fracturing. U. S. Atomic Energy Commission Report No. ORNL 2994, 1960, 128-136.
DONALDSON, E. C.	Subsurface disposal of industrial wastes in the United States. U. S. Bureau of Mines Inf. Circ. 8212, 1964, 34 p. 85-88.
ELLIOT, A. M.	Subsurface waste disposal problems. Water and Sewage Works, 1968, (Nov 29), R203-R212.

GRAVES, B.S.   Underground disposal of industrial wastes in
                               Louisiana.
                               Paper presented at Soc. of Mining Engineers,
                               1964.

HENKEL, H.O.   Deep-well disposal for chemical wastes.
                               Chem. Engng Progress, 1955, 51, (12),
                               551-554.

HOLLAND, H.R. and CLARK, F.R. A disposal well for spent sulphuric acid from
                               alkylating isobutane and butylenes.
                               Proc. Nineteenth Industrial Waste Conference,
                               Purdue University, 1965, (1), 195-199.

HUNDLEY, C.L. and MATULIS, J.T. Deep-well disposal of industrial waste by
                               F.M.C. Corporation.
                               Industrial Wat. Wastes, 1962, (Sept-Oct),
                               128-131.

KAUFMAN, W.J. and EWING, B.B. Disposal of radioactive wastes into deep
                               geologic formation.
                               J. Wat. Poll. Contr. Fed., 1961, 33, 73-84.

LANCING, A.C. and HEWETT, P.S. Disposal of phenolic waste to underground
                               formations.
                               Proc. Ninth Industrial Waste Conference,
                               Purdue, 1954, Purdue Univ., 1955,
                               Ext. Ser. No. 87, 184-194.

LEVEQUE, P.C., and others   Hydrodynamic criteria to guarantee
                               perenniality for the surface geology storage
                               of radioactive waste.
                               Conference, Tokyo, 1971, 1248-III-71-5.

LIEBERMAN, J.A. and   Practices and problems in disposal of
SIMPSON, E.S.   radioactive wastes into the ground.
                               Int. Ass. Sci. Hydrol. Helsinki Symposium,
                               Comm. Subterranean Waste, Publ. 52, 1960,
                               581-591.

LUM, D.   Preliminary report on geohydrological
                               exploration for deep well disposal of effluent.
                               Waimanalo Sewage Treatment Plant, Waimanalo
                               Oahu, Honolulu, Hawaii Division of Water and
                               Land Development Circular No. C54, 1969.

MACLEOD, I.C.   Disposal of spent caustic and phenolic water
                               in deep wells.
                               Proc. Eighth Industrial Waste Conference,
                               Ontario, 1961, 49-50.

MARSH, J.H.   Design of waste disposal wells.
                               Ground Water, 1968, 6, (2), 4-8.

MECHEM, O. E. and GARRETT, J. H. Deep injection disposal well for liquid toxic wastes.
Proc. Am. Soc. Civ. Engnrs, 1963, 89, (CO2), 111-121.

MOKHA, J. S. Report on subsurface disposal of liquid wastes and dispersal of oil-field brines.
Dept. of Petroleum Engng, University of Tulsa, Okla., May 1972, 72 p.

PARADISO, S. J. Disposal of fine chemical wastes.
Proc. Tenth Industrial Waste Conference, Purdue 1956.
Purdue Univ., 1956, Ext. Ser. No. 89, 49-60.

POWERS, T. J. and QUERIO, C. W. Fundamentals of deep well disposal of wastes.
Proc. Ohio Water Pollution Control Conference, Cleveland, Ohio, 1961, 11-12.

QUERIO, C. W. and POWERS, T. J. Deep-well disposal of industrial waste.
J. Wat. Pollut. Control Fed., 1962, 34, 136-144.

ROEDDER, E. Problems in the disposal of acid-aluminium high level radioactive waste solutions by injection into deep lying permeable formations.
Washington, USGPO, US Geol. Survey Bull. No. 1088, 1959, 65 p.

ROSE, J. L. Injection of treated waste water into aquifers.
Water and Wastes Engineering, 1968, 5, (10), 40-44.

SELM, R. F. and HULSE, B. T. Deep well disposal of industrial wastes.
Chem. Engng Progress, 1960, 56, (5), 138-144.

SHELDRICK, M. G. Deep well disposal. Are safeguards being ignored?
Chemical Engineering, 1969, 76 (Apr), 74-76 and 78.

SLAGLE, K. A. and STOGNER, J. M. Oil fields yield new deep-well disposal technique.
Wastes and Sewage Works, 1969, 116, (6), 238-244.

TALBOT, J. S. Some basic factors on the consideration and installation of deep well disposal systems.
Water and Sewage Works, 1968, Nov. 29th, R213-R219.

TALBOT, J. S. and BEARDON, P. The deep well method of industrial waste disposal.
Chem. Engng Progress, 1964, 60, (1), 49-52.

THEIS, C. V.   Geologic and hydrological factors in ground
                                      disposal of waste.
                                      U. S. Atomic Energy Commission Report No.
                                      WASH 275, 1954, 261-283.

UNITED STATES   Proc. 2nd Atomic Energy Commission
Department of Commerce   Working Meeting - Groundwater disposal of
                                      radioactive wastes - Chalk River, Canada.
                                      TID 7628, Washington, DC, 1962.

VAN EVERDINGEN, A. F.   Fluid mechanics of deep well disposals.
                                      Am. Ass. Petroleum Geologists, Memoir 10,
                                      Tulsa, 1968, 32-42.

VEIR, B. B.   Deep-well disposal pays off at Celanese
                                      Chemical Plant.
                                      Water and Sewage Works, 1969, **116**, (5),
                                      IW21-IW24.

VLADIMIRSKY, V. I.   Hydrogeological criteria for the delimitation
                                      of zones of sanitary protection of underground
                                      water sources (in French).
                                      Prosp. et Prot. Sous Sol., 1962, (8), 43-49.

WARNER, D. L.   Deep-well injection of liquid waste - a review
                                      of existing knowledge and an evaluation of
                                      research needs.
                                      U. S. Public Health Service, Division of Water
                                      Supply and Pollution Control, Cincinnati,
                                      Ohio, 1965, 55 p.

WARNER, D. L.   Subsurface disposal of liquid industrial wastes
                                      by deep-well injection.
                                      Am. Ass. Petroleum Geologists, Memoir 10,
                                      Tulsa, 1968, 11-20.

WARNER, D. L.   Deep-well disposal of industrial wastes.
                                      Chemical Eng., 1965, **72** (Jan 4), 73-78.

WALKER, W. R. and STEWART, R. C.   Deep-well disposal of wastes.
                                      Proc. Am. Soc. Civ. Engrs, J. Sanitary
                                      Engng Div., 1968, **94**, 945-968.

                                      *   *   *

                                      Spent pulping liquors to be discharged under-
                                      ground.
                                      Chem. Engng News, 1963 (Sept 2nd), 128-129.

                                      Industrial wastes forum subsurface disposal.
                                      Sewage Ind. Wastes, 1953, **25**, (6), 715-720.

6. SURFACE WASTE DISPOSAL

APGAR, A. and LANGMUIR, D.   Groundwater pollution potential of a landfill
                                      above the water table.
                                      Ground Water, 1971, **6**, (9), 76-96.

BHIDE, A.D., and
MULEY, V.U.

Studies on pollution of ground water by solid
wastes.
Central Public Health Engineering Research
Institute, Nagpur, India, 1973, 8 p.
(reprint from the Proceedings of a Symposium
on Environmental Pollution).

COE, J.J.

Effect of Solid Waste Disposal on Groundwater
Quality.
J. Am. Wat. Wks Ass., 1970, 62, (12),
776-783.

DAVISON, A.S.

The effect of tipped domestic refuse on
groundwater quality.
Proc. Soc. Wat. Treat. Exam., 1969, 18,
(1), 35-41.

FARKASDI, G., and others

Microbiological and sanitary investigations of
groundwater pollutants in the underflow of
waste tips (in German).
Städtehygiene, 1969, 20, 25-31.

FREEZE, R.A.

Subsurface hydrology at waste disposal sites.
IBM Journal of Research and Development,
1972, 16, (2), 117-129.

FUNGAROLI, A.A. and STEINER, R.L.

Laboratory study of the behaviour of a
sanitary landfill.
J. Wat. Poll. Contr. Fed., 1971, 43, (2),
252-267.

GOLWER, A. and MATTHESS, G.

Research on groundwater contamination by
deposits of solid waste.
Int. Ass. Sci. Hydrol., 1968, 78, 129-133.

GOLWER, A. and MATTHESS, G.

Quality deterioration in groundwater caused
by waste (in German).
Proc. Deutsch. Gewasserk. Meeting, 1968.
Deutsche Gewasserk. Mitt, 1969, 51-55.

GOLWER, A., and others

The influence of waste tips on groundwater
(in German).
Der Städtetag, 1971, (2), 1-5.

HOLDEN, W.S.

The effect of tipped domestic refuse on
groundwater quality.
Proc. Soc. Wat. Treat. Exam., 1969, 18,
42-69.

HOPKINS, G.J. and POPALISKY, J.R.

Influence of an industrial waste landfill
operation on a public water supply.
J. Wat. Poll. Contr. Fed., 1970, 42, (3),
431-436.

HOUSING AND LOCAL GOVERNMENT
MINISTRY OF

Pollution of water by tipped refuse. Report
of the Technical Committee on the experimental disposal of house refuse in wet and dry tips.
London, HMSO, 1961, 141 p.

HOUSING AND LOCAL GOVERNMENT
MINISTRY OF

Disposal of solid toxic wastes. Report of the
Technical Committee on the disposal of toxic
solid wastes.
London, HMSO, 1970, 107 p.

ILLINOIS STATE GEOLOGICAL SURVEY	Selection of Refuse Disposal Sites in Northeastern Illinois. G. M. Hughes, EGN 17, 1967.
ILLINOIS STATE GEOLOGICAL SURVEY	Hydrogeologic Data from Four Landfills in Northeastern Illinois. G. M. Hughes, R. A. Landon, and R. N. Farvolden, EGN 26, 1969.
ILLINOIS STATE GEOLOGICAL SURVEY	Evaluating Sanitary Landfill Sites in Illinois. Keros Cartwright and F. B. Sherman, EGN 27, 1969.
ILLINOIS STATE GEOLOGICAL SURVEY	Summary of Findings on Solid Waste Disposal Sites in Northeastern Illinois. G. M. Hughes, R. A. Landon and R. N. Farvolden, EGN 45, 1971.
ILLINOIS STATE GEOLOGICAL SURVEY	Hydrogeologic Considerations in the Siting and Design of Landfills. G. M. Hughes, EGN 51, 1972.
LANGER, W.	Means of protection of ground water against pollution from refuse tips (in German). Schr. Reihe. Ver. Wass. Boden und Lufthyg. 1961, 19, 113-130.
LeGRAND, H. E.	System for evaluation of contamination potential of some waste disposal sites. J. Am. Wat. Wks Ass., 1964, 56, 959-974.
MacLEAN, R. D.	The effect of tipped domestic refuse on groundwater quality. A survey in North Kent. Proc. Soc. Wat. Treat. Exam., 1969, 18, (1), 18-34.
MATSCHAK, H. and WALDE, M.	Investigation of infiltration and flow components in the regeneration of the level of groundwater in open pit dumps (in German). Bergbautechnik, 1970, 20, 561-569.
NÖRING, F., and others	The decomposition processes of groundwater pollutants in the underflow of waste tips (in German). Gas und Wasserfach (Wasser/Abwasser), 1968, 109, (6), 137-142.
NÖRING, F., and others	The effects of industrial and domestic waste on groundwater (in German). Mem. Congs. Int Ass Hydrogeol, 1967, 7, 165-171.
OTTO, F.	Containment of water contaminants in refuse tips. Galvanotechnik, 1966, 57, (8), 528-529.

QUASIM, S.R. and BURCHINAL, J.C.   Leaching of pollutants from refuse beds.
Proc. Am. Soc. Civ. Engrs, 1970, 96, (SA1),
49-59.

QUASIM, S.R. and BURCHINAL, J.C.   Leaching from simulated landfills.
J. Wat. Poll. Contr. Fed., 1970, 42, (3),
371-379.

ROBERTSON, J.S., and others   Pollution of underground water by pea silage.
Monthly Bull. Min. Health and Public Health
Laboratory Service, 1966, 25, 172-179.

ROESSLER, B.   Pollution of groundwater by refuse tips.
Vom Wasser, 1951, 18, 43-60.

SACKMAN, L.A.   Aquifer protection: control of the Rhine watertable aquifer.
La Houille Blanche, 1971, 8, 717-722.

SALVATO, J.A., and others   Sanitary landfill leaching - prevention and
control.
J. Wat. Poll. Contr. Fed., 1971, 43, (10),
2084-2100.

SCHINZEL, A. and BENGER, H.   Pollution of waters by wastes from cellulose
and cellulose wool factories (in German).
Arch. Hyg. Berl., 1960, 144, 320-344.
Z. Bl. Bakt 1961, 181, 158-159.

STONE, R.   Land disposal of sewage and industrial wastes.
Sewage and Wastes, 1953, 25, 406-418.

STONE, R. and GARBER, W.F.   Sewage reclamation by spreading basin
infiltration.
Proc. Am. Soc. Civ. Engrs, 1951, 77, 117-125.

US ENVIRONMENTAL CONTROL
ADMINISTRATION Bureau of Solid
Waste Management   Hydrogeology of solid waste disposal sites in
North eastern Illinois - an interim report on a
solid waste demonstration grant project.
Cincinnati, The Bureau, 1969, 137 p.

US ENVIRONMENTAL PROTECTION
AGENCY   Hydrogeology of solid waste disposal sites in
North eastern Illinois.
Washington, USGPO, 1971, 154 p.

WALTZ, J.P.   Methods of geological evaluation of pollution
potential at mountain homesites.
Ground Water, 1972, 10, (1), 42-47
(discussion 47-49).

WATERTON, T.   The effect of tipped domestic refuse on ground
water quality.
J. Wat. Treat. Exam., 1969, 18, 15-17.

WRIGHT, R.M.   Hydrogeological criteria for evaluating solid-
waste disposal sites in Jamaica.
J. Sci. Res. Counc. Jamaica, 1972, 3,
59-90.

ZANOUI, A.E.   Groundwater pollution and sanitary landfills -
a critical review.
Ground Water, 1972, 10, (1), 3-13
(discussion 14-16).

7. SPECIFIC POLLUTANTS

7.1. Salt water

BRUNOTTE, and others	A study of groundwater pollution by salt. Advances in Water Polln. Research Proc. 5th Internat. Conf. San Francisco and Hawaii, 1970, pub. Pergamon 1971, 1, Papers 1-34.
CHARMONMAN, S., and others	A fresh-water canal as a barrier to salt-water intrusion. Int. Assoc. Scient. Hydrol., General Assembly of Berne, 1967, Publ. 77, 343-352.
FRIED, J.J., GARNIER, J.L. and UNGEMACH, P.	Quantitative study of aquifer pollution - The salinity of a phreatic aquifer in the province of Haut-Rhin (in French). Bull. du B.R.G.M. (2nd series), 1971, III (1), 105-115.
HULING, E.E. and HULLOCHTER, T.C.	Groundwater contamination by road salt; steady state concentrations in East Central Massachusetts. Science, 1972, 176, 288-290.
SCHMORSK, S., and others	Salt water encroachment. Proc. Symposium of Haifa/Gentbrugge 1967 Int. Ass. Sci. Hydrol. Publ. No. 72, 1967.

7.2. Heavy metals

ALVARADO, A.	Chronic arsenic poisoning in the Lake Region. Salud. Pub. Mex., 1964, 6, 375-420.
BECKSMANN, E.	The behaviour of oil derivatives in groundwater (in German). Gas und Wasserfach (Wasser/Abwasser), 1963, 104, 689-694.
DAVIDS, H.W. and LIEBER, M.	Underground water contamination by chromium wastes. Water and Sewage Works, 1951, 98, 528.
DENSHAM, A.B. and BEALE, P.A.A.	Physical and chemical aspects of underground storage (of gas). London, Inst. Gas Engrs, 1963, 14 p.
DRACOS, Th.	The behaviour and movement of immiscible fluids in the subsoil (in German). M. bull. schweiz. Ver. Gas und Wass-fachm. 1968, 48, (10), 293.
DRACOS, Th.	Experimental investigations on the migration of oil products in unconfined aquifers. De Ingenieur, 1969, 81, 51-52.
GAS COUNCIL (London Research Station)	Experience with underground storage in aquifers. London, The Station, 1962, var. pag.

GOLEVA, G.A., and others

Distribution and migration of lead in groundwaters.
Geochemistry International (Washington), 1970, 7, 256-268.

GREENE, L.A.

The problems of iron and manganese in water supplies.
Brit. Waterworks Assoc. Jubilee Travelling Scholarship Report 1968-9, 45 p.

HAFFNER, A. and HOPTON, G.V.

Underground storage of gas.
London, Parliamentary and Scientific Committee, 1963, 9 p.

JONES, SIR HENRY

Storage of gas underground in aquifers.
Proc. Instn Civ. Engrs, 1963, 26, 317-330.

KAUFMAN, W.J.

Inorganic chemical contamination of groundwater.
Proc. Ground Water Contamination Symposium, Cincinnati 1961. Sanit. Engng Center Tech. Rept. W 61-5, 1961, 43-50.

LIEBER, M. and WELSCH, F.W.

Contamination of groundwater by cadmium.
J. Am. Wat. Wks Ass., 1954, 46, 541-547.

LIEBER, M., and others

Cadmium and hexavalent chromium in Nassau County ground water.
J. Am. Wat. Wks Ass., 1964, 56, (6), 739-747.

MERRITT, G.C. and EMERICH, G.H.

The need for hydrogeologic evaluation in a mine drainage abatement programme.
Proc. Third Symposium on Coal Mine Discharge Research, Pittsburgh, Pa., 1970.

MINDEN, L.

Iron and manganese in groundwaters (in German).
Schweizerische Zeitschrift fur Hydrologie, 1960, 22, 228-241.

MINK, L.L., and others

Effect of early day mining operations on present day water quality.
Ground Water, 1972, 10, (1), 17-26.

PERLMUTTER, N.M., LIEBER, M. and FRAUENTHAL, H.L.

Waterborne cadmium and hexavalent chromium wastes in South Farmingdale, Nassau County.
USGS Professional Paper 475C, 1963, C179-C184.

POWER, MINISTRY OF

The underground storage of gas; questions and answers.
London, The Ministry, 1965, 15 p.

SAUER, K.   Poisoning of groundwater by chemical wastes
                                        (in German).
                                        Gas und Wasserfach (Wasser/Abwasser),
                                        1961, 102, 998.

SCHWILLE, F.   The migration of mineral oil in porous media
                                        (in German).
                                        Gas und Wasserfach (Wasser/Abwasser),
                                        1971, 112, 307-311, 331-339 and 465-472.

SEMMLER, W.   Groundwater contamination by iron
                                        (in German).
                                        Bohrtechnik, Brunnenbau, Rohrleitungsbau,
                                        1970, 21, (7), 247-251.

VAN DAM, J.   Hydrocarbon migration in a water-bearing deposit
                                        Eau (Asnières), 1969, 56, (12) 581-587.

WALSH, F. and MITCELL, R.   Biological control of acid mine pollution.
                                        J. Wat. Poll. Contr. Fed., 1972, 44, (5),
                                        763-768.

7.3. Cyanides

EFFENBERGER, E.   Pollution of groundwater by cyanide.
                                        Arch. Hyg. Bakt., 1964, 148, 271-287.

SCHEWE, L.D.   Groundwater contaminated by cyanides from
                                        blast furnace wastes (in German).
                                        Gas und Wasserfach (Wasser/Abwasser),
                                        1969, 110, (26), 702-706.

7.4. Hydrocarbons

AMERICAN PETROLEUM   The migration of petroleum products in soil
INSTITUTE   and groundwater.   Principles and
                                        countermeasures.,
                                        Institute's Committee on Environmental
                                        Affairs Publication No. 4149.
                                        Washington DC, December 1972.

BEYNON, L.R., and   Prevention of contamination of water resources
OLDHAM, G.F.   by oil: Proceedings of the Eighth World
                                        Petroleum Congress, Moscow, 1971, 77-83.

BUYDENS, R.   The pollution of water by petroleum products
                                        (in French).
                                        Tribune du Cebedeau, 1960, 13, (111), 90-100.

DIETZ, D.N.   Pollution of permeable strata by oil compo-
                                        nents in water pollution by oil.
                                        Proceedings of Aviemore Seminar, May 1970,
                                        London, Institute of Petroleum, 1971.

DRACOS, Th.   Survey and results of model trials of protective
                                        and restorative measures for oil spillages:
                                        plan No. 3.
                                        Technical University, Zürich, 1968.

GERMANY, FEDERAL REPUBLIC   Evaluation and treatment of oil spill
                                        accidents on land with a view to the
                                        protection of water resources.
                                        Working Group "Water and Petroleum" of the
                                        Federal Ministry of the Interior, Bonn,
                                        December 1970.

GLEBIN, V.E. Advanced techniques and methods of
protection against the failure of storage tanks,
pipelines and means of transport for crude
oil and oil products. Proc. Seminar on
protection of ground and surface water from
crude oil and oil products.
Economic Commission for Europe (Geneva
1-5 Dec 1969), New York, United Nations,
1970.

HEPPLE, P. (Ed) Water pollution by oil.
I. W. P. C. and Inst. Petroleum Seminar,
Aviemore, May 1970.
Institute of Petroleum and The Elsevier
Publishing Co. Ltd., 1971, 393 p.

INESON, J. and PACKHAM, R.F. Contamination of water by petroleum
products.
Proc. Symp. Joint Problems of the Oil and
Water Industries, Brighton, 1967.
London, Institution of Water Engrs, 12 p.

KASSECKER, F. Pollution of groundwater by escaping motor
fuel after accident to tank wagons (in German).
Gas Wass. Wärme, 1961, 15, 249-252.

KÖLLE, W. and SONTHEIMER, H. The problem of groundwater contamination
by mineral oils (in German).
Brennstoff Chem., 1969, 50, (4), 123-129.

LAUFER, W. Saarbrucken defeats oil danger at a works
saved by luck and action (in German).
Ztg. Komm. Wirtschaft, 1960, 70, 15.
Lit. Ber. Wass. Abwass. Luft und Boden,
1960, 9, 91-92.

McKEE, J.E., and others Gasoline in groundwater.
J. Wat. Poll. Contr. Fed., 1972, 44, (2),
293-302.

MOLLWEIDE, H.U. Dispersion of mineral oils in aquifers
(in German).
Wasserwirtschaft-Wassertechnik., 1972, 22,
(1), 16-19.

NYDEGGER, H. Protective measures for groundwater in the
construction of the railway shunting yard at
Zurich-Limmattal.
Wasser-Boden-Luft. Special Publication
No. 16, May 1969.

RINCKE, G. Damage to waters by mineral oil products
(in German).
Schr. Reihe Ver. Dtsch. Gewässerschutz,
1961, No. 7, 24 p.

SCHAACK, E.A., WEYER, T. and
LORZING, H. Accident to a tank wagon near a water works
(in German).
Monatsbull Schweiz Ver. Geo und
Wasserfachmannern, 1961, 41, 262-265.

SCHLATFER, P. Prevention of oil pollution of surface and
groundwater.
Monatsbull Schweiz Ver. Gas und Wassersachs
1959, 39, 33-34.

SCHMASSMANN, H.

Oil seepage in ground-water areas and examples of measures taken from practical experience.
Gas. Wass. Wärme, 1969, 49, 65-76.

SCHWILLE, F.

Petroleum contamination of the subsoil, a hydrological problem. Proc. Symp. Joint problems of the oil and water industries, Brighton 1967.
London, Institution of Water Engineers, Paper 2.

SMITH, H.E.

Water polluted by aviation fuel.
Paper 5, Symp. Poll. groundwater.
Soc. of Wat. Treat. Exam., 1962, 11, 34-37.

UNITED NATIONS ECONOMIC COMMISSION FOR EUROPE

Protection of ground and surface waters against pollution by crude oil and oil products.
Seminar held at Geneva, December 1969.
United Nations, New York, 1970, 2 volumes, 166 p. and 224 p.

VAN DAM, J.

The migration of hydrocarbons in a water bearing stratum. Proc. Symp. The joint problems of the oil and water industries, Brighton 1967.
London, The Institution of Water Engineers, Paper No. 3.

VEREININGUNG DEUTSCHER GEWÄSSERSCHUTZ

The importance of modern energy sources in water hygiene; dangers and disasters through oil (in German).
Bad Godesburg VDG Schriftenreiche No. 11, 1963, 44 p.

ZIMMERMANN, W.

Effect of mineral oil on ground water (in German).
Gewasserschutz-Wasser-Abwasser 1970, 3, 23-32.

ZIMMERMANN, W.

Problems of pollution of surface and ground-waters by mineral oils.
Paper 9, Proc. International Congress PROAQUA, Basle 1961, 137-153.

ZIMMERMANN, W., and others

Water supply and car products.
Zbl. Bakt., 1, Orig., 1959, 174, 155-164.

ZIMMERMANN, W., and others

Experimental inquiry into the contamination of groundwater by mineral oil products (in German).
Gas. und Wasserfach. (Wasser/Abwasser), 1964, 105, 1089-1092.

*   *   *

Measures for the protection of waters in the event of oil spillages.
Gesundheitswesen, Godesberg, 1969, 1-137.

7.5. Phenols

DEUTSCH, M.

Phenol contamination of an artesian glacial drift aquifer at Alma, Michigan, USA. Proc. Soc. Wat. Treat. Exam., 1962, 11, 94-100.

LINGELBACH, H., and others

Pollution in water supply plants by gas works waste waters (in German). Z. Hyg. Grenzgeb, 1962, 8, 761-767.

STANIER, G.

Phenol in the water supply of a large city (in German). Wasserw-Wass. Techn., 1958, 8, 20.

WOOD, E.C.

Pollution of groundwater by gas works waste. Proc. Soc. Wat. Treat. Exam., 1962, 11, 32-33.

7.6. Detergents

CALIFORNIA. STATE WATER QUALITY CONTROL BOARD

Dispersion and persistence of synthetic detergents in groundwater San Bernardino and Riverside Counties. Sacramento, The Board, Publ. No. 30, 1965, 114 p.

CAMPENNI, L.G.

Synthetic detergents in groundwaters. Wat. and Sewage Wks, 1961, 108, 188-191 and 210-213.

HENCHAN, E. and MAYER, J.

Investigation of underground and surface water pollution by waste water from generators. Coll. Sci. Rep. 1952-1955. Inst. Hyg. Prague, 1957, 73-74.

NETSCH, F.

Pollution of groundwater by detergent residues from laundry (in German). Zietschrift fur die Gesamte Hygiene und ihre Grenzgebiete, 1959, 5, 207-212.

NEWELL, I.L. and ALMQUIST, F.O.A.

Contamination of groundwater by synthetic detergents. J. New Eng. Wat. Wks Assn, 1960, 79, 61-77.

OLIVIER, G.E.

ABS in Michigan supplies. J. Am. Wat. Wks Assn, 1961, 53, 301-302.

*   *   *

Detergents foul wells, halts home building Engng News Record, 1958, 161, (10), 25.

7.7. Pesticides

ALDOUS, J.R.

2, 4-D residue in water following aerial spraying in a Scottish forest.
Weed Res., 1967, 7, (3), 239-241.

ALY, O.M., and EL DIB, M.A.

Studies on the persistence of some carbamate insecticides in the aquatic environment - hydrolysis of Sevin, Baygon, Pyrolan and Dimetilan in waters.
Water Research, 1971, 5, 1191-1205.

BAUER, U.

Pesticides and water supply.
Report No. 100 of the Dortmunder Stadtwerke AG (in German), 1971, 16 p.

BAYS, L.R.

Pesticide pollution and the effects on the biota of Chew Valley Lake.
Proc. Soc. Wat. Treat. Exam., 1969, 18, 295-326.

BERAU, F. and GUTH, J.A.

Organic insecticides in various soils with particular reference to possible groundwater pollution.
Pflanzenschutz Ber., 1965, 33 (5-8), 65-117.

BONDE, E.K. and URONE, P.

Plant toxicants in underground water in Adams County, Colorado.
Soil Science, 1962, 93, 353-356.

BOUMAN, M.C., and others

Behaviour of chlorinated insecticides in a broad spectrum of soil types.
J. Agric. Food Chem., 1965, 113, 360-365.

COHEN, J.M., and others

Effect of fish poisons on water supplies, Part III, Field study at Dickinson.
J. Am. Wat. Wks Ass., 1961, 53, 233-246.

CROLL, B.T.

Organo-chlorine insecticides in water.
Part I.
Proc. Soc. Wat. Treat. Exam., 1969, 18, 255-274.

DUBEY, H.D. and FREEMAN, J.F.

Leaching of linuron and diphenamid.
Weed Sci., 1965, 13, 360-362.

EDWARDS, C.A.

Insecticide residues in soils.
Residue Reviews, 1966, 13, 83-132.

EICHELBERGER, J.W. and LICHTENBERG, J.J.

Persistence of pesticides in river water.
Environ. Sci. Technology, 1971, 5, (6), 541-544.

EYE, J.D.

Aqueous transport of dieldrin residues in soil.
J. Wat. Poll. Contr. Fed., 1968, 40, (8), R316-R332.

FAUST, S.D.

Pollution of the water environment by organic pesticides.
Clinical Pharmacology and Therapeutics, 1964, 5, 677-686.

FRANK, P.A., and others

Herbicides in irrigation water following canal-bank treatment for weed control.
Weed Sci., 1970, 18, (6), 687-692.

GIASSO, C., and others

Detection of pesticides in underground waters.
Annali dellasamita publica, 1968, 29, (4), 1029-1032.

GRUNTER, F.A., and others

Reported solubilities of 738 pesticide chemicals in water.
Residue Reviews, 1968, 20, 1-148.

JE HINAR, H.M.

Pollution of groundwater by substances difficult to decompose (in German).
Osterreichische Wasserwirtschaft, 1957, 7, 56.

JONES, K.H., and others

Acute toxicity data for pesticides (1968).
World Rev. Pest Control 1968, 7, (3), 135-143.

KOREN, E., and others

Absorption, volatility and migration of thiocarbamate herbicide in soil.
Weed Sci., 1969, 17, (2), 148-153.

LEWALLEN, M.J.

Pesticide contamination of a shallow bored well in Southeastern coastal plains.
Groundwater, 1971, 9, (6), 45-49.

LICHTENBERG, J.J., and others

Pesticides in surface waters of United States. 5 year summary, 1964-1968.
Pesticide Monit. J., 1970, 4, (2), 71-86.

LOWDEN, G.F., SAUNDERS, C.L. and EDWARDS, R.W.

Organo-chlorine insecticides in water - Part II.
Proc. Soc. Wat. Treat. Exam., 1969, 18, 275-287.

McCARTY, P.L. and KING, P.H.

The movement of pesticides in soils.
Proc. 21st Indust. Wastes Conf.
Purdue Univ., 1966, Engng Extn. Ser. No. 121, 156-171.

TARRENT, K.R. and TATTON, J. O'G.

Organo-chlorine pesticides in the British Isles.
Nature, 1968, 219, 725-727.

TSAPKO, V.V., and others

Possible penetration of Hexachloran and Heptachlor into groundwaters (in Russian).
Gig. Nasennykh Mest. (Hygiene in Residential Areas), 1967, 93-95.

7.8. Sewage and agriculture

BOGAN, R.H.

Problems arising from groundwater contamination by sewage lagoons at Tieton, Washington.
US Dept. Health Educ. and Welfare Cincinnati. Robert A. Taft Sanit. Engng Center, Tech. Rep. W61-5, 1961, 83-87.

BORNEFF, J.

Carcinogens in water and soil.
XV Survey of investigations.
Arch. Hyg., 1964, 148, 1-11.

BORNEFF, J. and KUNTE, H.

Carcinogens in water and soil.
XXVI Method of measurement of polycyclic aromatics in water.
Arch. Hyg., 1969, 153, 220-229.

BOUWER, H.

Water quality aspects of intermittent infiltration systems using secondary sewage effluent.
Proc. Artificial Groundwater Recharge Conf. Reading, 1970.
Medmenham, Water Research Assoc., 1970, 1, 199-221.

CALDWELL, E.L.

Pollution from a pit latrine when an impervious stratum closely underlies the flow.
J. Infect. Disease, 1937, 61, 270.

CALDWELL, E.L.

Pollution flow from a pit latrine when permeable soils of considerable depth exist below the pit.
J. Infect. Disease, 1938, 62, 125.

CALDWELL, E.L. and PARR

Groundwater pollution and the bored hole latrine.
J. Infect. Disease, 1937, 61, 148.

COOK, G.T. and SMITH, A.J.

Enterovirus excretors in residential nurseries.
Monthly Bull. Min. Health and Public Health Laboratory Services, 1962, 21, 47.

COOKE, G.W. and WILLIAMS, R.J.B.

Losses of nitrogen and phosphorus from agricultural land.
Proc. Soc. Wat. Treat. Exam., 1970, 19, (13), 253-276.

DAVEY, K.H.

An investigation into the nitrate pollution of the chalk borehole water supplies.
Scunthorpe, North Lindsey Water Board, 1970, 167 p.

DE JONG, B.

Contamination of groundwater in the catchments.
Vom Wasser, 1971, 38, 141-156.

GILLHAM, R. W. and WEBBER, L. R.

Nitrogen contamination of groundwater by barnyard leachates.
J. Wat. Poll. Contr. Fed., 1969, $\underline{41}$, (10), 1752-1762.

GREENE, L. A. and WALKER, P.

Nitrate pollution of chalk water.
Proc. Soc. Wat. Treat. Exam., 1970, $\underline{19}$, 169-182.

LOEHR, R. C.

Drainage and pollution from beef cattle feedlots.
Proc. Am. Soc. Civ. Engrs. J. Sanit. Engng Div., 1970, $\underline{96}$, (SA6), 1295-1311.

NAVONE, R., and others

Nitrogen content of ground water in Southern California.
J. Am. Wat. Wks Ass., 1963, $\underline{55}$, 615-618.

PREOBRASCHENSKAJA, A. S.

Pollution of soil and ground water by soakaways.
Hyg. and Sanit. (Moscow), 1950, (12), 4-12.

SCHMIDT, K. D.

Nitrate in ground water of the Fresno-Clovis area.
Ground Water, 1972, $\underline{10}$, (1), 50-61 (and discussion, 61-64).

SCHROILLE, F.

Nitrate in groundwater.
Gerassenkundliche Mitteilungen (Deutsche), 1961, $\underline{6}$, (2), 25-32.

SCHTERNIK, C.

Groundwater pollution by nitrogen compounds in Israel.
Proc. Sixth Int. Wat. Poll. Res. Conf. Jerusalem, 1972, Paper B. 30, to be published by Pergamon Press.

SCHWILLE, F.

High nitrate levels in well waters of the Mosel Valley between Trier and Koblenz.
Gas. und Wasserfach. (Wasser/Abwasser), 1969, $\underline{110}$, 35-44.

SHEPHERD, J. M.

Pollution of groundwater supplies by sewage.
Proc. Soc. Wat. Treat. Exam., 1962, $\underline{11}$, 12-16.

TAYLOR, R. G., and BIGBEE, P. D.

Fluctuations in nitrate concentrations utilized as an assessment of agricultural contamination to an aquifer of a semi-arid climatic region.
Water Research, $\underline{7}$, 1973, 1155-1161.

WAGNER, G. H.

Changes in nitrate-N in field plot profiles as measured by porous cup technique.
Soil Science, 1965, $\underline{100}$, (6), 397-402.

7.9. Micro-organisms

BARTOS, D., and others

The survival of enteric organisms in well water.
Z. Hyg. Infektionskrankh., 1947, $\underline{127}$, 247-254.

BRIGHTON, W. D.

Pollution of chalk boreholes by filamentous organisms.
Proc. Soc. Wat. Treat. Exam., 1958, 7, 144-156.

CLARKE, N. A. and others

Survival of Coxsackie virus in water and sewage.
J. Am. Wat. Wks Ass., 1956, 48, 677-682.

CLARKE, N. A. and others

Human enteric viruses in water source: Source survival and removability.
Proc. First International Conference on Water Pollution Research, London, 1962.
Oxford, Pergamon, 1964, 523-542.

DREWRY, W. A. and ELIASSEN, R.

Virus movement in groundwater.
J. Wat. Poll. Contr. Fed., 1967, 40, R257-R271.

ELIASSEN, R., and others

Studies on the movement of viruses in groundwater.
Ann. Reports. Water Quality Control Research Lab. Stanford University, Stanford, California, 1964-1967.

ELIASSEN, R. and CUMMINGS, R. H.

Analysis of waterborne outbreaks 1938-1945.
J. Am. Wat. Wks Ass., 1948, 40, 509-528.

FARQUHAR, J. D., STOKES, J. and SCHNACK, W. D.

Epidemic of viral hepatitis apparently spread by drinking water and by contact.
J. Am. Med. Ass., 1952, 149, 991 and abstracted in J. Am. Wat. Wks Ass., 1953, 45, (5), P & R 84.

GILCREAS, F. W. and KELLY, S. M.

Relation of coliform-organism test to enteric virus pollution.
J. Am. Wat. Wks Ass., 1955, 47, (7), 683-694.

GLANTZ, P. J. and KIADEL, D. C.

Escherischia coli serogroup 115 isolated from animals: isolation from natural cases of disease.
J. Am. Vet. Res., 1967, 28, 1891-1895.

GREEN, D. M., and others

Waterborne outbreak of viral gastroenteritis and Sonne dysentery.
J. Hyg., 1968, 66, 383-392.

HALL, H. E. and HAUSER, G. H.

Examination of faeces from food handlers for Salmonella shigella, Enteropathogenic Escherichia coli and Clostridium perfringens.
Appl. Microbiol., 1966, 14, 928-933.

MacDONALD, K. B.

The transport of waterborne viruses in soil.
University of Guelph, Ph. D. Thesis, 1971.

MacLEAN, R. D.   Imported intestinal parasites.
            Medical Officer, 1958, <u>99</u>, 61.

McCOY, J. H.   The presence and importance of <u>Salmonella</u>
            in sewage.
            Proc. Soc. Wat. Treat. Exam., 1957, <u>6</u>,
            88-89.

MOSELEY, J. W.   Transmission of viral diseases in drinking
            water.
            Transmission of viruses by the water route
            G. Berg (ed).
            New York, Interscience, 1967, 5-23.

POYNTER, S. F. B.   The problem of viruses in water.
            Proc. Soc. Wat. Treat. Exam., 1968, <u>17</u>,
            (3), 187-198.

ROMERO, J. C.   The movement of bacteria and viruses through
            porous media.
            Groundwater, 1970, <u>8</u>, (2), 37-48.

PUBLIC HEALTH IN LABORATORY Salmonella in pigs and animal feeding stuffs
SERVICE <u>and</u> SKOVGAARD, N. <u>and</u> in England and Wales and in Denmark.
NIELSEN, B. B.   J. Hyg., 1972, <u>70</u>, 127-141.

STILES, C. W. <u>and</u> CROWHURST, H. R. The principles underlying the movement of
            B. coli in groundwater with resultant
            pollution of wells.
            Public Health Reports, 1923, <u>38</u>, 1350.

VECCHIOLI, J., <u>and others</u> Travel of pollution-indicator bacteria through
            the Magothy aquifer, Long Island, N. Y.
            Geol. Survey Research, 1972, U. S. G. S. Prof.
            Paper 800-B, 237-252.

VOGT, J. E.   Infectious hepatitis epidemic at Posen, Mich.
            J. Am. Wat. Wks Ass., 1961, <u>53</u>, 1238-1242.

VOGT, J. E.   Infectious hepatitis outbreak in Posen, Michigan.
            Cincinnati, Robert A. Taft Sanit. Engng Center,
            Tech. Report W61-5, 1961, 87-91.

WATER RESEARCH ASSOCIATION Notes on virus in water: Factors relating to
            their survival and removal.
            Technical Inquiry Report, TIR 192,
            Medmenham, The Association, 1969, 9 p.

YOW, M. D., <u>and others</u>   Association of viruses and bacteria with
            infantile diarrhoea.
            J. Am. Epidemiol, 1970, <u>92</u>, 33-39.

8. PURIFICATION PROCESSES

ALY, O. M. and FAUST, S. D.

Removal of 2, 4-Dichlorophenoxyacetic acid derivations from natural waters.
J. Am. Wat. Wks Ass., 1965, 57, (2), 221-230.

CECIL, L. K.

'Injection water' treating problems.
World Oil, 1950, 131, (3), 176-180.

DEUTSCH, M.

Natural controls involved in shallow aquifer contamination.
Groundwater, 1965, 3, (3), 37-41.

DITTHORN, F. and LUERSSEN, A.

Experiments on the passage of bacteria through soil.
Engng Record, 1909, 60, (23), 642.

FRANK, W. H.

Fundamental variations in the water quality with percolation in infiltration basins.
Proc. Artificial Groundwater Recharge Conf., Reading 1970.
Medmenham, Water Research Assoc., 1970, 1, 169-197.

GASSER, J. K. R.

Some processes affecting nitrogen in soil.
Min. Ag. Fish and Food, HMSO, 1969, Tech. Bull. No. 15, 15-29.

GOLWER, K., and others

Self purification processes in aerobic and anaerobic groundwater (in German).
Vom Wasser, 1970, 36, 64-92.

GUINOVART, J. M. C. and CUSTODIO, E.

The influence of impounding reservoirs on the quality and quantity of underground water.
Vienna Congress, IWSA, 1969.
London, Int. Wat. Supply Assn, 1969, 36 p.

HOLZMACHER, R. G.

Nitrate removal from a groundwater supply.
Water and Sewage Works, 1971, 118, (7), 210-213.

KRETZCHMAR, R.

Investigation on the reactions of ammonium nitrate, chloride and sulphate on the lower terraces of the Rhine at Bonn and its modification in the underlying stratum.
Dissertation, Bonn Univ., 1964.

MERKLE, L. N., ELLIS, J. R., SWANSON, N. P., LORMIOR, J. C. and McCALLA, T. M.

Relationship of agriculture to soil and water pollution.
Cornell Univ. Conf. Agric. Wastes Management, 1970, 31-40.

McGAUHEY, P. H. and KRONE, R. B.   Soil mantle as a wastewater treatment system. Berkeley, California Univ. Sanitary Engng Research Lab. Report No. 67-11, 1967, 201 p.

ROBECK, G. G. and others   Effectiveness of water treatment processes in pesticide removal. J. Am. Wat. Wks Ass., 1965, 57, 181-199.

SMITH, J. M., and others   Nitrogen removal from municipal waste water by columnar denitrification. Env. Sci. and Tech., 1972, 6, (3), 260-267.

UNIVERSITY OF CALIFORNIA   Studies on water reclamation. Sanit. Engng Laboratory Univ. California Tech. Bull. No. 13, 1955, 43 p.

WARE, G. C. and PAINTER, H. A.   Bacterial utilization of cyanides. Nature (London), 1955, 175, (4464), 900.

WIJLER, J. and DELWICHE, C. C.   Investigations on the denitrifying processes in soil. Plant and Soil, 1954, 5, (2), 155-169.

WRIGHT, S. J. L.   Degradation of herbicides by soil micro-organisms. Microbial aspects of pollution, ed. G. Sykes and F. A. Skinner. London, Academic Press, 1971, 233-254.

*   *   *

A study of sewage effluent purification by filtration through natural sands and gravels at Sycamore Canyon at Santee. California Bureau of Sanit. Engng Dept. Public Health, 1965.

EXTRACTS RELATING TO INSTANCES OF GROUNDWATER POLLUTION

Compiled by J. L. Stringer, Hertfordshire County Council

The following instances of groundwater pollution have been extracted by J. L. Stringer, County Health Inspector of Hertfordshire County Council. The two reference numbers preceding the extract refer to the Water Pollution Abstract number and the respective volume of Water Pollution Abstracts from which it was obtained. The extracts have been arranged in the following order.

1.   BACTERIA

160;<u>36.</u>   <u>Groundwater contamination in Indiana.</u>   Jordan, E. J.   Journal of the
American Water Works Assoc., 1962, 54, pages 1213-1220.   The author describes
various instances of pollution of groundwater in Indiana by bacterial contamination from
septic tank effluents, oil field brines, sulphates, chromates and phenols.   The need for
controlling disposal of wastes to prevent such contamination, which may affect groundwater for many years, is stressed.

2.   CADMIUM AND CHROMIUM

1731;<u>38.</u>   <u>Cadmium and hexavalent chromium in Nassau County groundwater.</u>
Leiber, M., Perlmutter, N. M., Frauenthal, H. L.   Journal of American Water Works
Association, 1964, 56, pages 739-747.   Previous surveys have shown that groundwater
in Nassau County, New York was polluted with chromium and cadmium from waste waters
from an aircraft manufacturing plant (see Water Pollution Abstracts, 1952, No. 25,
1953 and 1955, No. 28 1195).   The plant installed facilities for treatment of the waste
waters before disposal in recharge basins, and in 1962, a further survey was carried
out to re-evaluate the effect of groundwater flow on the concentration of contaminants.
It was found that the slug of contaminated water is elongated in the direction of flow and
now extends south from the recharge basins to the Valley of Massapequa Creek where it
is moving slowly into and under the seam.   The maximal concentration of hexavalent
chromium detected was 14 mg/litre at the southern end of the slug and the maximal
concentration of cadmium was 3. 7 mg/litre at a well a short distance south of the recharge
basins.   Although the concentrations of the contaminants in most parts of the slug were
generally much lower than in previous years, the overall shape and dimensions of the
slug had not changed appreciably since 1949.   For the first time, the contaminants were
detected in the waters of Messapequa Creek, the maximal concentration of chromium in
the creek being 2. 1 mg/litre and the cadmium concentration rate being from 0 to
0. 07 mg/litre.   It is planned to improve waste treatment at the plating plant and this may
result in further reduction and complete elimination of most of the contaminants at
present discharged to the groundwater;   however, even if the treatment was entirely
effective, in view of the low rate of flow of the groundwater and the slowness of dilution
by fresh groundwater recharge, <u>it might be ten to fifteen years before the contaminated</u>
<u>water is completely flushed out under natural conditions</u>.   The possible toxic effects of
ingestion of cadmium and chromium are discussed and the standards of these elements
adopted by various authorities are indicated.

3.   CYANIDE

1551;<u>38.</u>   <u>Pollution of groundwater by cyanide.</u>   Effenberger, E.   Arch. Hyg.
Bakt., 1964, 148, pages 271-287.   Bulletin, Hygiene, 1964, 38, 947.   The dumping of

waste material containing cyanide in a gravel pit near Cologne caused pollution of the groundwater with a potassium ferricyanide.   Although the maximum concentration of cyanide detected was about 0.5 mg/litre, use of the water as a source of supply was continued, since the experiments had shown that the complex salt was relatively resistant to decomposition and was therefore fairly harmless.

4.   DETERGENTS

1922;38.   Pollution of groundwater by detergent residues from a laundry. Netsch, F.Z. Hyg. Grenzgeb., Berl., 1959, 5, 207-212; Bull. Hyg. Lond., 1960, 35, 192. Waste water from a laundry was discharged to a pond with no outflow, and this has caused the pollution of well waters in the surrounding districts in one year.   The water foams when pumped, and has an unpleasant taste.   Pollution extends over a radius of 200 m, and it is considered that in the present circumstances it will spread still further. Investigation indicated that the detergents in the waste water cannot be broken down, and can be expected to remain unaltered in the groundwater for years.

5.   FERTILIZERS

133;38.   Contaminants and their effects on groundwater supplies.   McCracken, R.A., Nickerson, H.D.   Sanitalk, 1963, 11, No. 1, 11-13 and page 27.   To stress the need for providing an adequate protected area around sources of groundwater supplies, the authors describe two forms of pollution of well water.   A gravel packed well installed on privately owned farm land at Bedford, Massachusetts, was polluted by fertilizer which was spread near the well.   Nitrates which were leached out of the fertilizer caused the nitrate nitrogen concentration to rise from 2 mg/litre to more than 34 mg/litre.   The well was pumped to waste over a long period before the nitrate concentration was reduced to 3 mg/litre.   A gravel packed well at Holbrook, Massachusetts, became polluted by phenolic wash water which was discharged by a nearby factory to a swampy area about 420 ft from the well.   It is probable that a clay seal which had previously protected the aquifer from surface pollutions was ruptured during the installation of a new drain. Although waste water is no longer discharged near the well and it has been pumped to waste for long periods, a strong phenolic odour remains in the water and the well cannot be used.

6.   HYDROCARBONS

2160;34.   Groundwater contamination in the Rocky Mountains Arsenal area, Denver, Colorado.   Walker, T.R.   Bulletin Geological Society America, 1961, 72, 489-494.   Since 1943, chemical factories at the Rocky Mountain Arsenal, near Denver, Colorado, have manufactured materials for chemical warfare and other chemical products.   Until 1957, wastes have become contaminated with chlorex, 2,4-D type

502.

compounds which render the water unfit for human consumption, injurious to livestock and plants. In 1957, a reservoir with an asphalt sealed bottom was constructed to receive all the waste waters, and it is planned to construct an injection well for disposal of the waste waters in deep bed-rock formations that cannot be used for water supply. The polluted groundwater has, however, <u>migrated five miles north-west from the waste lagoon</u>, and continued migration can be expected. Eventually the concentration of contaminants in the groundwater will be reduced to non-toxic level as a result of mixing with uncontaminated water and recharged from uncontaminated surface sources such as rainfall and irrigation ditches, but it is expected that it will be many years before all the water in the affected area will be consistently safe for use.

2233;<u>35.</u>   <u>Tastes and odours caused by hydrocarbons.</u>   Beneden, G. V.   Bull. mens. Centre Belge Et. Document Eaux, 1960, 111, pages 95-100.   The intensity of the taste or odour of a sample of water can be measured by the amount of dilution required to reach the threshold of perception.   Difficult cases of taste and odour are considered; water contaminated by hydrocarbons such as crude oil, petroleum, petrol and lubricating oil generally has odour but no taste.   These hydrocarbons may be present in such small concentrations that they are not detected by standard methods of analysis.   The effects of hydrocarbon contamination on streams, open reservoirs and groundwater are discussed, and their toxic effects are described.   As groundwater polluted by petrol is known to lack nitrates, the progress of pollution can be followed by recording the nitrate concentration. Treatment by activated carbon is the best method found for removing taste and odour but is not entirely satisfactory.

7.   <u>INDUSTRIAL</u>

1562;<u>33.</u>   <u>New York State Department of Health, Annual Report of the Division</u> <u>of Laboratories and Research</u>, 1958. ..... In connection with water pollution control, the survey required to classify the streams in the Delaware and Chemung Watersheds have been completed.   Reports are also included on the following investigations: contamination of groundwater supplies by waste waters discharged by the Philbrick Starch Company, Suffolk County, into a percolating lagoon; the use of sodium arsenite to control aquatic weeds and spectrographic determination of barium and strontium in water from wells contaminated by the waste waters from the Westinghouse Electric Company, Horseheads. ......

1919;<u>33.</u>   <u>Pollution of groundwater by waste materials.</u>   Schinzel, A.   Wass. u Abwass, 1957, 2, 192-217.   The author first discusses the conditions which are of importance in influencing the access of polluting material deposited in or on the ground to the groundwater.   These include the condition, type and amount of the deposited material, the structure and chemical composition of the soil, and the depth of the groundwater. In the second part of the article, the author describes the number of instances investigated by him and illustrating the inferences discussed. In acute cases of pollution by sewage from leaking sewers and cesspools, by liquid manure, by flooding of refuse dumps, by waste waters from fish preserving works, milk industry, cellulose factories and other industries, and by oil and tar acids.

2160;34. (See under HYDROCARBONS)

804;35. Pollution of waters by wastes from cellulose and cellulose wool factories. Pollution of groundwater by waste waters of the cellulose and cellulose wool industry. Refers to a Paper by Schinzel, A. and Benger, H., (Arch. Hyg. Berl. 1960) 144, pages 320-344; Z. Bl. Bakt., Ref, 1961 181, pages 158-159. Groundwater in a river valley below a cellulose and cellulose wool factory was examined. The water was affected both by seepage and the position of the stream receiving the waste waters. The most polluting constituents were sodium sulphate in the cellulose wool waste water and lignin sulphate in the cellulose waste waters.

8. METALS

698;33. Crash programme for expensive water. Flynn, J. L. Proceedings Thirteenth Industrial Waste Conference, Purdue Univ. Engineering Extension Fer. 1958, 253-259. At the Belvedere Illinois plant of the Standard Brass Corporation, plating waste waters were disposed of in a large lagoon but when water wells in the vicinity showed increasing concentrations of copper, nickel and chromium and traces of cyanide, it became necessary to treat the waste waters. Chlorination of the waste water to destroy cyanide was begun as a temporary measure and calcium hypochlorite was added to the lagoon to destroy cyanides already present there. As soon as possible, plant was installed to provide complete treatment of the waste waters so that they could be re-used. The treatment comprises chlorination for destruction of cyanides and reduction of hexavalent chromium with sulphuric acid, coagulation with sodium bisulphate and lime to precipitate metal ions and sand filtration. Sludge is dumped in the old lagoon. As there tends to be a build up of sulphate, chloride and calcium in the treated effluent, ion exchange units are now being installed to de-mineralize the water before re-use. The Municipal Authorities have led mains to connect the affected area with the municipal water supply and until this was done the residents in this area were provided with bottled water.

1562;33. (See under INDUSTRIAL)

504.

2059;38.　　Chronic arsenic poisoning.　　Salud. Pub. Mex. 1964, 6, 375-420.

(1) Epidemiological study in areas of Torreon, Coahulia.　　Alvarado, A., Vinigra, G., Gracia, R. E. and Acevedo, J. A.　　In 1962, an extensive outbreak of illness in parts of Torreon City, Mexico, was found to be the result of chronic arsenic poisoning.　　The water supply is obtained from wells and was found to contain 3.98 mg/litre of arsenic. Investigations showed that running near the wells is a flume carrying drainage from the laundry, sedimentation and storage tanks of a metal works where, among other substances, pure arsenious oxide is stored in the open.　　The water in the flume contained 0.534 mg/litre of arsenic.　　Soil near the well was also found to be heavily contaminated to a depth of 5.5 m.　　It was concluded that the groundwater of a large area had been contaminated with arsenic from the leaky storage tanks and with dust from the large spoil banks which also contained small amounts of arsenic.

(4) Treatment of the drinking water.　　Contamination had spread over a very wide area, even outside the city; 62% of the samples examined contained more than the maximum permissible concentration of 0.05 mg/litre of arsenic.　　Experiments showed that the arsenic could be removed by precipitation with ferric hydroxide, and a simple apparatus was devised for domestic use containing sand and ferrous sulphate for precipitation and filtration.

660;36.　　Groundwater contamination.　　Proceedings at the 1961 Symposium, April 5th-7th, 1961, Cincinnati, Ohio.　　U. S. Department, Sanitary Engineering Center Technical Report W. 61-5 1961, 220 pages.　　This Report comprises papers and summaries of the discussions presented to the Symposium on Groundwater Contamination, Cincinnati, Ohio, and includes sessions on the hydrogeological aspects, types of contaminants, etc.

(7) Inorganic chemical contamination of groundwater.　　Kaufman, W. J. (pages 43-50). This review dealt with the contamination of groundwater with inorganic chemicals which, compared with organic and biological contaminants are indestructible causing persistent pollution which is difficult and costly to abate.

(15) Incidents of chromium contamination of groundwater in Michigan.　　Deutsch, M. (pages 98-104).　　Chromium, especially hexavalent chromium compounds, are one of the most serious sources of contamination of groundwater in Michigan and a number of instances of such contamination were reported and discussed including those caused by percolation from ponds or infiltration pits receiving electroplating waste waters, by leaching from the land surface after uncontrolled spilling or spreading (for example in using chromium treated salts to melt snow) and by the introduction of spilt or airborne chromium dust which is washed into the ground by rainfall.　　Schematic diagrams were included showing the possible mode of entry of chromium into the aquifer in each case.

(18) Two cases of organic pollution of groundwaters.　　(pages 115-117).　　Instances given of pollution of well water within a radius of about 300 yards from a plant producing

pyridine compounds where the waste was discharged to a lagoon on porous gravel soil. The second example given concerned contamination of an isolated well in a rural district caused by pollution by leakage from a nearby fuel oil tank.

9.   MINE DRAINAGE

1353;36.   Symposium on pollution of groundwater.   Proceedings of the Society of Water Treatment Exam., 1962, 11, 11-49 and 76-105.   At a symposium on the pollution of groundwater organized by the Society for Water Treatment and Examination, March 1962, various papers were presented.

Paper No. 4 - Pollution of groundwater by gasworks waste.   E.C. Wood (pages 32-33) says that in 1950 to 1951 a borehole was sunk in the chalk at Norwich to provide an industrial water supply; the water was of good quality but the yield was insufficient and an adit was driven at right angles to the bore at a depth of 150 ft to increase the yield. Soon after, the water developed a strong tarry taste and became unusable; analysis showed the presence of about 0.2 mg/litre of total phenols.   Examination of the adit showed black tarry material exuding from the roof and this material was found to contain a proportion of volatile matter.   It was subsequently discovered that the bore had been sunk on the site of a gas works in 1815 to produce gas by the destructive distillation of whale oil; this gas works had been closed down in 1830 and the tarry material must have remained in the chalk for at least 120 years.

Paper No. 5 - A water supply polluted by aviation fuel.   H.E. Smith (pages 34-37). The water supply to Stamford, Lincs., is obtained from five main groundwater sources. One of these is near a large airfield where oil traps have been installed round a services area to prevent contaminants entering the groundwater.   The water is chlorinated before distribution.   In mid-January, 1960, an unusual odour developed in the water; an alternative supply had to be provided and the water from this source was allowed to run to waste.   It was finally discovered that early in December, 1959, a large volume of aviation fuel of petroleum/kerosene mixture had been accidentally spilled on the edge of the airfield; it was sprayed with foam and disappeared rapidly, but had by-passed the interceptors.   No warning was given to the waterworks of this incident and in the forty-one days between then and the first complaint of taint in the water, the pollution had travelled 450 to 500 yards.   The quality of the water gradually improved again but was not fully restored to normal until mid-April.   This incident indicates the importance of watching gathering grounds for possible pollution, particularly in areas such as this where the Lincolnshire limestone provides large amounts of water supplies.

Paper No. 8 - Notes on pollution of groundwater supplies in Queensland, Australia. (Simmonds, M.A. - pages 76-83).   The author reviewed briefly occurrences of ground-water contamination at various places in Queensland, Australia, by nitrate and nitrogen compounds, iron and iron bacteria, turbidity and fluorides.

Paper No. 9 - The Montebello Incident. (Swenson, H.A. - pages 84-88). In 1945, a plant at Alhambra, California, manufacturing 2,4-D weedkiller, dumped a batch of chemicals which had not reacted properly into the municipal sewers. The effluent from the activated sludge plant was discharged after chlorination to the River Hondo in an area of high underground water levels, and the chemicals in the effluent seeped into the aquifers which supplied the municipal water wells for the town of Montebello, causing medicinal tastes and odours. The water had to be treated with chlorine dioxide for two years and tastes and odours persisted in the water for four to five years although it was estimated that the original batch of chemicals discharged had been diluted in the ratio 1 part in 10 000 000.

Paper No. 11 - Phenol contamination of an artesian glacial drift aquifer at Alma, Michigan, U.S.A. (Deutsch, M. - pages 94-). The author described an occurrence of pollution of some of the municipal water wells at Alma, Michigan, by phenolic waste waters from an oil refinery which were discharged into a pit near the Line River. One well had to be abandoned and another is used only in summer when the demands are high. The geology of the area was discussed, the incident indicated that some chemicals can penetrate readily through clayey sediments into underlying artesian aquifers. The possibility of using scavenger wells in such cases to intercept the polluted waters and localize the areas of contamination were discussed; the water from such scavenger wells might be discharged into the river, or used for cooling and other industrial purposes where the quality would not be important.

Paper No. 12 - Disposal of drainage water from coal mines into the chalk in Kent. (Buchanan, F. - pages 101-105). Tabulated results were given of the analyses of ground-water in the chalk in East Kent to estimate the effect of mine drainage water which is allowed to soak into the ground in this area. The results indicated that a mixture of fresh and saline water was being spread over an area of about 10 square miles in the direction of groundwater flow, and extending to the natural outlets from which groundwater discharges.

10. OIL AND PETROLEUM

1335;32. Oil and tar products in groundwater and water supplies. (Zimmermann, Korresp. Abwass., 1958, May 2nd to 3rd). A summary is given of a paper presented to a meeting of the Zweckverband für Kanalisations Sörderung in Cologne in April, 1958. The author dealt with the sources of pollution by motor fuel and car products. Their effects on soil and groundwater, and the extent of this spread in soil and groundwater; period for which the effect has been found to last; precautions required in the construction of storage tanks and drains and legal provisions for protection.

2340;32. Prevention of oil pollution of surface and groundwaters. Schlatfer, P.,
Monatsbull Schweiz. Ver. Gas - U. Wassersachs. 1959, 39, 33-34. In January 1959,
a symposium was held in Baden-Baden by the European Federation of Water Protection
and Special Risks on the pollution of surface and groundwaters. Papers and discussions
dealt with the results of oil pollution, long distance oil pipes and surface and underground
storage, industrial and communal waste waters, inland navigation and the precautions
and international agreements required.

177;35. Saarbrucken defeats oil danger at a works saved by luck and action.
Laufer, W., Ztg. komm. Wirtschaft, 1960, 70, 15. In November 1959, 27 m^3 of
heating oil leached from an underground container 1200 m from the nearest well of the
waterworks at Saarbrucken. The tank was raised and the oil-soaked soil was removed
from below it down to groundwater level. The water was pumped from the surface of the
groundwater until the level had been lowered. Borings around the pit gave no large
amounts of oil. Digging showed that the sand retained the oil and released it only on
agitation. Borings up to 350 m from the point of leakage showed oil pollution in the
groundwaters. From the soil and water only 8.5 m^3 meters of oil were recovered and
more soil had to be removed. No tastes or odours were observed in the groundwater
drawn from the supply.

2232;35. The pollution of water by petroleum products. Buydens, R.,
Bulletinmens. Centre Belge. Et. Document Eaux, 1960, 111, pages 90-95. Pollution of
water by hydrocarbon compounds has become a serious problem; recent cases of pollution
of surface and groundwaters in Germany, Sweden and the USA are reviewed. The
creation of protective zones, where the construction of underground fuel storage tanks
would be prohibited and the routine inspection of all equipment used in the transport and
storage of fuel would help to avoid pollution. An accurate method of detecting small
quantities of hydrocarbons in water is still required. Fluorescence can be measured by
spectrophotometry but the results tend to vary with the origin and nature of the product.
Treatment with granular activated carbon has been found the most effective method of
purification of oil contaminated water. A bibliography of twenty-one references is
appended.

2233;35. (See under HYDROCARBONS)

1186;36. Accident to a tank wagon near a waterworks. Schaack, E.A., Weyer, T.,
and Lotzing, H., Monatsb. Schweiz. Ver.G/U. Wassersachm., 1961, 41, 262-265.
In October 1960, through an accident to a tank wagon, 10 000 litres of heating oil flooded
over and drained into the soil within the groundwater intake district of the Westhoven

508.

Waterworks of Cologne. Immediate action included removal and storage of polluted soil until it could be conveyed to a power works for incineration. From information available at the waterworks, the course of the possibly polluted groundwater was determined and an emergency well was sunk to divert polluted waters from the works and pump it into the river.

1187;36. Pollution of groundwater by escaping motor fuel after accident to tank wagons. Kassecker, F., Wass. Warme, 1961, 15, 249-252. After an accident to a tank wagon near Graz, it was found that 12 000 lites of diesel oil which escaped were retained in the 6 m layer of humus and sand above the groundwater from which it would later be washed into the groundwater by rain. Immediate steps would thus be possible to prevent pollution of groundwater. The author then described a discussion among various authorities with proposals for traffic control to prevent such accidents and for immediate steps to be taken on the occurrence of accidents, especially for the removal and safe disposal of polluted soil. The action taken after a later accident, also near Graz, and the cost of the necessary precautions are described.

1475;36. Water management and pollution control. Proceedings of the International Congress PRO AQUA 1961 in Basel. 448 pages. (Printed by R. Oldenbourg, Munich, 1962). This presents a full text of papers given at the International Congress in Basel in October, 1961.
Paper 9 - Problems of pollution of surface and groundwaters by mineral oils.
Zimmermann, W., pages 137-153. The author reviews work on the pollution of ground- and surface waters by oil giving a list of 43 references. The aspects covered include leakage of oil from household storage tanks, affecting both ground- and surface waters; causes and effects of oil pollution in rivers; and particularly the pollution of groundwater, including its effects on plants and soils, the effects and possible accumulation of carcino-genic substances, experiments on the decomposition of mineral oils and their migration through the soil (particularly studies Franzius Institute, Hanover and Institute of Hygiene, Hamburg). New methods of analysis.

2068;36. The behaviour of oil derivatives in groundwater. Becksmann, E., Wass-u. Wassersach, 1963, 104, 689-694. This report deals with the flow of oil derivatives in groundwater and the various effects of movements of oils together with permeability factors with particular reference to pollution of groundwater by oil. A flow diagram showing the various stages of oil percolating into the groundwater and illustrated data are supplied.

861;36. Damage to waters by mineral oil products. Rincke, G., SchrReihe Ver. Dtsch. Gewasserschetz, 1961, 7, 24 pages. Deals with oil pollution and contains a list of occurrences of accidental oil pollution in recent years and their causes.

11. PHENOL

40;32. Four years of practical experience with chlorine dioxide. Widemann, O., Vom Wasser, 1957, 24, 50-70. Part of the water supply of Basel is drawn from a groundwater stream running almost parallel to the course of the River Wiese. In 1950, pollution by phenol from a damaged sewer carrying waste waters from a gas works appeared and spread to several of the wells. Further pollution was stopped but it was not until 1955 that all the wells could again be brought into use. Experiments are described on the treatment of the polluted water with ozone and chlorine dioxide. Experiments with ozone showed that the difficulties were great and the amount of ozone required to remove the phenolic taste and odour was high. Parallel experiments with chlorine dioxide were more satisfactory and the ozone experiments were stopped. Treatment of water containing 30 µg/litre of phenol with 0.07 mg/litre of chlorine dioxide produced a treated water in which phenol was not detectable by analysis or by taste or odour. The process and reversible nature of the reaction is discussed. A water containing 20 µg/litre of phenol was treated satisfactorily with chlorine dioxide for two years. Experiments on the removal of tastes and odours due to other causes are then described. Works are planned in the neighbourhood of Basel in which infiltration of treated surface water is to be used to dam back from the intake a groundwater stream which is almost free from oxygen and polluted by oil and chemicals. Oxidation of this water altered but did not remove the taste. Treatment with active carbon proved satisfactory and is to be used until the infiltration plant is in action. Experiments have also been made on the treatments of Rhine water. Treatment by super chlorination, ozone, and chlorine dioxide gave variable results, but in each case, subsequent treatment with active carbon reduce the permanganate demand to less than 6.0 mg/litre, which is the maximum permissible concentration in Switzerland. Experiments on the bacteriological action of chlorine dioxide showed that heavy bacterial pollution was completely removed within one minute by the addition of 0.06 mg/litre of chlorine dioxide; 0.12 mg/litre of chlorine achieved the same effect only after twenty minutes and the result was not so reliable. The author then described the principal reactions of chlorine dioxide treatment, the plant used, and the methods of analysis and control of the process.

1744;32. Report of a meeting of the Austrian Association for Microbiology and Hygiene, 24th to 27th February, 1958. This includes a reference by W. Zimmermann to a report on pollution of groundwater by car products. The author describes pollution of the groundwater supply of a small community by a car works. Water samples showed an increase in the permanganate demand and considerable amounts of volatile phenols. A new source of supply had to be found. (Water Supply and Car Products, Zimmermann, W., Rheinforth, H. and Schwille, F. Zbl. Bakt., 1, Orig., 1959, 174, 155-164).

184;32. Investigation of underground and surface water pollution by waste water from generators. Hluchan, E. and Mayer, J. Coll. sic. Rep. 1952 to 1955, Inst. Hyg. Prague 1957, 73-74. Waste waters from the generator at a glass works in Czechoslovakia containing phenolic substances and tar, are discharged into a local brook. This discharge has affected the fish, geese and ducks in the area, directly, and has soaked into the groundwater. The discharge of waste waters to filtering pits has caused pollution of groundwater. The state of the water in the brook was examined and the extent of infiltration of waste waters was investigated by examining the water in local wells. The results indicated that the discharge of phenolic waste waters to filtering pits and surface streams can cause serious pollution. Methods of purifying the waste waters were therefore considered and a domestic water supply system had to be proposed in which water could be drawn from sources not contaminated by infiltration of phenolic substances into the ground.

987;32. Phenol in the water supply of a large city. Stanier, G., Wasserw-Wass. Techn., 1958, 8, 20; Zbl. Bakt., 1 Ref. 1958, 168, 450-451. The author described the pollution of a water supply by the discharge of gas liquor on land. The effect on the water supply of concentrations of phenol below the toxic limit are discussed.

1384;36. Pollution in water supply plants by gas works waste waters. This reference refers to the damage which can be done to a water supply by even a very low concentration of phenolic substance. It refers to tests of several polluted wells which showed that after eight months they were still unsuitable for use. The authors; Lingelbach, H., Faalbreite, R. and Kuhn, H. (Z. Hyg. Grenzgeb, 1962, 8, 761-767).

133;38. (See under FERTILIZERS)

1353;36. (See under MINE DRAINAGE)

12. REFUSE

2211;35. Means of protection of groundwater against pollution from refuse tips. Langer, W., Svhr. Reihe. Ver. Wass. Boden u Lufthyg., 1961, 19, pages 113-130. After a description of the composition of town and industrial refuse and of investigations into the amount and type of matter which can be dissolved in and carried by groundwater, the author describes investigations into two occurrences of pollution of groundwater from refuse tips. As removal of a refuse tip once deposited is generally impossible, there are only two methods of protection of the groundwater, diversion of the groundwater stream to by-pass the tip, and the boring of wells between the tip and supply wells to draw off the water containing concentrated polluting matter leaching from the tip. An account is then given of the precautions necessary in the choice of new tipping areas. The condition and use of the surrounding land, the geological and hydrological conditions and the chemistry and bacteriology of the groundwater must be investigated. The protection of the neighbourhood from dust and odour by the planting of the green zone, the drainage of the land, the laying of an impermeable foundation and the control and operation of the tip are described.

161;36. **Problem of refuse dumps and their connection with water supplies.** Nietsch, B., Steirlische Gemeinde-Nachr. 1961, 14 (9) pages 16-19. The author describes the possibilities of pollution of groundwater by matter from refuse dumps and gives accounts of occurrences of damage to well waters. The importance of control of the disposal of refuse is emphasized. (See German Water Law of 1960).

13. SEWAGE

183;32. **Pollution of underground water.** Carter, G., Water and Water Engineering, 1958, Vol. 62, pages 244-246. In a paper presented at a meeting of the Institution of Public Health Engineers in London, April, 1958, pollution of groundwater supplies was reviewed, including instances of groundwater pollution in various parts of England. Three cases of direct pollution of groundwater with sewage are described. (Journal of the Institution of Public Health Engineers, 1958, 57, 150-160).

1919;33. (See under INDUSTRIAL)

14. GENERAL ASPECTS OF GROUNDWATER POLLUTION

1561;32. **Water management in town developments.** Popel, F., SchrReihe Ver Dtsch. Gerwaserschutz, 1957, (No. 2), pages 62-79. The author discusses the future requirements of world population and the risk of pollution of surface and groundwater by domestic, industrial and solid wastes. Also the co-operation between towns and rural communities in water supply and disposal of wastes.

1738;32. **Pollution of water and protection of reserves.** Sanzot. E., Bulletin Centre. Belge. Et. Document Eaux, 1958, 40, pages 155-158. In a paper presented at a meeting in Liege in 1958, the author discussed the causes and nature of the pollution of surface and especially of groundwater, giving various examples of the occurrence of pollution due to trade waste waters and heavy rainfall and described legislative and technical methods for preventing pollution of groundwater.

512.

940;33. Pollution of water: causes, effects and methods of control.
Schmassmann, H., Reprint from Eau et Sante, 1958, 6, 12 pages, 4 plates. Discusses
causes and effects of pollution of surface and groundwater, etc., in Switzerland.

380;34. New York State Department of Health. Annual Report of the Position of
Laboratories and Research for 1959. This report includes the Report of the Laboratories
for Sanitary and Analytical Chemistry (pages 96-103). Pollution of numerous private
wells in one area was found to be caused by the discharge to a lagoon of waste waters
from a fertilizer factory; the lagoon is no longer in use and the use of the wells has
been discontinued until the normal chemical balance is restored.

2138;34. The pollution of surface and underground waters. Taylor, E. W.,
Journal of British Water Works Association, 1960, 42, 582-603. In the British National
Report presented at the 1961 International Water Supply Congress in Berlin, the author
answered questions concerning the pollution of surface and groundwaters in Britain.
The following subjects were discussed - problems encountered as a result of pollution of
ground and river waters by domestic sewage or water, waste waters from industry,
agricultural enterprises, fertilizers, plant protectives, garbage deposits, storage of oil,
etc., together with technical measures taken to control the quality of water supplies
derived from rivers; legislation to control pollution of ground and surface waters;
protected areas and other special measures; and radioactive contamination of water.

367;35. International Water Supply Congress and Exhibition, 29th May - 3rd June,
1961. Subject No. 2. Pollution of surface and underground waters. International
Water Supply Association, 1961, 92 pages. Based on the answers to a questionnaire
submitted by rapporteurs of fourteen countries, information is summarized on the
difficulties caused by pollution of drinking water supplies obtained from both surface and
groundwater sources, with particular reference to pollution caused by oil, fuel and
radio-activity, and the measures taken to eliminate or overcome these difficulties. The
data collected were examined from the technical, economic and legal points of view.

80;39. A review of the literature of 1964 on waste water and water pollution control.
The United States Water Pollution Control Federation. The Journal of Water Pollution
Control Federation, 1965, 37, pages 587-646, 735-799, 887-979. The report covers a
great many subjects on water pollution including effects of pollution on water supplies,
pollution of groundwater and polluting effects of groundwater recharge. The Abstract
contains no details.

23;39. Problems in groundwater pollution. Walton, G., United States Public Health Service. Department of Health and Education and Welfare, Robert A. Taft Sanitary Engineering Centre, Cincinnati, Ohio. Technical Report W. 62-25, 1962, 12 pages. With increasing water demand in the USA it was considered that more use will have to be made of groundwater resources and information is required on methods of protecting these resources from pollution. The author reviews some examples of groundwater pollution to illustrate the insidious nature of such pollution and the time that may elapse before it is detected. It also reviews the prolonged period before the water may regain its original quality.

1214;40. Contamination of rivers and other water supply sources. Roberts, F. W., Municipal Engineering, London, 1966, (143, 678, 683 and 685). The author discusses some of the causes of pollution of rivers and groundwaters, and various factors affecting the safety and taste of public water supplies.

1638;40. The review of the literature of 1966 on waste waters and water pollution control. The United States Water Pollution Control Federation. Journal of Water Pollution Control Federation, 1967, 39, pages 689-749, 867-945 and 1049-1154. This is a review of literature published during 1966 on waste treatment and pollution control and includes effects of pollution on water supplies, pollution of groundwater, eutrophication of natural waters, etc.

1353;36. (See under OIL AND PETROLEUM)

1556;34. Survey of groundwater contamination and waste disposal practices. American Water Works Association, Task Group 245OR The Journal of American Water Works Association, 1960, 52, pages 1211-1219. Continuing the survey of groundwater contamination in the United States (See Water Pollution Abstract 1958, 31, Abstract No. 646) the information obtained from a questionnaire survey in 1960 is summarized in a table and discussed. Individual consideration is given to the situation in the States of Michigan, Massachusetts, New Mexico, Missouri, Virginia, Minnesota, New York and Rhode Island, showing the types and seriousness of groundwater contamination experienced. It is emphasized that further action is necessary to eliminate undesirable waste disposal practice and control the construction of wells supplying water for human consumption. Contamination of groundwater by sewage can be detected by the presence of synthetic detergents which produce foam even when present in a concentration of less than 1 mg/litre and can be detected at a much lower concentration by chemical analysis. Contamination caused by the increasing use of sewage lagoons is also discussed.

2398;35. Effect of local conditions on artesian well water quality. Vinogradova, M. I.,
Trubmikova, A. S., Bolkahovitinova, M. N., Hygiene and Sanitation, Moscow, 1959.
Vol. 24, No. 6, pages 46 to 48. USSR Lit. Wat. Supply Pollution Control, 1961, Vol. 1,
pages 231-233. Artesian wells on the banks of the Moscow River near Lyublinsk Combine,
USSR were found to be showing evidence of pollution. Although the water was of
satisfactory bacteriological quality, it contained excessive amounts of chlorides, ammonium
salts, iron and hydrogen sulphide. It appeared that the geological strata below the waste
treatment plant serving the combine had gradually become permeable to water allowing
chemical impurities to pass into the groundwater.

1368;36. Research for clean water. The US Department of Health, Education and
Welfare. US Publication Health Service, Robert A. Taft, Sanitary Engineering Centre,
Tech. Rep. W. 62-2, 1962, 36 pages. This publication, constituting a report on the
activities of the research branch of the Taft Sanitary Engineering Centre, Cincinnati,
Ohio, contains brief descriptions of the major investigations at present being carried out
and a list of publications being issued during the past two years.

1921;39. A review of the literature of 1965 on waste water and water pollution
control. The US Water Pollution Control Federation. This contains more details of
water pollution work in the USA including surveys of polluted water.

15. LEGAL ASPECTS OF GROUNDWATER POLLUTION

392;33. The public health aspect of the struggle against pollution of waters:
Delecourt. A paper presented to a meeting of the Committee on Public Administration
of the Union of Western Europe, 1959. 14 pages. The author discusses the French
Laws controlling the hygienic quality of mineral water and the public water supplies and
prohibiting the pollution of groundwater and surface waters, including individual laws
relating to radio-activity, cemeteries, shell-fish beds, bathing places, fisheries and
trade waste waters.

1738;32. (See under GENERAL ASPECTS OF GROUNDWATER POLLUTION)

1349;35. On the provisions concerning waste water in our new water law.
Airaksinen, K., Vesitalous, 1961, Vol. 2, No. 1, pages 24-26 (English summary page 32).
The author discusses the provision of the new Finnish Water Law governing the disposal
of sewage and waste waters, to prevent pollution of streams and groundwater. Sewage
must receive treatment before discharge. Other waste waters may be required to be
treated before discharge to the ground and into open ditches; in such cases, permission
is required before discharge.

24;39. <u>Groundwater contamination and legal control in Michigan</u>. Deutsch, M.,
US Geological Survey Water Supply, Pap. 1961, United States Government Printing Office,
Washington DC, 1963, 86 pages. In Michigan, about 90% of the rural population and 70%
of the total population are served by groundwater supplies. Many cases of pollution of
groundwater have occurred in the State; the author reviews various examples. Typical
sources of pollution are industrial and domestic wastes, septic tanks, leaking sewers,
flood waters and other poor quality surface waters, mine waters, solids stored or spread
on the surface and airborne wastes; in addition, natural recurring saline waters have
entered fresh water aquifers, etc.

1419;39. <u>Water pollution: development and remedies</u>. Schroeyers, G.A.,
Trib. CEBEDEAU, 1963, 16, pages 384-388. Increasing pollution of surface and
groundwaters in Belgium and its prevention are discussed. Details are given of standards
of surface waters recommended by the International Water Supply Association and Belgian
Legislation for the discharge of sewage and trade waste waters is outlined.

16. <u>GROUNDWATER TRACING</u>

1317;33. <u>Colour and salt experiments in groundwaters in South West Germany</u>.
Reference to work of Schulz who carried out a large number of experiments in tracing
groundwaters in South West Germany. Tests showed that the direction of flow in non-
compact rocks is generally easily placed; in fissured and especially cast formations it
is more difficult. Sewage seepage can be considerable in several kilometers. Rates
of flow vary considerably .... effects of pollution by poison, oil and tar are discussed.

LIST OF CONFERENCE DELEGATES

Addresses refer to the United Kingdom, unless stated otherwise, and relate to October 1972. Certain institutional changes are detailed in the footnote on p. 523.

ABRAHAM, P.J.   Hydrologist, Greater London Council

ACKROYD, F.   Fisheries and Pollution Officer, Lincolnshire River Authority

AHLSTRAND, C.   Superintendent, Uppsala Kommuns Gatukontor, Sweden

ALLCROFT, J.B.   Chief Chemist, Mersey and Weaver River Authority

ALLEN, Dr. R.G., OBE   Director, The Water Research Association

ANDERSEN, L.J.   Chief Geologist, Geological Survey of Denmark

ASHFORD, P.L.   Hydrologist, East Suffolk and Norfolk River Authority

ASPINWALL, R.   Consulting Hydrogeologist, Burnham on Crouch, Essex

ATKINSON, Dr. T.C.   Research Fellow, Department of Geography, University of Bristol

AVERY, M.T.   Chief Assistant Engineer, Cumberland River Authority

BAKER, G.P.   Technical Liaison Officer, The Water Research Association

BEAL, D.   Mechanical and Electrical Engineer, West Wilts Water Board

BEARD, M.J.   Hampshire River Authority

BENALI, A.   Université Catholique de Louvain, Belgium.

BERRY, N.   Project Engineer, Babtie Shaw and Monton, Glasgow

BLACKFORD, M.   Works Engineer, Imperial Chemical Industries - Plant Protection Ltd.

BLACKSTOCK, J.   Project Engineer, Babtie Shaw and Morton, Glasgow.

BLOXHAM, Mrs. S.   Chief Chemist and Bacteriologist, Wolverhampton County Borough Council

BORLAND, J.R.   Assistant Pollution Prevention Officer (Chemist), Dee and Clwydd River Authority

BOWLER, G.K.   Analyst, London Brick Company Ltd.

BRASSINGTON, F.C.   Geologist, Severn River Authority

BREWIN, D.   Trent River Authority

BROMLEY, A.B.   New Works Engineer, Wolverhampton County Borough Council

BROWN, Dr. J.   Head of Chemistry Section, Central Electricity Generating Board

BUCHANAN, D.   Principal Assistant Inspector, Clyde River Purification Board

BURFORD, Dr. J.R.   Soil Science Department, University of Reading.

CALCUTT, T.   Technical Liaison Officer, The Water Research Association

CHAPMAN, Dr. B.T.   Technical Officer, Severn River Authority

CHARLESWORTH, D.L.   University of Aston, Birmingham

CHILLINGSWORTH, P. C. H. Hydrologist, Kent River Authority

CLARK, A. J. L. Principal Assistant Engineer, Thanet Water Board

COLE, J. A. Chief Hydrologist, The Water Research Association

COLLINGE, V. K. Assistant Director (Technology), Water Resources Board

COLLINSON, G. County Health Inspector, Lindsey County Council

COLLINSON, P. A. J. Hydrological Engineer, Mersey and Weaver River Authority

COLLYER, M. L. Senior Assistant Engineer, Thames Conservancy

COTTON, C. J. N. Water Resources Engineer, Kent River Authority

CREASE, R. I. Deputy Engineer, East Yorkshire (Wolds Area) Water Board

CROLL, Dr. B. T. Resources Group, The Water Research Association

CROW, M. Ready Mixed Concrete (UK) Ltd, Feltham, Middlesex

CUSTODIO, E. Comisario de Aquas del Pirinea Oriental, Barcelona, Spain

DAVISON, A. S. Chief Chemist, The East Surrey Water Company

DAY, J. B. W. Principal Scientific Officer, Institute of Geological Sciences

DICHTL, L. Hydrogeologist, Empresa Nacional Adaro, Madrid, Spain

DORFMEIJER, R. Environmental Health Engineer, Provincial Water Board, Arnhem, Netherlands

DOWNING, R. A. Senior Engineer, Water Resources Board

DOWSE, L. H. Geologist, Trent River Authority

DRACOS, Prof. Dr. Th. Eidgenössiche Technische Höchschule, Zürich, Switzerland

DREW, E. A., OBE Chief Engineer, Middle Lee Regional Drainage Scheme

DROST, Dr. G. Duinwaterleiding van Den Haag, Netherlands

DUDGEON, C. R. Resources Group, The Water Research Association

EDEN, G. E. Assistant Director, Water Pollution Research Laboratory

ELLIOTT, E. Director, Stainless Steel Development Association

ELPHICK, A. Assistant Chief Chemist, Wallace & Tiernan Ltd.

ELVY, B. D. Chemist, Redland Purle Ltd.

EXLER, Dr. H. J. Bayerisches Geologisches Landesamt, München, West Germany

FLETCHER, J. Effluent Officer, Imperial Chemical Industries Ltd., Agricultural Division, Billingham

de FLEURY, A. Geologist, Amey Group, Oxford

FORD, D. B. Chemist and Bacteriologist, Bucks Water Board

FOSTER, S. S. D. Senior Scientific Officer, Institute of Geological Sciences

FOX, R. A. Manager of Exploration Department, Ready Mixed Concrete (UK) Ltd., Feltham, Middlesex

FRANKLIN, R. W. D. Senior Assistant Engineer, Water Resources, Yorkshire River Authority.

de FREITAS, M. H.	Lecturer, Imperial College, London
FRIED, Dr. J. J.	Ecole Nationale Superieure des Mines de Paris, France.
FUNNELL, Prof. B. M.	Dean of School of Environmental Sciences, University of East Anglia
GASKARTH, J. W.	Lecturer in Geology, University of Aston, Birmingham
GILLIS, A. W.	Division Water Chemist, Imperial Chemical Industries Ltd., Mond Division, Runcorn
GLOVER, H. G.	Scientist, National Coal Board
GOLDTHORP, G. D.	Hydrologist, Great Ouse River Authority
GRAY, D. A.	Chief Hydrogeologist, Institute of Geological Sciences
GREAVES, G. F.	The Water Research Association
GREEN, B. M.	Distribution Engineer, Rugby Joint Water Board
GREEN, M. J.	Resources Group, The Water Research Association
GRIFFITHS, Dr. G. H.	Head of Water Examination, Bristol Waterworks Company
GUILLEN, J. C.	Chief, Groundwater Division Institute of Geology and Mines, UN Food and Agricultural Organization, Madrid, Spain.
de HAAN, F. A. M.	Senior Res. Officer on Soil Pollution, Agricultural University, Wageningen, Netherlands
HALE, H. T.	Partner, J. D. & D. M. Watson, London
HALL, E. S.	Distribution Group, The Water Research Association
HANSEN, H. K.	Civil Engineer, The Danish Geotechnical Institute
HARDCASTLE, B. J.	Deputy Chief Engineer, Thames Conservancy
HARRISON, R.	Chief Engineer, East Shropshire Water Board
HAWES, F. B.	Senior Assistant Engineer (Planning), Central Electricity Generating Board
HAWKINS, K.	Divisional Engineer, Hales Containers (RMC)
HAWKINS, R.	Company Lawyer, Redland Purle Ltd., Rayleigh, Essex
HEADWORTH, H.	Hydrologist, Hampshire River Authority
HERBERT, Dr. R.	Senior Geohydrologist, Ercon Ltd.
HERZBERG, Dr. S.	Head, Environmental Conservation, Koninklijke/Shell-Laboratorium, Amsterdam, Netherlands
HOLMES, R. E.	Director, Pollution Prevention (Consultants) Ltd., Crawley, Sussex
HOOD, A. E. M.	Crop Production Agronomist, Imperial Chemical Industries Ltd., Agricultural Division, Bracknell, Berks.
HOPKIN, D. V.	Great Ouse River Authority
HOWSAM, P.	Research Student, Manchester University
HILLYER, M. A.	Senior Engineer, Wallace Evans & Partners
HIRSCH, H.	Resources Group, The Water Research Association
HUNTER BLAIR, A.	Resources Group, The Water Research Association
HUTCHINSON, Dr. M.	Resources Group, The Water Research Association

IRELAND, R. J.	Assistant Hydrologist (Groundwater) Mersey and Weaver River Authority
JAMES, T. E.	Area Chief Scientist, National Coal Board, Doncaster
JEFFCOATE, P. R.	Partner, Sandford, Fawcett, Wilton & Bell, London
JOBLING, T. W.	Chief Chemist and Bacteriologist, Portsmouth Water Company
JOHNSON, K. S.	Experimental Officer, Imperial Chemical Industries Plant Protection Ltd., Yalding
JOHNSON, P. F.	South East of Scotland Water Board
JOHNSTON, F. A. S.	Supplies and New Works Engineer, Doncaster and District Water Board
JOLLIFFE, A. W.	Assistant Hydrologist, Devon River Authority
JONES, G. P.	Senior Lecturer in Hydrogeology, University College, London
JONES, H. H.	Senior Hydrologist, Essex River Authority
KEAY, G. R.	Deputy Chief Chemist (Water), City Laboratories Service, Coventry
KENYON, W. G.	Deputy Chemist and Bacteriologist, Bedforshire Water Board
KERSHAW, G. M.	Assistant Hydrologist, Northumbrian River Authority
KNIGHT, D. O.	Pollution Control Officer, Bristol Avon River Authority
KNIGHT, H. W.	Chief Civil Engineer, Essex Water Company
KNILL, Dr. J. L.	Reader in Engineering Geology, Imperial College, London
LAMBERT, P. W.	Chief Technical Assistant, Water Quality Department, Lancashire River Authority
Le ROUX, N. W.	Warren Spring Laboratory, Stevenage
LEWIS, G. E.	Deputy Chief Engineer, Middle Thames Water Board
LINDHARD, J.	Senior Research Officer, Government Experiment Station on Plant Nutrition, Denmark
MACHON, F. J.	Manager, Member Services, The Water Research Association
MAGINESS, T. O.	Engineer and Manager, Armagh and Dungannon Waterworks Joint Board, Northern Ireland
MALCOMSON, N.	H. B. Berridge and Partners, Chelmsford, Essex
MARSHALL, J. K.	Partner, Willcox, Raikes & Marshall, Birmingham
MARSTRAND, Mrs. P. K.	Research Fellow, University of Sussex
MAWER, Dr. P. A.	Chief Economist, The Water Research Association
MERCER, D.	Assistant Director, Directorate General Water Engineering, Department of the Environment
MILNE, D. M.	Senior Engineer, Sir M. MacDonald & Partners, London

MILTON, Mrs. V.A.	Hydrogeologist, Sir William Halcrow & Partners, London
MOLES, J.	Research Engineer, Société Degremont, France.
MONK, R.B.	Deputy Engineer and Manager, Folkestone and District Water Company
MONTGOMERY, Dr. H.A.C.	Water Pollution Research Laboratory, Stevenage
MOORE, Dr. B.	Public Health Laboratory Service, Exeter
MULDER, Dr. F.G.	Geohydrologist, N.V. Waterleidingmij Oostelijk Gelderland, Netherlands
NEWMAN, Dr. A.T.	Senior Engineer, Howard Humphreys & Sons, Reading
NICHOLLS, Dr. G.D.	Consultant Geochemist, Edgar Morton & Partner, Adlington
NICOLSON, N.J.	Deputy Chief Purification Officer, Thames Conservancy
ORR, W.E.	Chief Assistant Engineer, Makerfield Water Board
OWEN, Dr. M.	Senior Assistant Engineer, Thames Conservancy
PACKHAM, Dr. R.F.	Chief Chemist, The Water Research Association
PAIN, B.F.	University of Reading
PARKER, D.E.	Pollution Prevention and Fisheries Manager, Sussex River Authority
PEARSON, C.R.	Assistant Manager, Imperial Chemical Industries Ltd., Brixham
PEEL, A.	Effluent Disposal Ltd., Walsall
POITRINAL, D.	Ecole Nationale Supérieure des Mines de Paris, France.
POTIE, L.	Ingénieur, Société des Eaux de Marseilles, France.
POWLING, Miss J.	Senior Biologist, Water Supply Commission of Victoria, Australia
PRENTICE, Prof. J.E.	Department of Geology, Kings College, London
RANSHAW, R.C.	Mill Chemist, Portals Ltd., Basingstoke
RENOLD, J.	Local Government Operation Research Unit, Manchester
RICHARDS, H.J.	Head, Resources Division, Water Resources Board
RILEY, R.W.	Senior Engineer, Binnie & Partners, London
ROBINSON, K.	Assistant Engineer, North Lindsey Water Board
ROFE, B.M.	Partner, Rofe, Kennard & Lapworth, London
ROGIER, Ph.A.	Development Engineer, Stichting Concawe, Netherlands
SATCHELL, R.L.H.	Senior Engineer, Water Resources Board
SAUTY, J.P.	Bureau des Recherches Géologiques et Minières, France.
SAYERS, D.R.	Senior Technical Officer, Northumbrian River Authority
SELBY, K.	Inspector, Trent River Authority

SEVEL, T.	Scientific Officer, The Danish Isotope Centre
SHARPE, R.	Chief Chemist and Bacteriologist, Makerfield Water Board
SIDENVALL, Dr. J.	Hydrogeologist, Uppsala Kommuns Gatukontor, Sweden
SMITH, D.B.	Project Manager, Atomic Energy Research Establishment, Harwell
SMITH, D.I.	Lecturer, Department of Geography, University of Bristol
SMITH, E.J.	Resident Engineer, Metropolitan Water Board
SMITH, H.S.	Chemist and Bacteriologist, Ipswich County Borough Council
SMITH, R.M.	New Works Engineer, St. Helens County Borough Council
SPEIGHT, H.	Director of Technical Services, Hampshire River Authority
STANTON, W.I.	Consultant Hydrogeologist, Bristol Avon River Authority
STEENVOORDEN, J.H.A.M.	Research Officer, Instituut Cultuurtechniek en Waterhuishouding, Wageningen, Netherlands
STOTT, Dr. D.A.	Chemist/Head of Laboratory, Lee Conservancy Catchment Board
STOW, G.R.S.	Director, George Stow & Co. Ltd., Henley on Thames
STREATFIELD, Dr. E.L.	Senior Consultant, Cremer & Warner, London
STRINGER, J.L.	County Health Inspector, Hertfordshire County Council
SWALES, G.M.	Chief Engineer, The East Surrey Water Company
SWEETING, F.	Chief Chemist, North West Sussex Water Board
TAYLOR, Miss J.	Research and Development Department, Degremont Laing Limited, London.
THORN, Mrs. B.M.	Resources Group, The Water Research Association
THURSTON, E.F.	Div. Assistant Environment Adviser, Imperial Chemical Industries Ltd., Mond Division, Winnington
TODD, Prof. D.K.	Professor of Civil Engineering, University of California, Berkeley, USA
TOFT, Mrs. H.	Geologist, Lee Conservancy Catchment Board.
TOMS, R.G.	Pollution Prevention Officer, Yorkshire River Authority
TRAC, N.Q.	United Nations Technical Officer, UN Food and Agricultural Organization, Madrid, Spain
TRUESDALE, Dr. V.W.	Aquatic Geochemist, Institute of Hydrology, Wallingford
TYLDESLEY, J.A.	Chief Engineer, North West Leicestershire Water Board
UNGEMACH, P.	Bureau de Recherches Géologiques et Minières, Strasbourg, France.
VLASBLOM, W.J.	Qual. Civil Engineer, North Holland Waterworks, Bloemendaal, Netherlands
van WAEGENINGH, H.G.	Geohydrologist, Government Institute for Water Supply, The Hague, Netherlands
WARREN, S.C.	Chemist and Bacteriologist, Brighton Water Department

WEST, J. T.	Technical Liaison Officer, The Water Research Association
WILSON, A. L.	Treatment Group, The Water Research Association
WITHERSPOON, Dr. P.A.	Professor of Geological Engineering, University of California, Berkeley, USA.
WOODHEAD, D.	Senior Water Scientist, Yorkshire River Authority
YOUNG, D.D.	Deputy River Conservator, Essex River Authority
ZOETEMAN, B.	Government Institute for Water Supply, The Hague, Netherlands

On 1st April 1974 the following changes took effect

Former organisation	Became
Water Pollution Research Laboratory	Water Research Centre, Stevenage Laboratory
The Water Research Association	Water Research Centre, Medmenham Laboratory
Water Resources Board	(most staff redeployed to Central Water Planning Unit and Water Data Unit, Reading or to Water Research Centre, Medmenham)
River Authorities, Water Boards and Sewage Treatment Authorities in England and Wales	10 Regional Water Authorities (see DEPARTMENT OF THE ENVIRONMENT and WELSH OFFICE, Background to water reorganization in England and Wales, London, HMSO, 1973, 37 p.)

Other changes of address or title are as follows:

ASPINWALL, R.	Consulting Hydrogeologist, Oswestry, Salop.
DUDGEON, C.R.	Senior Lecturer, Department of Civil Engineering, University of New South Wales, Australia.
ELLIOTT, E.	British Steel Corporation, Special Steels Division, Sheffield.
KNILL, Dr. J.L.	Professor of Engineering Geology, Imperial College, London.

PREFACE
TO
INDEX

This covers the principal topics mentioned in the preceding text, together with the various contributing authors and individuals named in the discussions. There is no attempt to indicate the more important references, in case of multiple entries.

References to the literature occur at the end of most papers: a handful of official agencies are indexed out of these, but individual authors of references are not indexed. None of the Bibliography (pp 465 - 516) is indexed. Towns and place names are given under the heading of country, county (UK) or state (USA).

The index has been set up to conform closely to text wording. Many synonyms occur in consequence; the more important of these are indicated via 'See also ....' instructions.

XYLENE 122